Sources and Studies
in the History of Mathematics and Physical Sciences

For further volumes:
www.springer.com/series/4142

Sources and Studies
in the History of Mathematics and
Physical Sciences

Managing Editor
J.Z. Buchwald

Associate Editors
J.L. Berggren and J. Lützen

Advisory Board
C. Fraser, T. Sauer, A. Shapiro

Dirk van Dalen

The Selected Correspondence of L.E.J. Brouwer

 Springer

Dirk van Dalen
Utrecht University
Utrecht, The Netherlands

Whilst we have made considerable efforts to contact all holders of copyright material contained in this book, we have failed to locate some of them. Should holders wish to contact the Publisher, we will make every effort to come to some arrangement with them

Additional material to this book can be downloaded from http://extras.springer.com
Password: [978-0-85729-527-9]

ISBN 978-1-4471-2691-1   ISBN 978-0-85729-537-8 (eBook)
DOI 10.1007/978-0-85729-537-8
Springer London Dordrecht Heidelberg New York

British Library Cataloguing in Publication Data
A catalogue record for this book is available from the British Library

Mathematics Subject Classification: 00A30, 01A60, 97Exx, 03F55, 32Fxx, 54F45

Cover design: VTeX UAB, Lithuania

Printed on acid-free paper

Springer is part of Springer Science+Business Media (www.springer.com)

Preface

The editing of the Brouwer correspondence has been a long journey with pleasant surprises and grave setbacks. It was, so to speak, the natural sequel to the publication of Brouwer's biography. When I embarked on the project, I did not foresee the problems and the adversities that lurked at the wayside—in spite of the warnings of our sorely missed colleague Karl Schumann. My retirement was largely filled by taking part in and supervising the project; although my role in the project can be viewed as a free donation to science and our national heritage, I was fortunate enough to obtain financial support for a number collaborators.

I have found goodwill and support from many sides, for which I am grateful indeed. Let me thank here the many institutions and persons that have enabled me to reach this journey's end.

Special thanks go to my wife Dokie, who, against her better judgement, bravely tolerated this seemingly interminable project.

Finally, the publisher Springer has kindly and efficiently supported me in the task of converting the texts into a book; in particular I am indebted to the help and encouragement of Karen Borthwick, Lauren Stoney and Donatas Akmanavičius.

Utrecht   Dirk van Dalen
July 2011

Contents

Chapter 1

Introduction

The Brouwer archive consists largely of material from Brouwer's estate, which was presented by his heir, Mrs. Cor Jongejan, to the *Wiskundig Genootschap*, the professional organization of Dutch mathematicians. H. Freudenthal and A. Heyting made use of the collection for the edition of the Collected Works, vol. 1 and 2. In 1976 it was passed on by the former chairman of the *Wiskundig Genootschap* to the present editor in order to set up an archive. The archive contained correspondence, scientific documents, and legal and business documents. In due time more original documents were made available by various correspondents. At first, efforts were concentrated on the collection and preservation of original material for the composition of Brouwer's biography. To this end, copies of letters were also obtained from a number of sources; mostly archives but also from individuals.

The present collection of originals and copies is by no means complete. There is no certainty that the collection that was donated to the *Wiskundig Genootschap* has indeed been preserved in its totality; moreover, during Brouwer's life, fire has destroyed an unknown number of documents in his files. Some of the documents in the archive are indeed partly singed.

The letters fall roughly into three categories: scientific, personal, and business. As is to be expected, these categories overlap here and there. The business part contains mostly material that deals with the pharmacy, real estate and legal topics. Only a few of these letters have been put into our collection. A brief explanation may be in order: Brouwer's wife ran a pharmacy in Amsterdam that was actually bought by Brouwer in 1905 from his mother-in-law. Brouwer handled all matters concerning the property, including negotiations with the city and appointments of staff. Furthermore,

D. van Dalen, *The Selected Correspondence of L.E.J. Brouwer*,
Sources and Studies in the History of Mathematics and Physical Sciences,
DOI 10.1007/978-0-85729-537-8_1, © Springer-Verlag London Limited 2011

Brouwer owned property in Amsterdam, Blaricum, and Laren, and also in Berlin, Harzburg, and in former East Prussia, now in Poland. All of this led to a massive correspondence, some of which has been preserved. Finally, Brouwer was in the 1930s involved in an investment conflict (the Sodalitas affair, concerning a Spa in Budapest); the resulting correspondence would have been enough to keep a couple of clerks in business. Brouwer reported that his Sodalitas file in the end contained at least 70 kg of letters, court orders, documents, memoranda, etc., of which only a small part has been preserved.

Background information on the content of the letters is, as a rule, to be found in the biographies of Brouwer, D. van Dalen. *Mystic, Geometer and Intuitionist: The Life of L.E.J. Brouwer. Vol. 1. The Dawning Revolution, Volume 2: Hope and Disillusion* Oxford University Press 1999, 2005. In Dutch[1]: D. van Dalen. *L.E.J. Brouwer. Een biografie. Het heldere licht van de wiskunde.* Bert Bakker. Amsterdam, 2001.

Selection criteria

The letters vary considerably in topic and in significance; numerous letters were selected for the translations that shed light on Brouwer's life, thoughts and mathematics. It would have been easy to double the amount of letters; however, the size of the book and range of topics forced us to refine our selection.

The early years of Brouwer were to a large extent influenced and shaped by his friend, Carel Adama van Scheltema, and his teacher Diederik Korteweg. Scheltema took the 16 year old freshman in hand, and acted as his guide and tutor in the exciting world of fraternities, literature and politics, and Korteweg taught him all the wisdom and techniques of late nineteenth century mathematics. No collection would be complete without a generous display of the exchange between the two friends[2] and between student and teacher[3]. In particular, the correspondence dealing with the dissertation is a illuminating supplement to the dissertation itself.

The second period of Brouwer's mathematical journey brings us to the extraordinarily rich topological period. The correspondence of these years (1909–1913) involves the great men of the period, such as Baire, Fricke, Hadamard, Hilbert, Klein, Koebe, Lebesgue, Poincaré, and Schoenflies.

---

1. A separate biography, not a translation.
2. *Droeve snaar, vriend van mij,* [Van Dalen 1984].
3. *Brouwer en de Grondslagen van de Wiskunde* [Van Dalen 2001].

When Brouwer entered into his mature intuitionism, there was a certain distancing from topology. However, in 1924 he was brought back into the center of the subject when the two young Russians Alexandrov and Urysohn took the stage, almost immediately followed by Karl Menger. A great deal of the correspondence of that period concerns dimension theory. In a later stage also Hans Hahn took part in the correspondence.

After 1930, the number of big topics dwindled. There are in effect two topics that played a prominent role later in Brouwer's life; the first one is the treatment he received from the authorities after the war, and the second one is the loss of his journal, *Compositio Mathematica*. In both cases he was treated less than fairly.

A large amount of correspondence has not been included in the present collection, nor in the online collection. The letters that deal with the pharmacy, the internal revenue service, the Sodalitas investments, real estate, general business matters, and legal matters have not been adopted. Furthermore, the correspondence with Henri Borel (the Sinologist), E. Dubois (neither had Brouwer's letters available), and L.S. Ornstein (permission withheld) are not in the collection. We have also chosen not to include a number of letters and cards with little or no relevance.

The original letters can be found on 'Springer Extras' and can be accessed and downloaded. To view the content on Springer Extras, please visit extras.springer.com and search for the book by its isbn. You will then be asked to enter a password, which is given on the copyright page of this printed book.

Editorial conventions

The letters are preceded by a header that contains the name of the recipient or the sender, the date, and the place of dispatch. When relevant, the letter head is included. The address of the sender is included when known from the document or envelope. In the case of Brouwer's letters the address, when in Blaricum or Laren, is suppressed, being invariably Torenlaan 70.

We have now and then provided the original salutations of the letters; these often convey more information than the translations.

The bottom line contains the standard information about the document: *handwritten, autograph, typescript, signed/unsigned, copy, carbon copy, printed document, partially burned,* etc. Some of the documents have already been published in the Collected Works; in those cases the location is indicated.

There are two kinds of footnotes: those of the author [1] and those of the editor [(1)].

Footnotes may also refer to letters that have not been adopted in the present volume. The reader can consult these in the online publication of the original versions.

As a rule we have stuck to the original format of the letters. Sometimes we have adopted a variation, indicated, when relevant, in a footnote.

Figures have almost always been redrafted; I am indebted to John Kuiper for his invaluable assistance in providing figures.

The correspondents are listed in a separate index.

References have been provided for most papers or books mentioned in the correspondence.

There is an online bibliography of Brouwer's publications; see *A bibliography of L.E.J. Brouwer,* `http://igitur-archive.library.uu.nl`, in which extra information, such as 'submitted by', 'published in', ' translated in', 'reviewed by', ... is indicated.

Major topics

One may distinguish in Brouwer's life a number of affairs or topics that belong to a particular period, and that have to be kept in mind to obtain a correct interpretation of certain letters in certain periods. All of these have been treated in the biography. Here we will restrict ourselves to a brief résumé of the main themes. The references are to the above mentioned biography.

Friendship with Adama van Scheltema, 1898–1924. Carel Adama van Scheltema (1877–1924) and Brouwer were members of the same fraternity. They became close friends; Scheltema was the older protector and instructor of Brouwer. Their friendship was a very private one, it has found its expression in a long exchange of letters. Scheltema became the leading Dutch socialist poet of his time. He was responsible for the transformation of Brouwer from a shy freshman into a ripe man of the world. [Bio 1.4]

The doctorate. From 1905 through 1907 Brouwer carried on an intensive correspondence with his Ph.D. adviser, D.J. Korteweg. This series of letters provides a great deal of extra insight into Brouwer's views

---

[1]author's footnote

[(1)]editor's footnote

on mathematics and on life. After Brouwer was appointed, Korteweg acted as his older and wiser guardian. [Bio Ch. 3, 6.6]

Lebesgue, topology and dimension. After Brouwer published his proof of the invariance of dimension, a protracted correspondence between Brouwer, Blumenthal, Hilbert, [4] Baire, Fréchet, and Lebesgue followed. Although Lebesgue did not claim priority on the grounds of his (defective) proof, he did not contribute toward a balanced historical evaluation. [Bio 5.1]

Koebe and automorphic functions. When Brouwer applied his invariance of domain theorem to the theory of automorphic functions and uniformization by salvaging Klein's Continuity Method (1912), Koebe did whatever he could to avoid acknowledging Brouwer's contribution. The episode counted a number of curious situations, which offended Brouwer's feeling for justice. [Bio 5.3]

Interbellum political frictions in science—the boycott of Germany. In 1918 a new international scientific union was founded, which had as one of its aims to exclude the former members of the axis from international scientific contacts. Brouwer took a staunch internationalist position and condemned the idea and practice of the Conseil International des Recherches. He carried his opposition to the Dutch Academy, the Dutch Mathematical Society, and finally to the Ministry of Education. In 1925 he successfully defended the German position in the board of the Mathematische Annalen at the occasion of the Riemann volume. In 1928 he acted again in the hope to secure a truly international mathematical congress in Bologna. Sticking to his overly strict position he overplayed his hand, but for all practical purposes the German boycott was terminated. [Bio 9.1 – 9.3]

Dimension theory. The discussion concerning dimension theory proper started in 1923, when Urysohn entered the subject. In 1924 Menger joined in. From then on there is a great deal of correspondence on the topic, slowly developing into the Brouwer–Menger conflict. The mathematicians involved are, among others, Alexandrov, Urysohn, Menger, Vietoris, Hahn, Sierpiński.

The Riemann volume. The Mathematische Annalen prepared a volume at the occasion of the centenary of the birth of Riemann (1826). Within the editorial board no unanimity could be reached on the issue of admitting French authors. [Bio – 13.3]

---

4. We point out that the Brouwer archive contains no letters from Hilbert.

The Grundlagenstreit. Although this was a major topic of discussion in the 1920s, there is hardly any correspondence in the Brouwer archive that deals with it.

The Bologna Conference. The opposition to the boycott of German scientists was gradually growing, when in 1928 the International Congress of Mathematicians in Bologna was being organized. Brouwer untiringly campaigned for a completely open admission policy. He succeeded in reaching a positive arrangement with the Italian organizers, but in the end was not convinced that a guarantee for future open international scientific relations was ensured. He advised against participation. [Bio – 15.2]

The War of the Frogs and the Mice (or 'the Mathematische Annalen conflict'). A combination of serious illness and an accumulation of conflicts brought Hilbert to an unfortunate act; he dismissed Brouwer from the board of the Mathematische Annalen, considering him a danger to the future of the journal. A great deal of correspondence was generated by this decision. In the end Brouwer's discharge was disguised as a wholesale revision of the board. [Bio – 15.3]

Compositio Mathematica. Soon after his discharge from the board of the Mathematische Annalen, Brouwer started negotiations for the founding of a new mathematics journal. The journal appeared in 1935; the political changes in Germany caused some serious problems. [Bio – 16.6]

Post-war conflicts. This deals mostly with Brouwer's suspension and faculty policies at the Mathematical Centre. After the war Brouwer for the second time was ousted from his prominent position in the editorial board of a mathematical journal. This time he was the victim of a coup in the board of Compositio Mathematica by his Dutch colleagues and the publisher. [Bio – 18.4]

Acknowledgements

It would have been impossible to collect the material in the present volume without the help and advice of a large group of colleagues, and of archives and institutions all over the world. I will, with disregard for the alphabet, or ordering principles in general, mention persons and organizations that have been helpful in any respect. Let me first mention the archives that have provided me with information and with copies of documents. The University Archive in Göttingen has most generously assisted me in finding the relevant material from its collection. I am most grateful to the archivist Dr. Helmuth Rohlfing. In view of the many connections of Brouwer with his Göttingen friends and colleagues, the project could not have been undertaken without his support.

A rich collection of letters and drafts in the Alexandrov archive has been made available to me by Albert Shiryaev, who has taken care of the original letters from Brouwer's personal archive, which were given to Alexandrov by Brouwer's heir. In addition, he provided me with additional material, which was crucial for the biography of Brouwer.

The *Letterkundig Museum* (Literary Museum) in The Hague at an early stage allowed us to make use of the archive of Carel Adama van Scheltema. Its archivist Sjoerd van Faassen most kindly assisted us in the intricate transcription of the letters exchanged between the two friends since their students days. The collection is most valuable, as it allows us a glimpse at an unguarded exchange of letters during the formative years of both men.

The *Centrum voor Wiskunde en Informatica* supported us by making its information available, and by allowing us to make use of the collections of correspondence of J.G. van der Corput and J.A. Schouten.

The collections of letters of F. van Eeden, D.J. Korteweg, and G. Mannoury were put at our disposal by the University Library of the University of Amsterdam. The *Noord-Hollands Archief* at Haarlem has an impressive collection of archives on the sciences. A large part of the correspondence with H.A. Lorentz and P. Zeeman used in the present collection was made available by the Noord-Hollands Archief. The archive also contains the correspondence of E.W. Beth, D. van Dantzig, H. Freudenthal and A. Heyting. H. Visser granted permission for the use of material from the Beth archive. Furthermore, the Noord-Hollands Archief contains the archive of the *Koninklijk Wiskundig Genootschap* (KWG, Royal Dutch Mathematical Society) and the *Koninklijke Nederlandse Akademie van Wetenschappen* (KNAW, Royal Dutch Academy of Sciences). Both archives have been most supportive. The *Archief van de Nederlandse Provincie der Jezuïeten* (Archive of the Dutch Jesuits) kindly allowed us to make use of the correspondence with Jac. van Ginneken in their archive. The archive of the University of Amsterdam is under the care of the *Stadsarchief Amsterdam* (Amsterdam City Archive); I am indebted for the permission to use the various documents relevant to Brouwer. The library of St John's College gave us permission to use the letters from the Max Newman collection. I am grateful to the archivist, Jonathan Harrison, who also directed me to the son of Newman, William Newman, who also gave permission for the publication. Vagn Lundsgaard Hansen kindly gave permission to use the letters from the Nielsen papers in the archive of the Mathematics Department of the University of Copenhagen. The Library of Leiden University granted permission for the adoption of the Bolland letters.

Financial support

The project has been generously supported by the *Nederlandse Organisatie voor Wetenschappelijk Onderzoek*, NWO (Netherlands Organisation for Scientific Research). I am indebted for the patience and understanding of the organization that enabled us to carry on with the research, transcription, annotation, editing, and preparation of the text. The project started in 1999 and was funded until 2007. The funding also supported the research for, and publication of, a Ph.D. thesis, *Ideas and Explorations. Brouwer's Road to Intuitionism*, by J.J.C. Kuiper in 2004. The topic of the thesis was the genesis of Brouwer's foundational views, as expressed in Brouwer's dissertation. The publication is available online: `http://igitur-archive.library.uu.nl`

In addition to the financial support of NWO, the Faculty of Philosophy at Utrecht, and the University of Amsterdam have generously helped out over a prolonged period, both in funding and in providing material support. The helpdesk of the Faculty of Philosophy deserves our gratitude for helping us out on many occasions. Thomas Müller and Nick Boerma have lent a hand in giving the text a well-ordered form. Their expertise in the art of TEX and UNIX has been most valuable. Various funds and associations have been kind enough to fund our enterprise; I am most indebted to the Zeeman Foundation (A. Kox), the Heyting Foundation (D.H. de Jongh, A. Visser), the Korteweg–de Vries Institute (J.J.O.O. Wiegerinck), the Institute for Logic, Language and Computation (F.J.M.M. Veltman), and the Universiteitsvereniging van de Universiteit van Amsterdam. Dr. P. Blok of the latter university has been very supportive and inventive in finding ways and means for keeping the project going under difficult circumstances.

The project has also been so fortunate to receive the financial support of three winners of the Spinoza Price Award: H.P. Barendregt, R. Dijkgraaf, and H.W. Lenstra. It is needless to say that I have deeply appreciated their support and involvement.

The following persons were supported by NWO for taking part in the Correspondence project: Dr. M.S.P.R. van Atten, Dr. M. Brandsma, Dr. K. de Leeuw, and Dr. P. van Ulsen. Dr. J.J.C. Kuiper took part as a volunteer. All of the above persons took part in the transcription process; Dr. van Atten in addition handled a large part of the TEX details involved in the

handling of the letters. Dr. Kuiper was most helpful by taking care of the
larger part of the illustrations. Dr. van Ulsen was involved in almost all
parts of the project; he was responsible for the layout and unification of
notation and comments. After the termination of the grant he was prepared
to continue his work on the project on a freelance basis. Without the help
and friendship of in particular the last three collaborators, the project would
have been an awkward burden.

A special grant was made available by NWO for the translations in the
present book. Dr. J.W. Nienhuys undertook the challenge of translating
Brouwer's letters, which already in their original form presented formidable
problems of transcription and composition. I am in his debt, not only for the
translations, but also for comments and suggestions concerning the original
letters. In the further process his translations have been scrutinized, revised,
and polished by myself, and by Prof. M. Knus and Prof. Th. Verbeek where
the original texts were in French. In the final stage the Springer corrector,
Peter Laurence, has carried out a thorough check.

Support and advice

I have been most fortunate to secure the cooperation of Ernst Specker
and Max-Albert Knus, both from the ETH, Zürich. They have checked
the major part of the French and German language letters for misspellings,
grammatical slips, etc. Their efforts have gone far beyond the call of duty;
indeed a number of transcription errors have been set right by them on the
basis of deduction and extrapolation.

Hans Freudenthal has supported the Brouwer project in the widest sense.
He had generously shared his considerable store of memories of "the old
days", when he joined Brouwer in Amsterdam, and in the subsequent years.
Furthermore, he made available his private correspondence archive, thus
adding important information that greatly contributed to the coherence and
depth of the correspondence collection.

The collection of the material had already started before the project
was even conceived. The biography could not have been written without
all the evidence and information buried in letters. During the whole prolonged search for letters and documents, I have received invaluable help and
encouragement from many colleagues, friends, relatives of correspondents,
archivists, etc. I express my gratitude toward all who have during the past
years helped me to find the right sources, and warned for the dangers that
the editor of correspondence faces. My attempt to provide a complete list
of all those who over the past years have been kind enough to assist me in

my self-chosen task will almost certainly fail, nonetheless let me mention my close and distant supporters:

C. van Aardenne (Ehrenfest); Dr. C. Adelmann (Fricke, Braunschweig); Dr. G. Alberts; Prof. K. van Berkel (Dijksterhuis, Groningen); U. Bieberbach (Munich); N. Boerma (Utrecht); M.F. Brouwer-Rueb; Prof. M. Duewell (Utrecht); Prof. R. Dyckhoff (St Andrews); Sigurd Elkjear (Copenhagen); Dr. M. Folkerts (Munich); P. Forman (National museum of American history); Prof. Hans Freudenthal; Dr. B. Glaus (ETH-Bibliothek); H. Hübner (Helen Ernst); Dr. U. Hashagen (von Dyck, Munich); Prof. M.A. Knus (ETH); Prof. A. Kox (Amsterdam); C.M. Los-van Kampen (Weesp); Prof. A.G.B. ter Meulen (Groningen); Ir. W.C. Mulder (Den Haag); Prof. T. Müller (Utrecht); Dr. J.W. Nienhuys (Eindhoven); Prof. H. Osswald (Munich); Prof. K. Parshall (University of Virginia, Charlottesville); Dr. V.R. Remmert (Mainz); Prof. D. Rowe (Mainz); Prof. Karl Schuhmann; Dr. C. Smorynski (Chicago); Prof. E. Specker (ETH); Prof. F. Veltman (Amsterdam); Prof. Th. Verbeek (Utrecht); Dr. M. Weyl; Prof. P.G. Ziche (Utrecht).

I am grateful to the following persons who granted publication permission for certain letters, or who were instrumental in the search for persons who could give permission — asking clemency for a possibly overlooked name:

C. van Aardenne; mrs. M.J. Aerts; Alonzo Church jr; mrs. B. Arden; mrs. H. Baer; mrs. M. Baier; T. Bastin; mrs. C. Becker; O. Bieberbach; R. van Blommestein; mrs. H. Blumenthal; H. Bohr; G.V. Born; J. Braithwaite; mrs. S.M.P. Brouwer-Euwe; N.G. de Bruijn; Leonard Bruno; S.S. Cairns; J.H. ten Cate; T. Corkery; mrs. S. Coxeter Thomas; mrs. A. Cullingford; E. Cuypers; mrs. M-M Daisley; R. van Dantzig; B. Denjoy; D. Devriese; L. van den Dries; L. van Duinen; R. Dyckhoff; B. Fraenkel; mrs. A. Frank; mrs. F. Greffe; V.L. Hansen; J. Harrison; mrs. B. Hertwich-Koppel; mrs. L. Heyting; H-P. Hillig; M.C van Hoorn; mrs. R. Jeltsch Fricker; A. Kanamori; mrs. J. Kneser; D. Kohnstamm; A. Kox; mrs. T. Kuit; mrs. R. van der Laan-van Heemert; S.V. Langeveld; mrs. M. Lederer; G. Lochak; A. Lubbers; mrs. K. McKee; mrs. E. Menger; H. Michels; K. Mothate; W. Newman; N. Peguiron; W. Pohlers; F. Poincare; K. Raine; mrs. B.I. Reimers; mrs. A. Sander; N. Schappacher; A. Shiryaev; Y. Sinai; mrs. J.R.E. Smit-Dijksterhuis; M. Stupperich; F. van de Velden; H. Visser; E. Vuijsje; mrs. U. Wajcen; mrs. M. Whitman; R. Winckworth; H. Zuidervaart.

The archive sources

Most of the letters are to be found in the Brouwer archive, which is at
the moment still at the Philosophy Department of Utrecht University. The
following list shows the archives where a number of original documents are
stored. What is listed as the Mathematische Annalen archive consists of a set
of copies of letters exchanged during the so-called Mathematische Annalen
conflict (or "The war of the frogs and the mice"). The originals were in the
possession of Courant, who made copies of the set and sent them to Van
der Waerden for a possible historical account. Van de Waerden then sent
various copies out to historians of mathematics, including Freudenthal. The
latter copy has been used for the present collection; it has been deposited
in the Brouwer archive.

The following abbreviations of archive names have been used at the end
of the letters:

Adama van Scheltema Nederlands Letterkundig Museum, Den Haag

Alexandrov Moscow (in charge of A. Shiryaev)

Baire Archive of the Académie des Sciences, Paris.

Bernays Eidgenössische Technische Hochschule, Zürich

Beth Noord-Hollands Archief, Haarlem

Bieberbach Deutsches Museum, Bibliotheksbau. Munich

Brouwer Philosophy Department, Utrecht University

Corput (van der) Centrum voor Wiskunde en Informatica, Amsterdam (Noord-Hollands Archief, Haarlem)

Courant Elmer Holmes Bobst Library

Dantzig (van) Noord-Hollands Archief, Haarlem

Doetsch Universität Freiburg

Ehrenfest Museum Boerhaave, Leiden

Einstein Hebrew University, Jerusalem

Fréchet Archives Académie des Sciences, Paris

Fraenkel Hebrew University, Jerusalem

Freudenthal Noord-Hollands Archief, Haarlem

Fricke Universitätsarchiv der Technischen Universität Braunschweig

GAA Stadsarchief Amsterdam

Heyting Noord-Hollands Archief, Haarlem

Hilbert Universitätsbibliothek, Göttingen

Hopf Eidgenössische Technische Hochschule, Zürich

Klein Universitätsbibliothek, Göttingen

KNAW Koninklijke Nederlandse Akademie van Wetenschappen, Noord-Hollands Archief, Haarlem

Kneser (Göttingen) Universitätsbibliothek, Göttingen

Kneser (Zürich) Eidgenössische Technische Hochschule, Zürich

Korteweg Universiteitsbibliotheek, Universiteit Amsterdam

KWG Koninklijk Wiskundig Genootschap, Noord-Hollands Archief, Haarlem

Lakwijk (van) private collection

MA Mathematische Annalen collection; various sets of copies (e.g. Brouwer archive)

Mannoury Universiteitsbibliotheek, Universiteit van Amsterdam

Menger Special Collections Department, William R. Perkins Library, Duke University / Mason Barnett

Ministerie van Onderwijs Zoetermeer (Den Haag)

Mises (von) Harvard University

Mittag-Leffler Djursholm (Stockholm), Swedish Academy

Ritter Utrecht University Library, Ritter Archives

Schoenflies Universitätsbibliothek, Göttingen

Schouten Centrum voor Wiskunde en Informatica, Amsterdam

Sommerfeld Deutsches Museum, Bibliotheksbau. Munich

Univ. Arch. Oslo University Archive, Oslo

Veblen Library of Congress, Washington

Zeeman Noord-Hollands Archief, Haarlem

Chapter 2

1900 – 1909

901-12-15

From C.S. Adama van Scheltema — 15.XII.1901   Amsterdam

Dear Bertus [Beste Bertus]

Your letter [1] is something of a small event on this first cold winter's day. I had asked Huet [2] whether I was allowed to visit you, [3] but he preferred to postpone that until you would ask for it by yourself,— and now your letter, which doesn't really ask for it, arrived, but which does show that you are recovering,— and also: *a new will to live.* Only with that too you will in fact be able to return to life, that just happens to push down the most brilliant man, when he primarily doesn't know how to live *intellectually* in a conscious direction. And when you really and consciously start all over again using your renewed vitality, a treasure will be open for your desires, and you will experience again the gratitude to live in just *this* wild and wonderful time,— but also a will to take part in it!

Sometimes I yearn for the quiet and endlessness of the country side, but I will have yet to remain for some time in the oppressing narrowness of city life, to breathe for a moment in the sad sphere of the time, to submerge for just a moment as 'die Wildente' [4] and after that to fly better and safer,— and foremost also to search for the simplest connection between the real and the ideal.— This summer I did in fact make something beautiful,— I hope

---

[1] *Brouwer to Scheltema, 5.XII.1901.*   [2] Brouwer's family doctor   [3] Brouwer invariably suffered from nervous breakdowns, complete with physical side effects, as a result of his military service. The effects often lasted for some time. In fact it seriously delayed his study.   [4] Refers to Ibsen's play *The Wild Duck*

D. van Dalen, *The Selected Correspondence of L.E.J. Brouwer,*
Sources and Studies in the History of Mathematics and Physical Sciences,
DOI 10.1007/978-0-85729-537-8_2, © Springer-Verlag London Limited 2011

to publish in April a new collection, [5] but fame won't yet be captured thus, I guess. Here I am sterile,— it is hard for me to reserve a couple of hours each day to lift my head above 'die Sumpfluft' [6] and ... there are no great minds around,—

returning to you I hope to find a smaller soul:— it's sometimes also a kind of art and sometimes it is courageous indeed to be a bit smaller,— for us men,— women are like that by nature;— more marrow is required to bear greatness.

Write to me then when you have recovered so far that you can see me, I long for that — the day is rather cold and so is my room, and hence maybe my letter, but all the warmer is my desire for your return, for your appearance, for your words,— if it is really to be a *revival* (— and you must want that) it can also be the *birth* of a new and closer relation:— I sometimes feel myself dying from the small desires of life,— the great ones I keep — because it is beautiful to *be* human,— I hope that you will have *become* so.

The addresses are:

Heelsum (Hupkes) f. 2,50 per day.

Ruurloo (H. v.d. Mey (baker)) f. 1,75

Doorn (Miss H.W. Boeschoten) f. 3,=

Bergen (near Alkmaar, this is very beautiful, the name I don't have near at hand, but if you want to go there I can easily send it, f. 3.=)

For the rest Huet will know others.— Send me then your new address.—

Your friend Carel

[Signed autograph – in Scheltema]

1903-05-23

To C.S. Adama van Scheltema — 23.V.1903   Amsterdam

'Il faut savoir se séparer' [7] (La Rochefoucauld).

'C'est le privilege des grands esprits, de ne pouvoir se brouiller' [8] (Voltaire).

Carel, my rich poet, I have finished your book, but now listen: In no realm there are two Kings, each must live in his own country of commoners;

---

[5] of poetry.   [6] the swamp air.   [7] One must know to separate   [8] It is the privilege of the great minds that they cannot quarrel

that solitude without peers, that is what they are Kings for. But once a year they visit each other, and behold their great contrasts, with nothing in common except their joint feeling of both being king, both standing in the direct grace of God; their intercourse cannot consist of anything but showing each other the powers and splendors of each other's realms.— And the only during charm that they can receive themselves from the knowledge of both being King, consists in showing each other prescribed courtesies and in the reporting of the outward *appearances* of their person and kingdom.

Carel, your realm is more summery than mine, and your people more pacified — both our countries are endowed by God with wondrous beauties.

Well, after our talk of Thursday evening I believe that you are right, but also that the best interpretation of the fact that we have to live apart is what I write here.

Let us get together every year on Ascension Day, and together solemnly bathe ourselves in the cool spring sunshine, and sup together and exchange what the past year has brought, and, by the way indiscernible for each other, feel ourselves joined by 'knowing each other to be King'.

So, brother, dost thou assent to this?
Then, hail to Thee and Thine Kingdom.— Until 1904.

Bertus

To be deposited for posteriority in the 'Archive of the Holy Twin Alliance', established Ascension Day 1903.

[Signed autograph – in Scheltema]

903-08-09

To C.S. Adama van Scheltema — 9.VIII.1903   Haarlem

Brother and King, [Broeder en Koning]

Be of good courage, be of good courage! - for your task, which is busy unwinding. It's a pity that also a King has a body, just as his subjects, a vestige indeed from an earlier stage of development, but he does have one and he has to guide it decently to the grave. That doesn't make the royal task any lighter; the subjects must not know worries of his body.

And this bodily concern weighs much heavier on us, yes is almost impossible. That body with its life of passion in a muddled brain needs the warmth of the soul, and man's soul warmth flows down; who gives some of it to the King?

That is the misery of purification, the warning misery which restrains the great soul that is rich at the bosom of its mother, the fair earth, and together with mankind, the laboring mankind, the freedom, seeking mankind, the naive tower building mankind at his feet — from too high a flight, and in the fear of God, of God whom he has to serve in the guarding of his children, asking nothing for himself — we are the elected — not in the world for our pleasure — we are the prophets, who, messengers between God and mankind, lead its development, its work, its growth and its blossoming and we inspire it with the dewdrops flowing from our fingers — you stride solemnly and stately through your garden, and you disperse them with a steady and knowledgeable gesture: I fly through my jungle, and they roll without me knowing — few though who find them, but for them all the more valuable.

But we may only turn a look of sadness to our mother, to nature, or to each other — not to the people — because if our tears fell there, they would burn themselves and they wouldn't understand. But if we direct our eyes upward, our tear will not fall, and it will eat into our face and leave the scars of ascesis.

That is our sad inner discord; that is the tragedy of the lonely resignation that we have known: it is heart-rending when among the hard workers beneath us there is one happening to look upward who then discerns his King sometimes even, clairvoyantly, sees his royal sadness, and wants to comfort him, but the King shakes his head: no comfort can reach him.

Brother, write, write every now and then, warmth doesn't flow horizontally either, but, to be seen, stimulates our vigor, even though it doesn't make our task more joyful.

Last week I talked a bit with your brother: [9] he is not at all like a poet, and he doesn't understand your divine side at all — but as a socialist he's one of the brightest that I have met, even if he has not made everything

---

[9] Frits Adama van Scheltema.

explicit to himself, he isn't trapped anywhere. I quite liked him. Addio, happiness and strength.

   your friend

   Bertus

[Signed autograph – in Scheltema]

.903-11-15a

To C.S. Adama van Scheltema — 15.XI.1903[a]   Amsterdam
Nieuwe Tolstraat 39

   Dear Carel, [Beste Carel,]

Of course you haven't minded that I have been silent for so long. I was busy: to return after two years of absence to my subject required some dedication, especially where all love for that subject was lacking. But now I have succeeded bit by bit and I am rowing in steady strokes to my 'doctoraal'.[10] My work proceeds without illusions, but with a feeling of cheerfulness about the activity as such. And that is growing big, broad and comprehensive. Always working on, reading and thinking — and harmonizing one's life ever more, borne by resignation and trust in God — that's bliss down here. My house is homely[11], striking and modestly cheering. And sacred to me — I could do no harm there nor have evil thoughts — here I am even friendly to everybody. If a boring person comes to me — in my house I don't find him annoying — and when I am out of doors the next day, I don't comprehend how I could have managed to tolerate him.

   That ultimate harmonization of our lives seems to be hardest and slowest and most cumbersome for people of our kind. It seems that in the line of progress of generations the eldest child of each parental couple must not be sent out into the big stream of 'strives and mates', but has to be offered as a opening sacrifice — as a flower that sprouts sideways without partaking in the upward striving of the stem — to the Gods of consciousness, a consciousness that is infertile in the worldly motions — and then the Gods forsake by way of compensation their rights to the other children. Let there-

---

[10] Final examination, comparable to 'masters'.   [11] Brouwer used the English term 'homely' (i.e. 'cosy') here.

fore those holy sacrificial animals be conscious of their role, and let them not be jealous of the coarse rye bread of the animals of the herd.

Ah, well, my dear companion and comrade in arms, the purest in ourselves, the resignation and dedication to our task, it lives best in our solitude. Associating with people is necessary, but as different from the jewels of our hearts, as rye bread – which is necessary as well, from ambrosia.

With someone else we descend from our holy tower, and we live in lower regions, which also have their own demands, which also want to be part of life. The two of us do not have such lower regions in common; blessed be the moment we have seen that.

But I do not value you less because of that — because you are far away I can see you better in your lonely splendor, of which I got the sensation the first time I saw you — the first moment I saw a person for whom I felt something. And that splendor of yours I have not observed in anyone else, but to perceive it, it's best to stand far away. Therefore, stay afar, know a warm friendship for you in me, and let us work with glad acquiescence on our task in this life, knowing that all sorrow is given by God and part of that task, and see each other

sub specie aeternitatis.

If you can maintain yourself on that level, and thus view the presently living herd animals, — and I believe you can — then you can indeed move on to tragedies. In the end of Dusseldorp [12] I saw an announcement of that — I am longing for their birth.

With a handshake from Christian King to Christian King

your Bertus

[Signed autograph – in Scheltema]

_____

1904-01-18

To C.S. Adama van Scheltema — 18.I.1904   **Amsterdam**

Dear Carel, [Beste Carel,]

Passing your house, I saw the green shelves of your book case being lowered down, and with this I associated the world, — which makes you

_____
[12] [Adama van Scheltema 1903].

look in vain for a quiet corner, where you might serve it without disturbance, wherein your rich soul is allowed to blossom. Say, do write me now and then, how you are, how you feel, and if you can find quietude in the dedication to your duty, leaving it to the Gods to give you the time to complete that duty.

I spent the Christmas holiday on the heath of the Gooi, [13] and sat there in wind and frost between the pines, and wept tears about the transience, when I for a moment remembered in a flash how ecstatic I used to be about such things five years ago. That dead life is worth crystal tears, and flowers on its grave.

Would that memory be alive in yet another person, but in my barren man's soul? If so, it would be in yours. Do some day hold a requiem mass for the dead too. What a lack of tenderness, artlessness, of dedication in the words I write; I know it, I would be ashamed of myself if I met myself as I was five years ago; but just as one cannot stop one's beard from growing, neither can one stop the growth of the philistine tissue in the soul. Then let me be great as philistine! And unfeelingly go my way through the dead stones, alone to the splendid End. And so leave my trace on the melancholic earth. That is, Ambition is born in me – perhaps. But in any case one that knows how to restrain itself, and to collect quietly building materials until its time has come! I will have to be obscure for some more years, and then my grasp shall be felt. Exactly because I feel the insignificance of all worldly affairs, no other ambition or fear will disturb my course.

My short lived socialist inclinations, of two years ago, have thus turned out not to be viable. And even yet I believe that in you they don't belong to your proper substance. Read that last volume of Gorter, [14] then you will also feel nothing but revulsion, I know you too well for that. In you too introversion is the main theme; take care that an overhasty ambition will not lead us, out of yearning for quick success, to assimilation and to consorting with low company.

I heard from a nobody, who had heard you read Dusseldorp, [15] and found it beautiful, and who apparently thought that there had to be a lot of serious stuff behind it; please don't do such things, or maybe you need that kind of caresses, you would know best; in that case, do.

Wednesday Riechers and Waldemar [16] play in the Paleis; [17] maybe they can soften my philistine tendencies; you will of course also go there; I remember Die Wildente as if it were yesterday evening.

---

[13] At the time a poor farming area, some 25 km southeast of Amsterdam. It had already been discovered by artists.   [14] Socialist poet.   [15] [Adama van Scheltema 1903]. [16] actors.   [17] *Paleis van Volksvlijt* (Palace of the People's Industriousness) in Amsterdam, the Dutch version of Chrystal Palace.

Please write to me your new address. I hope you will find there solitary quiet; surround yourself with books of your equals [18] and people you find sympathetic. I live with Pascal, Emerson, Madame Gimon and Montaigne.

And if you don't know it yet, do me a favor, read 'Journal de Marie Bachkirtoff' (Paris, Charpentier). She resembles us both, she is halfway between us. [19]

Carel, my writing is tough, I am nowadays hardened, but underneath these words I feel my long soft hand of yore.

Your friend Bertus

[Signed autograph – in Scheltema]

1904-07-04

To C.S. Adama van Scheltema — 4.VII.1904

Sunday evening

Dear Carel, [Beste Carel,]

I haven't left yet; the long duration of the vibrations of restlessness, which always precede my departure, may indicate that my absence will be long and far. Anyway, I need it very much, not so much because of my physical health, which doesn't leave much to be desired, but rather for regaining the pure relation in which I have to place myself with respect to the various people and institutions within my narrow social horizon, so as not to be distracted from nurturing my abilities and the development of my clairvoyance in the service of God.— Finding the purities of human relations is rather a comprehensive task — nowadays there is more interaction between the unconscious lives of me and my environment than before. But even if the equilibrium isn't reached, work in that direction is the happy task in our lives.

Coming winter I will be in Blaricum — where a small house is being built for me [20] — working on a philosophical confession, which is going to

---

[18] In text 'Ebenbürtige' ('of equal birth').   [19] Marie Bachkirtoff (Bashkirtseff) was a much admired genius artist-intellectual, who died prematurely   [20] This is the so-called 'hut', designed by his friend Ru Mauve.

be the prologue of my work, or in London, in the great British Library, for my dissertation: 'The value of Mathematics' with the motto

$$Ο\dot{υ}δεις \ \dot{α}γεομετρικος \ εἰσιτω ^{(21)}$$

I thank you for your well-meant admonishment to me at the gate of the paradise of freedom. If I were looking for kingship on earth, it might be good to wall myself in mathematics, and have myself crowned like a pope in the Vatican, a prisoner on his throne. But I covet a Kingship in better regions, where not the goal but the motive of the heart is of primary importance.

We are not on earth for our pleasure, but with a mission that we have to render account for. And a small kingdom by the Grace of God is better than a large one by the will of the people.

But all this are tough thoughts of the heavy work of steering the ship, which after all floats on the clear lakes, protected by sun and lucky stars. And this heavy steering work is the direct punishment for a lack of confidence to surrender oneself to unknown powers that after all would the best for us to sail by.

Anyway, such a trust is the essential thing, and whether one directs oneself in good faith or allow oneself to be directed is more a matter of a name, which mainly depends on the nature of your own character, which is after all what you are trustingly relying on.

In this manner we both are looking for what is strongest and permanent in ourselves, hence true, and we want to make ourselves independent of the weak, the changeable, what is false, where we will never find support. However — we must wittingly every now and then let our hair down, because without paying every once in a while a small tribute, we cannot get rid of the devil. Letting our hair down we do with the necessary contempt, because we don't want to become chummy with the devil.

Carel, I hope that you can read under all these words a current which is the eternal content of the letter.

Greetings from your friend,

Bertus

[Signed autograph – in Scheltema]

---

(21) Let no one who is not a geometer enter.

1905-05-13

From D.J. Korteweg — 13.V.1905   **Hilversum**

Dear Brouwer [Waarde Brouwer]

You are certainly not mistaken that I take much interest in you, and
therefore I appreciate you sending me your booklet. [22] Whether I will read
it? I leafed through it, but it is not the kind of reading I wish for, or that is
good for me. It is true that right next to us there are those unfathomable
abysses, but I don't like walking along their edges. It makes me dizzy and
less able for what I have to do. Whether it is good for *you*, I doubt. That
much is certain, that I rather like seeing you walk different paths, even if it is
difficult to follow you there as well, where you dig so deep into fundamental
matters.

With cordial greetings

Your
D.J. Korteweg

[Signed autograph – in Brouwer]

1906-09-07a

To D.J. Korteweg — 7.IX.1906[a]   **Blaricum**
                                                                Torenlaan

Professor,

For some time I have been in Blaricum now; where I can more easily
devote all my time to my work. I have stopped reading others, and I am
now busy ordering my notes and arranging them into chapters.

I feel the more strengthened in my convictions, now that I observe that
I can fully stand by my notes of roughly two years ago, even now, after all
my reading of the intermediate period.

---

[22] [Brouwer 1905A], *Leven, Kunst en Mystiek/Art, life, and Mysticism.*

But now I can support them better with mathematical elaborations than at the time. I already have a publisher, [23] and to coerce myself, I have agreed with him that he can start printing in the beginning of October. Before that time I'll drop by you, to hear whether you wouldn't rather see the copy before it's printed; then I can still change as much as I want; and once they are page proofs, I am of course much more restricted.

Hoping you have had an agreeable summer, and looking forward to see you again soon.

With polite greetings

L.E.J. Brouwer

[Signed autograph – in Korteweg]

906-09-07b

To C.S. Adama van Scheltema — 7.IX.1906[b]   Blaricum

Dear Carel, [Beste Carel]

For quite some time we haven't heard from each other; Sunday afternoon I was in town, and saw you at half past 4 at the Leidscheplein [24] in the streetcar; I whistled, but in vain; I'm working hard on my dissertation; but for 5 weeks I have been suffering from a terrible root canal-toothache, from which I'm free only the last 14 days; it was almost unbearable; the last sign of you were those German professors eating coconuts on an island; [25] the island is very good, the food is still wrong, and those Germans will probably eat a lot too. Enclosed here I send you two English translations, [26] by the end of September there is again a new thing in the Academy, [27] and by the end of October the dissertation, many 'deeds', isn't it, speaking in your language. Oh, if they knew how little energy I have, and how afraid I am of all those doers of deeds.

Life is a magic garden. With wondrous softly shining flowers, but between the flowers there are the little gnomes, they frighten me so much, they stand on their heads, and the worst is, they call out to me that I should also stand on my head, every once in a while I try, and I die of embarrassment;

---

[23] Maas & van Suchtelen.   [24] Square in front of the City Playhouse in Amsterdam.
[25] A newspaper cutting.   [26] [Brouwer 1906A2], [Brouwer 1906b].   [27] [Brouwer 1906c].

but sometimes the gnomes shout that I am doing very well, and that I'm indeed a real gnome myself after all. But on no account I will ever fall for that.

Would you like tomorrow (Saturday) to go swimming with me in the sea? Not in Obelt [28]? Then I will arrive about ten o'clock in the morning at your place, and we walk there from Overveen all the way through the dunes. If you don't join me, that's all right too, but by all means you should do it.

Bye, old chap,

Bertus

[Signed autograph – in Scheltema]

1906-10-16

To D.J. Korteweg — 16.X.1906   **Blaricum**

Professor,

I have subdivided the material for the dissertation, as it is now before me, into 6 chapters.

1°) The construction [29] of mathematics.

2°) Its genesis in relation to experience.

3°) Its philosophical meaning.

4°) Its founding on axioms.

5°) Its value for society.

6°) Its value for the individual.

The survey in the first chapter I sent you serves mainly as a support for the next chapters, and to be able to refer to it, moreover to display various investigations in the foundations of mathematics from one single point of view, namely that of their meaning for the constructing mathematics. [30] A couple of things I have worked out a bit further, for instance the research of Hamel on the straight line as a minimal curve, because I need that in

---

[28] Public swimming pool in Amsterdam.   [29] Brouwer used a more colorful terminology here "opbouw", i.e. "erection" or "building".   [30] Brouwer uses the term "building mathematics". To translate this with "constructive mathematics" would twist Brouwer's intentions. He indeed referred to a mathematics that constructs its own objects.

principle in chapter 3 for opposing Russell; furthermore the construction of the group of fundamental operations on the continuum, because I want to present the construction of groups independent of differentiability as an essential part in the construction of mathematics; and then the deduction of the non-Euclidean arc element by means of the calculus of variations because I can't find that anywhere, and because it seems to me the only way to make that arc element *also for n dimensions* appear directly from what has been deduced for 2 dimensions. (The ordinary way is on the basis of formulas for geodetic curvatures according to the investigations of Christoffel and Lipschitz in Crelle 1870 and the following years, which become very complicated in n dimensions.)

I plan to come to Amsterdam tomorrow or the day after tomorrow in the morning, and to hear what you think of it.

Perhaps the chapter can be typeset at the end of the week, after I have gone through it once more.

Meanwhile I'm working on the next part, and I hope to send it to you as soon as possible.

With polite greetings,

L.E.J. Brouwer

[Signed autograph – in Korteweg]

906-11-05a

To D.J. Korteweg — 5.XI.1906^a [31]

Allow me to send you the enclosed volume of the Göttinger Nachrichten, in which Hilbert's Paris lecture, 'Mathematical Problems' is printed. Then you will see that in the first chapter of my dissertation I have given a complete treatment of n°. 1, ('Cantor's problem of the power of the continuum'), [32] and indeed by going back to the intuitive construction that has to exist for all of mathematics.

N°. 2 ('Consistency of the axioms of arithmetic') [33] is discussed in the last chapter, in so far there the solution of Hilbert himself, as given at the Heidelberg Congress [34] is rejected, and that for the one and only solution

------

[31] Letter without salutation.   [32] Cantor's *Problem von der Mächtigkeit des Continuums*.   [33] *Widerspruchslosigkeit der arithmetischen Axiomen*.   [34] Heidelberg Congress

one is referred back to the construction of the arithmetic on the continuum, as it is given in the first chapter by characterizing addition and multiplication as *the* twofold group.

I have also solved n°. 5 (Lie's notion of continuous transformation groups without the assumption of differentiability') [35] for a simple case (the two parameter linear group). Hilbert himself has treated another case (the three parameter group of plane motions) in the Mathematische Annalen 56. [36]

I am sending you this book, because I seemed to note that you doubted somewhat whether the subjects in my dissertation were really worth the effort.

Then, regarding your remark that the name of Kant doesn't belong in a mathematical dissertation: you will see that the 'Foundations' of Russell deal repeatedly with Kant, and that 'The principles of mathematics' [37] of Couturat are completed with an Appendix of over 100 pages about Kant. And when you compare the Transcendental Aesthetics [38] of Kant to these, you'll see that he speaks about exactly the same things as Russell and Couturat. And Poincaré points out that the present struggle about the foundations is a continuation of the old mathematical-philosophical controversy between Kant and Leibniz.

Even though the name of Kant can be avoided here, his subjects are touched upon; is it then necessary to avoid his name because he is known as a philosopher? You can't really qualify the books of Russell and Couturat as outside of mathematics? Virtually all mathematical periodicals with a bibliographic section have always reviewed them.

As to my words that you find so absurd, namely that astronomy is nothing but a convenient summary of causal sequences of readings on our measuring instruments; Poincaré says something of comparable intent (even though I haven't copied him) in 'Science et Hypothèse'. There we find: 'The earth rotates' has no other meaning than: 'To order several phenomena in a convenient way, it is very useful to assume that the earth rotates.' And I think that such a thing is far from being absurd, on the contrary it immediately convinces anybody who happens to read it. The system of celestial bodies is indeed nothing but a mathematical system freely built by ourselves; of which people are so proud, only because it serves to control the phenomena.

And also such propositions belong indeed to the subject, at least nobody will deny 'La Science et l'Hypothèse' a place within the faculty of mathe-

---

[35] *Lie's Begriff der continuirlichen Transformationsgruppen ohne Annahme der Differenzierbarkeit.*   [36] [Hilbert 1902].   [37] *Les Principes des Mathématiques.*
[38] *Tranzendentale Esthetik.*

matics and physics. Incidentally, various congress talks of Klein, Cantor, Boltzmann, and others treat this kind of subjects.

Finally, Sunday you said you weren't sure at all that I had studied Kant thoroughly enough to be able to make a judgment. Of course I cannot give you such certainty, but I can tell you that I read the 'Kritik der reinen Vernunft' [39] in its entirety, and that I have studied many parts (among which those that bear on my dissertation) repeatedly and seriously.

That my work is unclear, and its structure unpolished and that it shows traces of having been edited in haste, will probably be true, and also that there are here and there inaccuracies, but that the thoughts in it are vague and that the preparatory study has been superficial, I emphatically deny.

I would like so much that it will not end up bargaining between you and me about what can remain and what must go, but that you would rather sense and acknowledge the fundamental ideas; in other words more the general than the specific what's written between the lines as it were, even though your fundamental thoughts are different, though you find mine absurd,— because I am a child of a different epoch than you are.

You will recall that when two years ago I chose my subject it was not because I wasn't able to handle a more 'ordinary' one, but only because I felt an urge to take on this subject: it evolved spontaneously in me. You agreed, 'if there remained enough mathematics in it', probably suspecting that it would drive me strongly into philosophy, which it did, to the extent that I sometimes lost sight of mathematics altogether. But what I brought you now treats exclusively *how mathematics is rooted in life* and *how the starting points of the theory therefore ought to be*, and all special subjects in it receive their meaning in relation to that fundamental proposition. Taken by themselves, some of these subjects remain of value (for example the solution to the three problems of Hilbert mentioned above), but others become, when torn from their context, rather trivial, for example the survey of physics.

For me the essential part of the work is the general spirit. That is why I would like to send it into the world as a dissertation, which is fitting because of old a dissertation has the character of taking a position. The doctor's degree would give me satisfaction only if that spirit will be appreciated by my thesis advisor. With polite greetings,

L.E.J. Brouwer

[Signed autograph – in Korteweg]

---

[39] *Critique of Pure Reason.*

1906-11-06

To D.J. Korteweg — 6.XI.1906

Professor,

I will be then with you on Saturday morning at about 10 o'clock, but I don't want to wait so long in answering your letter.

When I had left you on Sunday, November 4, I did not feel upset at all, but in the next days the particulars of our conversation were ever stronger brought back to me, and they brought me more and more into a state of dejection. I believe that it was mainly the recollection of that paragraph, which you thought so absurd that you even cut off my words that tried to give a further explanation. Besides, in my imagination the parts that you wanted to delete were perhaps larger than they were in reality. — for I really was under the impression that I wasn't allowed to speak about Kant, because I thought I recalled you saying that you weren't sure that I had sufficiently acquainted myself with the literature about Kant. But the main thing was the first one; if you found something too absurd even to discuss it at all with me, then I probably have consciously associated with this the idea that you doubted the earnestness of the writing, and that therefore you doubted how the honesty and the thoroughness of the reflections that had led to it. That probably made me defend myself in that respect in my last letter, which originally was intended to consist of only a few words to go with the book I sent to you, but which, under the influence of the thoughts that haunted me those days, involuntarily expanded into what it became.

Even now I still would appreciate it that you would not think that paragraph to be 'too absurd for words'; so allow me to elaborate on it for just a moment. To be rehabilitated in your eyes matters more to me than to keep it at all in my dissertation, if it can be removed completely without damage to the whole — and in that respect I certainly believe it can.

You think (this in reference to that what I mistakenly thought to remember as having been judged absurd) that the general law of attraction has very little to do with the instruments that led to its discovery; but are laws anything but inductive summaries of phenomena, means to control the phenomena, and existing nowhere but in the human mind? Taken by itself the law of attraction only exists in reference to Euclidean space, and the latter only exists by a suitable but arbitrary extension of the domain of motion of solid bodies here on earth. Without solid bodies on earth the law of attraction couldn't exist, and the connection between the two is made

by astronomical measuring tools. The law of attraction exists with respect
to astronomical phenomena in the same way as molecules with respect to
the state equation; both turn out to be suitable for summarizing a group
of phenomena, and to be effective as a means of prediction; but the law of
attraction just prevails over the molecular theory with respect to simplicity.
But once more: the law of attraction is a hypothesis; the distance from earth
to the sun is just as well a hypothesis.

Now I would like to say something about the main issue, namely that a
similarity of laws in physics is to be expected on the basis of a similarity of
the instruments used, and I would like to start with the remark:

Projected on our measuring instruments, there is no distinction between
the electromagnetic field of a Leclancher element and a Daniell element;
but if we look at it with an open mind, we must expect that both fields
differ as much as copper sulphate and ammonium chloride; [40] only on our
counting- and measuring instinct, working with certain selected instruments,
they act identically there it appears that the same mathematical system can
be applied to both, but it is merely the lack of suitable instruments that
has so far stopped us from finding other mathematical systems that can be
applied to one field but not to the other.

In each phase of the development of physics the measuring instruments
that 'have been found suitable' remain a restricted collection, with respect
to the totality of measuring instruments that 'might be found suitable to
control all kinds of other yet unknown phenomena'; parallel with this 'the
mathematical systems that have already been applied to nature' form a
restricted whole compared to the totality of mathematics which 'would be
applicable to nature if only physics would have expanded sufficiently.'—
And since every group of mathematical systems has its invariants, it is to
be expected that every restricted group of phenomena of nature has its
invariants, precisely because of those restrictions, namely in the form of
laws or principles that are valid for all phenomena of that group.

Now someone could say: 'But why should we expect invariants for the
whole of present day physics; as this physics doesn't make any *specific* restriction at all, but it arbitrarily includes the most heterogeneous things in
its scope?'

This could be answered as follows: 'There actually is a *specific* restriction, because, after all, the mathematical laws that have been observed in
nature don't express anything but relations between measures, which all are

---

[40] The electrolytes in these two types of batteries.

taken from the group of rigid motions; only the *influences* to which those rigid measures are exposed, are freely varied. The other physical quantities are only auxiliary quantities that are suitably chosen for certain influences of the measures, and which through their introduction as coordinates, give a simple form to the equations of state. The physical quantities are indeed never measured themselves, only the rigid measures are, in the fictive context by means of which they have been introduced; for example, one didn't measure magnetic forces and currents, but torsion angles of silk threads, and the angle-measure is based on the group of rigid motions.— And also: speaking about equivalent things, or about circumstances without influence, we always mean: with respect to our readings on measuring instruments. is only one thing that can be stated as an empirical truth by itself, namely: the group of motions of solid bodies has roughly such and such properties, and those remain roughly constant in time.'

— 'But we do measure after all things other than rigid measures; for example amounts of electricity; can't we for example give a conductor consecutively equal charges, by discharging a charged globule on it that has been charged twice in exactly the same manner, and don't we know then that the charge after the second discharge is double the amount after the first discharge?'

— 'No; because to what extent can we speak about *quantities of electricity*, in other words, to what extent can the effects of consecutively applied equal charges be superposed as equal effects? For example to the extent that they give cumulative effects on the torsion balance of Coulomb. But in how far can we superposition forces that give equal torsions there? In as far as they balance equal copper weights. But to what extent can we superpose the weights of equal pieces of copper? To the extent that the accelerations that they give rise to in the same body(for example in the Atwood machine) can be superposed. But those accelerations are only observed in solid bodies; for, both velocities and accelerations are observed in the rigid group. And this remains the case for weights of fluids, we measure them either by volume — and that is measured on the basis of the rigid group — or the weights are transferred as forces to a solid body, for example a balance or a piston.'

In this manner every physical measurement is in the end reduced to a measurement in the rigid group; and in fact *the laws of these measurements are examined in all kinds of different circumstances.* So we can really expect a *specific* restriction on physically applicable mathematical systems, and the existence of invariant principles shouldn't surprise us. Just as an organ pipe refuses to resonate with other than specific notes, we may expect that the rigid group refuses to *resonate* with other phenomena than those which satisfy the principles of energy, action and thermodynamics.— The more generally unknown, which lies outside, could still manifest itself in the physical laws as all kinds of 'contingent' constants, as unexplained atomic weights, dielectric constants, frequencies, specific weights etc., and also the 'accidental' fact that the laws are the way they are and not different.

Maybe you find in these arguments a weak spot, but in any case, they show that my statement is more than a vague feeling, and not merely founded in a pessimistic outlook. — To conclude, please regard this letter, and also the previous one, as inspired only by the apprehension that I might have to give up the empathy with you with respect to the subject, and by the deeply felt wish to preserve that as much as possible, also in all its parts.

With polite greetings,

L.E.J. Brouwer

[Signed autograph – in Korteweg; draft in Brouwer]

906-11-11

From D.J. Korteweg — 11.XI.1906 [41]

Dear Brouwer, [Waarde Brouwer]

I have now also made myself acquainted with your third chapter. The result is very satisfactory. I find a lot of beautiful things in it. I would prefer rather that some things would be expressed a bit less crudely, as this can only bring a note of passion where it doesn't belong and that a few statements be expressed somewhat less absolute.

For example it seems to me that you can't object that strongly against the logical figures by themselves as attempts, outside of mathematics, to analyze and classify the way people reason, and that, if you do, you are going beyond your subject. But all of that concerns just a few sentences or even words. For the rest see later on about the final part.

---

[41] The archive contains some drafts and notes for this letter. One carries a note 'eene andere redactie, waarschijnlijk meer overeenkomend met het verzondenen.' (another version, probably more conform the letter that was sent).

Concerning the first chapter 'the construction', you know that I would wish here only clarification, which you have declared yourself to be prepared to make an effort.

So only the second chapter remains.

After receiving your letter I have again considered whether I could accept it as it is before me. But really Brouwer, this won't do. A kind of pessimistic and mystic philosophy of life has been woven into it, that is no longer mathematics, and has also nothing to do with the foundations of mathematics. It may here and there have coalesced in your mind with mathematics, but that is wholly subjective. One can in *that* respect totally differ with you, and yet completely share your views on the foundations of mathematics. I am convinced that every supervisor, young or old, sharing or not sharing your philosophy of life, would object to its incorporation in a mathematical dissertation.

In my opinion your dissertation can only gain by removing it. It now gives it a character of bizarreness which can only harm it. It doesn't come back in the third chapter, except on a single page at the end, which therefore of course also should be deleted because it wouldn't remain comprehensible any longer.

I have tried to indicate how it could be removed from chapter 2. Take this in at your leisure, and try along these lines to make something out of it that *you* too find worth preserving.

I would regret it if this were impossible, because I find much that is good and to the point in some of your expositions and in your treatment of Russell's book, including in the conclusions you draw about Kant's considerations about the aprioristic in mathematics.

I guess that you now better understand my objection. Your last letter was a great disappointment because it shows all kinds of misunderstandings. This pained me all the more, as I was under the impression that we understood each other quite well last Sunday.

You inform me of all sorts of matters which could not possibly be unknown to me, as a regular reviewer [42] of the Revue de Métaphysique et de Morale, as if they were things that I would not know. You thought to have understood that you were not allowed to use the name of Kant, even where it concerned opinions of Kant on mathematics, and you thought that I found the view 'that astronomy is nothing but a convenient summary of causal sequences in the reading of our measuring instruments' absurd. No, not *that* view; I admit that one can present the matter in that way, although in my

---

[42] for the *Revue semestrielle des publications mathématiques.*

opinion the general law of gravity has indeed little to do with our measuring instruments which led to its discovery, than that these make measurement possible at all; but that the similarity of the laws which are valid in very different parts of physics would find its origin in the similarity of the used instruments; it was that claim that appeared absurd to me.

You also thought that I suspected that your preparatory studies had been superficial. This can only be caused by the explanations I asked from you (and which I most likely will ask every now and then). But the aim of these were, apart from enlightening myself, that I would be able to state (and that might become necessary) that I have repeatedly asked you for explanations, and that I have found each time that you had solid. Personally I don't doubt that at all.

Enough now! I am very busy this week and I prefer to see you on coming Saturday November 17. I'll keep the entire morning free for you, and I want to ask you some more particulars. Hopefully your revision of the first part is ready then.

Greeting,

Your
D.J. Korteweg

[Signed autograph – in Brouwer.]

907-01-10a

From D.J. Korteweg — 10.I.1907[a]   Amsterdam

Dear Brouwer, [Waarde Brouwer]

On the enclosed sheet I have only a couple of very insignificant corrections to propose, as you will see.

Meanwhile the question came up with me, can the proof on page 86 be given for higher differential quotients?

For example, let φ be the coordinate of an ordinary Weierstrass curve without differential quotient $\varphi = C + \sum_0^\infty b_n cos(a^n x\pi)$, then C can be taken such that φ is always positive, which makes it easier for me.

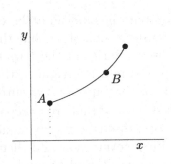

Now consider the curve that starts from an arbitrary point A and for which $\frac{dy}{dx} = \int \varphi dx$. That integral *exists*, at least Klein says that every 'continuous function' [43] admits such an integral. The curve that is thus obtained in this fashion certainly has certainly no second differential quotient, but doesn't it satisfy your requirement? In my opinion it does, because the differential quotient steadily increases within boundaries that become smaller as one considers a smaller part of the curve. In my view one has here indeed for nearby argument points (like A and B) approximately equal values for all [first] [44] differential quotients because they all 'continuously' approximate the differential quotients [45] when the increase becomes smaller.

But maybe I haven't understood you well.

Greetings,

Your
D.J. Korteweg

[Signed autograph – in Brouwer]

_____

1907-01-10b

To D.J. Korteweg — 10.I.1907[b]

Professor,

The curve you indicate is really one for which the proof of page 86ff. can be given with respect to the first differential quotient, and not with

_____

[43] '*stetige Function*' in letter.   [44] Korteweg's brackets.   [45] '*stetig*' in letter.

respect to the higher differential quotients. But it does not fully satisfy the requirements I formulated.

'In nearby argument points roughly equal behavior' means that all functions of the independent variable determined by the curve should be *continuous* [46] (this is indeed expressed unambiguously by my words); now I know about a continuous curve that its difference quotients *exist*, they are functions that are determined by the curve, hence by my assumption *continuous*.

The proof of page 86ff. shows that the first differential quotient *exists*, hence it is a function determined by the original curve and therefore it is (by my assumption) *continuous*, hence is has *existing* difference quotients, which now are also functions determined by the original curve, hence (by my assumption) continuous, and according to the proof of page 86ff. the *existence* of the *second* differential quotient is deduced. We can continue in this way; first it follows from the *assumption* on page 86 that the second differential quotient and its difference quotients are *continuous*. Then from the *proof* of page 86ff. the third differential quotient *exists*, etc.

The function you indicate doesn't satisfy my postulate; for there *exist* functions determined by the curve (difference quotients of the differential quotient) that are not continuous. If these functions were continuous, then the second differential quotient would exist too.

With polite greetings

L.E.J. Brouwer

[Signed autograph – in Korteweg]

907-01-11

To D.J. Korteweg — 11.I.1907 (in the morning)

Professor,

It occurs to me that my letter to you of last evening was somewhat incomplete, and that the question could be raised why I formulated my requirement on page 86 as 'in nearby points roughly equal behavior' and not as 'continuity of functions determined by the curve'.

---

[46] Everywhere in this letter Brouwer uses '*stetig*' for continuous.

The former does mean the latter, but one thinks then only of functions determined by physical measurements (or continuous operations on the results of those measurements), so *observed* functions. To these belong among others difference quotients and the various differential quotients, which, if they exist, can be approximated by measurements of $\Delta f, \Delta^2 f, \Delta^3 f$, etc.

However, in the latter phrasing one might include all kinds of arbitrary *mathematical* functions that I might construct from the ordinate in my mathematical imagination, and of course in that sense the postulate can never be satisfied. *Each mathematical* curve certainly determines *mathematical* discontinuous functions. I thought I expressed correctly what I meant with the word *behavior* of physically observed quantities, without the need for further elaborations.

All the more, because it was a vague feeling in people, that I pointed out, of which they have not made for themselves an outline of the precise mathematical purport I use in my proof will *indeed* lie inside that not sharply delineated domain.

Of course much can be said in addition to the short indication I gave; this is also true for quite a few other subjects that are treated in the second chapter. Maybe the reason is that in my head they were only accessory offshoots of a unifying fundamental idea (which isn't in the dissertation anymore), hence they only had secondary importance.

After their sudden appearance on the foreground as replacement of their former leader, it wasn't on the spur of the moment possible to dress them all up so that, left to themselves, could together save the whole performance.

At least that is what I occasionally feel when I take a look at the chapter. On the other hand, I more and more understand that thoughts, in the form I wrote them down at first, would have completely interfered with the mathematical tone in a mathematical dissertation, and I have tried to write about the connection between mathematics and experience as thoroughly and non-trivially as possible, while excluding these thoughts completely.

Now this letter has become somewhat longer than I thought.

The printer seems to be dawdling again; the last few days I haven't received anything. But maybe the printing of the first sheets takes up his time.

With polite greetings,

L.E.J. Brouwer

[Signed autograph – in Korteweg]

907-01-18

To D.J. Korteweg — 18.I.1907

Professor,

After consideration of your remarks about page 128, I also think that there the mathematical tone is obstructed, and I have deleted the sentences that implied a 'judgement'.

Also my remark that theoretical logic is not oriented towards the external world wasn't felicitous; by saying 'directed to the external world, to control or to oppose it' I roughly meant 'having practical applications in the external world', but one does not gather that from it.

From your characterization of theoretical logic as part of psychology I gathered that I had expressed myself rather vaguely, because it was actually my intention to show that theoretical logic on no account has a psychological meaning, even though it is a science.

I have therefore provided a few additions to page 128, and also reworked the last lines of page 127, so in the end a completely new and more extensive presentation of the subject resulted, and I will send you the proof again when I get it back, together with the proof of sheet 9. As I sent it yesterday evening to Nijmegen, I expect to have it back again tomorrow night, so I hope you will have it on Sunday morning.

The matter of differentiability of physical functions on sheet 6 has received an – adequate, I hope – supplement.

With polite greetings

L.E.J. Brouwer

[Signed autograph – in Korteweg]

907-01-23

To D.J. Korteweg — 23.I.1907

Professor,

In order to lose no more time, I will, with your permission, make no more changes in the text, but I would just like to answer your objections and try to bring you a little more on my side in the hitherto questionable subject.

In the beginning of the chapter [47] I show that mathematical reasoning is *not* logical reasoning, and that it is just out of poverty of language that mathematical reasoning uses the connectives of logical reasoning; and thereby perhaps will keep linguistic accompaniment of the logical arguments alive, long after the human intellect will have outgrown logical reasoning. Far from it being 'queer folk' that does not argue logically, I do believe that it's just a phenomenon of inertia that the corresponding words still exist in modern languages. Pure usage of those words only rarely occurs, and impurely they are used in daily life, where they have led to all kinds of misunderstandings and dogmatism, and in mathematics to the misconceptions of set theory. [48] Those misconceptions did *not* arise through insufficient mathematical insight, but because mathematics, lacking a pure language, *has to make do with the language of logical reasonings*; whereas *the thoughts of mathematics don't reason logically, but mathematically*, which is something completely different.

The theorem: *If a triangle is isosceles, it is acute-angled* [49] is used as a logical theorem — the predicate *isosceles* is considered to *imply* for triangles the predicate acute-angled, in other words, one imagines all triangles (of a flat surface [50] for example) pictured as points of an R_6, and then one sees that the region of R_6 that represents the isosceles triangles lies inside the region that represents the acute-angled ones. This is in this case really true, so the logical formulations of the logical language can be safely used here.

But the mathematician who, because of the poverty of language, phrases the above mentioned theorem as a logical theorem, thinks something different from the logical interpretation just mentioned. He imagines that he starts to construct an isosceles triangle, and then that after the construction either the angles will turn out to be acute, or the construction doesn't *succeed* if a right or obtuse angle is postulated. In other words, he gives the theorem in his mind a mathematical, not a logical interpretation. It is precisely the main content of the 3rd chapter to show that the naive use of a logical language rather than a mathematical one has led parts of mathematics astray.

---

[47] Chapter 3.   [48] '*Mengenlehre*' in the letter.   [49] To be precise: the angles opposite the equal sides are acute. Perhaps Brouwer had equilateral in mind. For the following the distinction is not important.   [50] The Euclidean plane will do.

Now let me briefly elucidate why I believe that logical language is obsolete; by the way, we discussed that the day before yesterday. The mathematical systems that are applied to the world, and that are thus the only ones that qualify for representation in language, will have to teach us something practical through their mathematical theory. But *the mathematics of whole and part* doesn't teach us through its theory anything new for applications. Once the system is applied to a part of the observational world, even a very mediocre intellect can immediately read off all consequences: no intermediary logical reasoning is necessary. Nowadays one knows very well that if one deduces something about the external world by logical reasoning that wasn't immediately clear a priori, then exactly because of that it is totally unreliable; because one doesn't believe anymore in the postulate on which it is based, that the world consists of an admittedly very large, but finite number of atoms, and that hence each word represents a (therefore also finite) group or group of groups of atoms. In other words, one knows very well that the world is not a logical system and one cannot argue logically about it; one knows very well that in the end every debate is hogwash and only can be decided for mathematical problems, but then *not by logical arguments* (even if it seems so in a deficient language; how false that appearance is, is shown in the case of the axiomatic foundations and the transfinite numbers), *but by mathematical arguments*.

Theoretical logic doesn't teach anything in the present day world, and people know this, at least the sensible people; it serves only lawyers and demagogues, not to instruct other people but to deceive them, and that is because the vulgar herd unconsciously reasons: that language with its logical figures is there, so it will be useful and so they meekly let themselves be deceived; just as I heard several people defend their habit of gin drinking with the words: 'What else is gin for?' Whoever has the illusion to improve the world, may just as well agitate against the language of logical reasoning as against alcohol; and just as little it would be a 'queer folk' that doesn't drink alcohol, it would be a 'queer folk' that doesn't argue logically; I believe though, that maybe no abuse is rooted deeper than what has grown together with the most popular parts of the language.

Your question about the word *exists* on page 141, line 5 from below is maybe answered with the example that is given 4 lines down. In the conditions is comprised that it is a *finite* number, i.e. a well known mathematical thing that can be searched for; but it isn't certain that the conditions can be fulfilled, in other words, that the mathematical thing *exists*.

Maybe I have after all stated my intentions in a clearer way in this somewhat wildly written letter than I succeeded in doing in my modest text. But perhaps after this letter the text will be seen in a different light. That would please me very much. With friendly greetings,

Sincerely,
L.E.J. Brouwer

[Signed autograph – in Korteweg]

1908-02-21

To C.S. Adama van Scheltema — 21.II.1908   Blaricum

Dear Carel, [Beste Carel]

There is always such a considerateness in you, to write to me on the anniversary of what may have been my last passage under the yoke of society. [51] But independent of that, I was happy with your call across the misty sea of life where our view is so limited, but where we feel so clearly the climate change, and see our lodestar ever more clearly and follow its guidance more patiently. Yet we will meet enough surprises, and the end and afterimage will remain exciting enough. What a marvelous sensation it will be to survey the whole episode in one's dying moment and to understand it as the purest novel. Sometimes I really long for it, when I've lost my ambition and mood for a while.

A month ago I thought almost that the time had come. I had skated to Rotterdam against the wind over bad ice, and back the next day. Three days afterwards I got a fever as I have never had in my life before. After a week it was gone, just like that. Meanwhile I was weakened so much that I'm still in bed the whole day, and I have no wish to get out of it ever. I also had my last will drawn up, but my state of mind was mainly affected by the impression of the pitiful lackey's role of the notary public in the whole social theatre, the only one of the players who isn't human, but a

---

[51] The public defense of the dissertation - 19.II.1907.

mere acting machine. The way he sat at my bedside gave me tears in my
eyes.

But now I believe that I shall and may have to behold and experience
various things, before I can say: 'at evening time it shall be light.' [52]

Do you think it a sign of weakness, that you can't maintain your mask?
What if that was caused by your continual urge to be honest, forced upon
you by your work?

What do you think of this little poem by De Génestet [53] for people who
want to be part of the movements of society

> Let each grow glad and quiet
> His happiness in the world
> If the rose adorns itself, it
> adorns the garden too. [54]

The 'Penseur' [55] is a beautiful thing. Only it is certain that a person
who sits like that will forever lose the balance in his head and never find
truth. He is much to active for that and he has too little self-confidence. He
has almost tumbled from the 'sublime' into the 'ridiculous'. He arrived here
when I was sick already, therefore you didn't hear anything about it. I also
got that crazy (but not disgustingly crazy) thing in Brusse's advertisement
booklet.

And the promise? [56] Well, Carel, I gave you my word that you will see
its fulfillment before you need it, and foi de gentilhomme, [57] I will keep it.
I don't need to be reminded of promises, didn't you ever notice that?

About or to you wife I have been silent, but now she has become such a
dear to you, all parts of the letter that are hers will flow to her. Bye now,
je t'aime toujours, [58]

Your Bertus

[Signed autograph – in Scheltema]

---

[52] *'Als het avond wordt, wordt het licht'*. Zechariah 14:7.   [53] P.A. de Génestet
(1829–1861), popular Dutch poet.   [54] *Rückert's egoïsme.*   [55] Rodin's Thinker. Probably a picture postcard sent by Scheltema.   [56] Cf. *Scheltema to Brouwer 11.I.1908*;
Scheltema, traumatized by the suffering of his father, feared a slow, lingering death; he
had asked Brouwer for a pill. See [Van Dalen 1999], p. 202.   [57] My word as a gentleman.
[58] I always love you.

1908-05-00b

To D.J. Korteweg — V.1908^{b} [59]

Professor

Do you have by any chance

U. Dini 'Grundzüge der theorie der Funktionen einer veränder-
lichen reellen Grösze'. [60]

I would very much like to consult it and probably quote it for the article for
the Academy. [61] I used to get it from the library, but now it has been lent,
and the man at the desk is very strict about the lending secrecy.

If you don't have it, could you, maybe Monday or so when you pass by
the library, ask in your capacity of Librarian of the Society [62] who has got
the book? Then I could ask that person, if we know him, to lend me the
book for a day.

I received the issues of Rennes; the next week Tuesday or Wednesday I
will have finished these, if nothing comes up. [63]

With polite greetings

L.E.J. Brouwer

[Signed autograph – in Korteweg]

————————

1908-06-08

To D.J. Korteweg — 8.VI.1908   Blaricum [64]

Professor,

After your last letter I started to hesitate dispatching the application
and finally I didn't do it at all, when I thought I had clearly understood
that there was no need whatsoever for me as a teacher, and that my being

---

[59] Undated; the month May is plausible as Brouwer needed the book for a paper that
was to be submitted on 30.V.   [60] Fundamentals of the theory of functions of one real
variable. At the time the basic text on the subject.   [61] [Brouwer 1908a], p. 59.   [62] Dutch
Mathematical Society.   [63] Concerns Brouwer's reviewing for the Revue semestrielle.
[64] The letter is erroneously dated 1907.

a privaatdocent [65] would be of service to nobody, it would only dissipate my strength. For, where I have dedicated myself to our mathematics, to achieve the best that my abilities will enable me, I have too much a sense of responsibility to be diverted from my work, when I cannot really be of any *service*. That I will, however, at all times be prepared to do the latter — I hope you will remain convinced of that.

I always have considered privaatdocents in optional subjects to be nothing but careerists, [66] and perhaps you know me well enough to know that I don't care about a social position; if you didn't know that side of me, I would set store by telling you so explicitly.

I'm sorry that you didn't get this impression about me right away this winter, but I have never had anything else in mind; and I also understood from you that it really was the intention to charge me with a few parts that actually belonged to the curriculum but that weren't done enough justice, as formerly De Vries and Coelingh. After the appointment of Van Laar I thought my privaatdocent position had certainly become superfluous, and I wondered about it when you subsequently brought it up again.

As regards my personal desires, I hope that I can stay away from teaching, at least as long as I am still such an immature mathematician, as I am now, and also afterwards if there is no need for me. Anyhow I will never be able to give courses, where I am dependent on the pleasure of the audience. I have consecutively attended such courses held by Van der Waals Jr., Mannoury and Kohnstamm, and I have always clearly felt incapable of something like that.

Of course I am well aware that you had nothing but my best interests in mind with your proposal, and I am grateful for that, but it would push me in the direction of a career that is alien to my desires; I sincerely hope that you won't hold my point of view against me, and that your interest in, and friendship for me will not be less on this account.

L.E.J. Brouwer.

[Signed autograph – in Korteweg]

---

[65] A *privaatdocent* was an academic teacher (as a rule with a PhD), who is allowed to give courses at the university for a nominal salary. It was usually a position for young, promising scholars, but it could equally well be used to get teachers for routine lectures. The German equivalent, *Privatdozent*, required a Habilitation, and as a rule had more status. [66] '*Strebers*' in the text.

1908-06-24

To C.S. Adama van Scheltema — 24.VI.1908   Blaricum

Dear Carel, [Beste Carel]

Being about to send you a mathematical reprint, I would at the same time just like to shake your hand. My trip to Paris was impossible until now because of various activities and worries that already now are tying me down in the world of my profession: looking back I notice how I am gradually encapsulated by them until I tear myself loose with a vigorous pull. Korteweg [67] and De Vries [68] want to make me privaatdocent after the vacation, I want to escape, I resist, I make conditions ... and in the end I'll give in perhaps: I love that subject, and why not serve it in society as well; what's a God without altars on earth? And if I might be more of a philosopher than a mathematician, it will break some day also out of that straightjacket.

The review that I will write about you in the Journal for Philosophy [69] is settled. I will get some eight pages, but it will depend on circumstances whether I will make it that long. The first weeks I will have no time for it: I didn't even finish your book yet. The main point is that I think it so beautiful and necessary of you to have written that book; what should a man be but a growing philosophy of life, as a tree full of lifeblood, that in the fall drops some fruits every now and then — his work. For the rest I completely agree with the main tenor, and I think it is a badly needed purge of the market, healthily simple and yet never said before. But you reject too much the doubting poetry, and what do you mean in your foreword where you identify specialized investigations with scientific investigation? You may choose your terms any way you wish, but really you don't want to accuse scientific instincts such as those of Goethe, Schopenhauer and Helmholtz, which mine, I believe, resembles, of a tendency to specialization or being subject to specialization, don't you? [70]

Maybe I can come now in about 14 days — an issue of a journal that I have to do quite a lot of paperwork for [71] must first appear—: I sometimes long so much for a summer evening with a chilly wind from the north in a Paris suburb, just cold enough for the girls in their white blouses to quicken

---

[67] Brouwer's Ph.D. advisor.   [68] Hk. de Vries, mathematics professor in Amsterdam. [69] *Tijdschrift voor Wijsbegeerte*; refers to [Adama van Scheltema 1907].   [70] Brouwer strongly objected to what he called "verbijzondering", specializing, in the sense that the attention was fixed on isolated details and phenomena, instead of keeping the full image in mind.   [71] Brouwer assisted Korteweg in editing the '*Revue semestrielle*'.

their step and get color on their cheeks. Also I long to see the Louvre again, and I hope to see a couple of decent statues there; did I tell you that Rome and Naples have almost only bad statues? And that it all was compensated by the Greek temple in Paestum, and that I wept there because of nostalgia, as when I read Aeschylus? Rome never had any sentiment. One can see that already from the landscape: in such a place only dogmatism and sensuality can grow, whereas in Tuscany a quiet humanity breathes, as in the past it must have been only in Attica; I am terribly curious whether *that* country will give me that intimate hallowed feeling if I get there later.

I hope I will find you strong of mind; your book sufficiently indicates so; I very much long to see you in your home with your wife who understands you and supports you. For the rest life is difficult, dreamlike and happy. Let us live it to the end in joy.

Bye Carel, and greetings to your wife too,

from Bertus

[Signed autograph – in Scheltema]

908-11-08

From D.J. Korteweg — 8.XI.1908

Amice

Thank you for showing me your correspondence with Schoute. [72] He told me more about his plan and you certainly have not committed an indiscretion.

I don't want at all to disapprove of your decision. What you had to take upon yourself is of such magnitude indeed, that it would interrupt your present study completely and it would draw you completely out of your present way of life. Whether this would contribute to your happiness or be of advantage to science, is impossible to judge.

There is just one aspect of your considerations I would like to react to. It concerns what you write about the 'loftiness' of the position of a professor. In my view the position of a professor is neither higher, or lower than any other. Just as for any other job, one has to ask for a man that will best

---

[72] Professor in mathematics in Groningen.

fill the role. It is clear that scientific merits and scientific insight are to be considered in the first place.

But that's not the *only* thing.

If we had just one large university instead of four, then matters would have been different. Then one would enjoy a luxury of professors where every highly qualified scientific man — because there aren't that many— can find his place.

Now we enjoy a luxury in a different respect, that is, there are four universities at relatively close distance, where similar courses are given.

And almost every professorship in mathematics has, which you correctly point out, a scope and meaning for all kinds of students and for the regular routines of the university, which forces [the university (ed.)] to make also other demands.

I am not saying that you would not satisfy those demands. But that they will be made seems unavoidable. I myself would think that one were to fail one's duty if one didn't make them.

Anyway, professor or not, a person's value depends on what one is as a human being and as a scientific man.

Greetings

Your
D.J. Korteweg

[Signed autograph – in Brouwer]

1909-03-01

To C.S. Adama van Scheltema — 1.III.1909   Amsterdam

Dear Carel, [Beste Carel]

Forgive me, that I forgot your birthday. I also forgot mine. [73] But I don't forget you, and I will not forget you; you have been too much a person in my life for that. Also my two promises to you are still in my head.

A while ago I stood at a crossroads: I was asked to take temporarily the place of a Groningen professor who had died; I did not qualify for a permanent position, as I never had taught. Thank God I had the strength to say no, which meant that at least temporarily I have almost cut off myself

---

[73] birthday Scheltema: 26 February; birthday Brouwer: 27 February.

from a social career. Actually, it wasn't a very attractive offer: after the course they would probably appoint a Delft professor in Groningen, and then I could have gone to Delft, where as professor you are something as a supervisor of drawing lessons.

Please remember that soundings for professors vacancies are confidential: except for my wife and brother you are the only one who knows this from me.

Lily's accident [74] has deeply moved me, she pulls through well, and she has gone to Paris with my brother; he is doing mineralogical research there, and when this is finished, they'll probably marry soon. [75]

Bye now, dear chap; live in happiness with your wife and your work; build up your view of life as a Cathedral; I can't, but on day I like to sit on its steps like a roaming pilgrim.

Will you inform me of your address, when you leave? Warmest greetings to Annie.

Your Bertus

[Signed autograph – in Scheltema]

909-03-16

To D.J. Korteweg — 16.III.1909

Professor,

May I once more ask for space in the Proceedings for the enclosed short article? [76]

I have been for a long time in doubt, whether it would be desirable to add figures; but as the text is now, I believe it doesn't need support, and in that case figures would make the ideas unnecessarily specific.

With respectful greetings

L.E.J. Brouwer

[Signed autograph – in Korteweg]

---

[74] The fiancee of Brouwer's brother Aldert lost both her legs in an accident.
[75] Marriage of Aldert Brouwer and Lily van der Spil, 2.VI.1909.   [76] Probably [Brouwer 1909e, Brouwer 1909d], the paper does contain a figure.

1909-05-27

From A. Schoenflies — 27.V.1909   Königsberg

Haarbrückerstrasse 12

Dear Doctor, [Sehr geehrter Herr Dr] [77]

First of all I would like to thank you most cordially for sending me a copy of your article submitted to the editors of the Annalen. [78] Please accept my apologies that I do so only now, I first wanted to think matters over thoroughly. I hardly need say that I am highly interested in your important results.

My joy that you have studied my article so thoroughly is however not without a bitter taste. Because I see that I overlooked one of the possible shapes, namely of domains that in my terminology do not have a closed curve as boundary. The error in the proof is on page 123, where I work with the set $\mathfrak{T}_{gh}$. This can be empty — as indeed it is in your Figure 1 — and then the conclusion that is based on that set fails.

I have addressed the question to what extent my results and methods therefore must be modified, in particular the presentation in § 3ff of Chapter V. With a few changes in the arrangement and the proofs they remain valid in the following manner:

1) First the theorems in § 3 must, as you stress yourself, be restricted to such closed curves etc. that can be decomposed into arcs without common points (except for the end points). It is even sufficient to prove them first for *polygons* etc. In that way I define the concept 'proper arc' — which is the only thing I need — from the outset in such a way that its complement is a single domain. Only that will be considered for my treatment.

I have changed it in the following way, by which it remains applicable to all cases.

---

[77] The letter looks more like a draft than a letter. There are many insertions, corrections, crossed out passages. The fact that a newcomer had found serious gaps and errors in the first exposition of point-set theory – topology, we would say – by the leading expert, must have had a devastating effect on Schoenflies. The document even lacks the obligatory polite closing sentences and a signature. In view of the fact that a man with Schoenflies' reputation would hardly send a letter of this sort, one might consider the possibility that these sheets were indeed his draft pages, and that for some reason Brouwer later obtained them from Schoenflies. This would not be a quite satisfactory explanation, for then Brouwer would doubtlessly have asked for (copies of) his own letters, which are not extant. Brouwer had sent the original letter of Schoenflies to Korteweg (see *Brouwer to Korteweg 18.VI.1909*); unfortunately it is not in the Korteweg archive. Both Brouwer and Schoenflies repeatedly appealed to Hilbert, sending him the letters of the other party; it is not clear what exactly happened to their letters.   [78] *Mathematische Annalen.*

One constructs in the domain $\mathfrak{H}'$ a polygon $\mathfrak{P}'$ that approximates its boundary $\mathfrak{T}'$ to within a distance ε. This divides the domain $\mathfrak{H}'$ into a ring shaped domain $\mathfrak{R}'$ and a complementary domain $\mathfrak{H}''$.

Then one concludes again that by continuity one can chose the polygon P so near to the given curve $\mathfrak{C}$ — which I assume to satisfy the more restricted sense — that the image set P' is completely contained in the ring domain $\mathfrak{R}'$. Now we only have to distinguish two cases, namely the remainder domain $\mathfrak{G}''$ is separated from $\mathfrak{T}'$ by P' or not.

In the first case the proof can proceed as in the article.

In the second case one obtains the contradiction by working with *two* points a'; for the sake of simplicity I use the images of two such points a_1 and a_2 of $\mathfrak{A}_1$ which are near to two points c_1 and c_2 of $\mathfrak{C}$ whose distance $\rho(c_1, c_2)$ is a maximum, and moreover I choose them such that 1) also the lines $a_1 c_1$ and $a_2 c_2$ are completely inside $\mathfrak{A}$, which fits the assumptions of § 2, and also 2) the image sets C_1' and C_2' of these lines have no point in common — which is possible because of continuity. That done, one chooses again ε such that both a_1' and a_2'' belong to the remainder domain $\mathfrak{H}''$. [79] The contradiction is then obtained, because both image sets C_1' and C_2' must on the one hand penetrate into the ring domain $\mathfrak{R}'$, but on the other hand they have no point in common; and they also may not contain a point of P'. From this it would follow that P' couldn't be a continuum. This remains true also when $\mathfrak{H}''$ is separated from $\mathfrak{T}'$ by P'.

The precise proof can anyway proceed by assuming first a_1 and a_2, then choosing ε, thereby determining the ring domain $\mathfrak{R}'$, and then one chooses P such that P' is in the ring domain, and then one argues as before. So one doesn't have to distinguish two cases at all.

The proofs of Theorems 1 and 2 on § 3 then still work — by the way, I see that on page 159 line 13/14 from below it should say: ... 'that C_1' and C_2' must be identical to, *or one a subset of* the other.'

However, Theorem V on page 160 can not be inferred at this point.

On the other hand, Theorem VI, which expresses the invariance of the order, still is valid, and also the general course of its proof. First a general remark about this proof. On page 160/161 it would be better to state the conclusion of the proof — to which you have objected — at the outset; then the basic idea of it will be clearer. Indeed, one always starts from a *specific* point, and then one chooses the quantities σ etc. namely as follows. (Line 3). Then immediately line 10ff.

---

[79] [In the margin, with the comment '*left out*']: 'i.e. $\varepsilon < \frac{\sigma_1}{3}$ and $\varepsilon < \frac{\sigma_2}{3}$ in case $\rho(a_1', \mathfrak{T}') = \sigma_1$ and $\rho(a_2', \mathfrak{T}') = \sigma_2$.'

This proof can be transferred without any problems to the more general interpretation that one has to give then to the image $\mathfrak{C}'$ of the curve $\mathfrak{C}$. Because from the auxiliary theorems of § 3 it doesn't only follow that $\mathfrak{C}'$ is a set that decomposes the plane $\mathfrak{E}'$ into *at least* two domains; but more isn't necessary for § 4. One of these is the exterior and one the interior of what is bounded by the border $\mathfrak{T}'$. One doesn't need to suppose or know more for the discussions of § 4. Of course Theorem VI only holds insofar as it relates to the respective single domains.

The theorem in § 5 can be proved as before, with the small addition indicated by you, to exclude that in the image plane there are domains that stay disjoint from the boundary curves $P_1', P_2' \ldots$.

With the invariance of the domain we have the basis for the rest, also for Theorem V which now can be concluded.

As you see, essentially one only has to avoid temporarily the general concept of a closed curve, and a few rearrangements suffice to preserve the building. The main thing, the actual conceptual construction, remains.

Concerning your further expositions and your own results, I permit myself today to make two remarks.

1) I don't understand how you get on the basis of your Figure 3 the division of domains where just the one 'curve' is simultaneously the boundary of three domains. Namely, the text and the picture have different letters. A more detailed explanation would be most welcome, and I kindly ask you for it.

2) Your admonishments at § 15/16 concerning reachability I don't understand at all; also I cannot interpret the figures. Maybe you overlooked that my theorems only refer to *simple* curves or arcs and their images. Also concerning this I would like to ask you for more detailed expositions.

I will send you shortly the above exposition about the changes to be made in §§ 3–5 of your article in the Annalen.

[Autograph – in Brouwer]

1909-06-18

To D.J. Korteweg — 18.VI.1909   **Amsterdam**

Professor,

Allow me to submit the enclosed article for the next Academy meeting. I am in town until tomorrow afternoon 3 o'clock. Then I will go with Lize to

> pension 'Sunny Home'
> nearby Ede (Gelderland)
> bus stop Doesburgerbuurt.

where I'll be until July 5.

So if next week you want to speak me about the article, write to that address.

Schoenflies has rather extensively gone into my latest article for the Annalen, of which I had sent him a copy. I actually pressed him hard, and my success is probably only due to that; all the same I'm glad that I at last had a 'bite', and something more than just a friendly card concerning my work.

I enclose the letter of Schoenflies, which I have answered just as extensively, convinced as I am that you like to stay informed about my scientific situation.

After your last post card saying that the next issue of the Nieuw Archief voor de Wiskunde is filled up already, I have postponed the French translation a bit. It seems to me that September 1 would be early enough for submissions? Maybe I would be allowed to append a translation of the enclosed article, which is closely related?

With polite greetings

L.E.J. Brouwer

[Signed autograph – in Korteweg; enclosures not extant.]

909-07-08

To D.J. Korteweg — 1 or 8.VII.1909   Blaricum [80]

Thursday evening

Professor,

I have heard that the position of curator of Teyler's physical Cabinet [81] and editor of the Archives du Musée Teyler is free. I would like to ask you:

---

[80] Undated, in view of *Brouwer to Korteweg 10.VII.1909* to be dated July 1 or July 8. Since the mentioned curator was still alive on June 24, the two Thursdays 1 and 8 July qualify.   [81] Part of the *Teylers Museum* in Haarlem (the oldest museum in the Netherlands).

would I overestimate myself to apply for that, and might I count on some support from you?

As a solution for my livelihood it would be an almost too beautiful piece of luck. From my earlier stay in Haarlem I seem to remember that Mr. v.d. Ven [82] had almost all his time available for his scientific work.

I didn't receive proofs of the little note I sent some time ago to the Nieuw Archief; do you know the reason for that? [83]

At the election bureau [84] there was very little for me to do.

with respectful greetings [85]
L.E.J. Brouwer

[Signed autograph – in Korteweg.]

1909-07-26

To D. Hilbert — 26.VII.1909   **Amsterdam**
                                                          Overtoom 565

Dear Mr. Geheimrat, [Sehr geehrter Herr Geheimrat]

I am sending you, as an enclosure, my note about Analysis Situs in slightly modified form. [86] May I kindly ask you to print it in this form, and to stick to the old figures?

I hope I have achieved in this way that Mr. Schoenflies (to whom I already sent a copy) can now approve of it in every respect.

The modifications consist of the following:

1°. I have given in to Mr. Schoenflies in so far, that I have removed the admonishments that only refer to the presentation, not to the proper contents of his theory. They actually only occupied a few

---

[82] dr. Elisa van der Ven, October 5, 1833 - June 27, 1909, curator since 1868.
[83] [Brouwer 1909f]. Korteweg must have taken action, for the editor Kluyver informed Korteweg that Brouwer would soon receive his proofs (Kluyver to Korteweg 12.VI.1909).
[84] The parliamentary election of June 11, 1909. Brouwer had offered his support to Korteweg.   [85] *Met beleefde groeten.*   [86] [Brouwer 1910h]; also in *CWII*, 352–366; cf. Freudenthal's remark's in *CWII*, 367–368.

lines, and maybe they were somewhat out of place in an article in the Annalen.

2°. At the end of the introduction I have inserted a note so that also for the reader my admiration for the achievements of Mr. Schoenflies will be beyond all doubt.

3°. My remark about Chapter V § 15 of Schoenflies' Report [87] is formulated a little differently, because of a communication of Mr. Schoenflies, according to which I here partially misunderstood the meaning of his text.

4°. At the end of the whole article I have added a summary for the sake of a better overview.

The date I have left the same, because nothing has been changed in the scientific contents of the article.

I have once more explained to Mr. Schoenflies in a letter why his deduction of the invariance of the closed curve and the ordering remains invalid even after the modifications he proposes in his addendum.

I expect now a copy of the definitive form of his addendum; we both hope that with that the matter can be considered as closed.

I will submit my second communication about finite continuous groups [88] to you in August. It will make use of a note in the Amsterdam Communications, [89] which I will dispatch to you today.

Most sincerely yours [90]
L.E.J. Brouwer.

P.S. Just now I receive your card; have you perhaps not received my letter of June 23 about the addendum that Mr. Schoenflies was going to send to you? That was the letter I referred to in my card of July 19.

In any case I will send my letter now by registered mail.

LEJ Brouwer

[Signed autograph – in Hilbert]

---

[87] Usually simply quoted as the *Bericht*, [Schoenflies 1913].   [88] [Brouwer 1910c].
[89] [Brouwer 1909g, Brouwer 1909a].   [90] *Mit ausgezeichneter Hochachtung.*

1909-08-08

From A. Schoenflies — 8.VIII.1909   **Königsberg**
                                                          Haarbrückerstr. 12

Dear Doctor, [Sehr geehrter Herr Dr]

Possible shapes, as contained in your letter, are available for *arbitrary*
polygons; that is not the case with approximating polygons. I ventured
to point this out already once before. Originally I did it as you did, I
worked with polygons of arbitrary shape, but I soon saw that then the
conclusion became [..?..], and have then subsequently, when I was editing the
Bericht [91] deduced a series of theorems about the shape an approximating
polygon can have.

As an example, forms as those reproduced here etc. are excluded, in
which the shaded part of the plane represents the polygon and the horizontal
edge has precisely the length ε, hence is a side of a square. It follows
immediately that for the *plane domain* of $\mathfrak{P}_\nu$ there are only such square
parts that themselves [..?..] next to the 8 [..?..] of $\mathfrak{T}$.

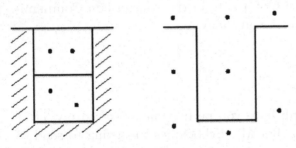

In the end I have left them
out too, in the conviction that
I could save the trouble of the
proofs of everything related to
polygons, except the little bit
in Chapter IV, § 1.

Now as concerns your ob-
jection, please consider the
following. Let p_m be the *last*
intersection point of the path with $\mathfrak{P}_m$, then it is impossible that $\mathfrak{P}_m$ — as
in your figure — penetrates once more into the circle, because

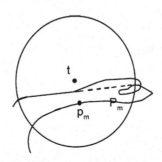

----
[91] [Schoenflies 1908].

I have drawn the circle around t in such a fashion that the whole part of the path from the circle around t is drawn in such a way that the whole part of the path from p_m to t is *inside* the circle. Hence, as the dotted line separates t from P_m, it will intersect $\mathfrak{P}_m$ once more.

Furthermore, as far as $\mathfrak{P}_{m+1}$ is concerned, your drawing is impossible for this polygon because of the theorems above. Also I refer to them on top of page 106 in the article. There I say that one should construct the polygons starting from a *fixed* division into squares. Now observe how as a consequence, the polygon $\mathfrak{P}_{m+1}$ is formed. It is best if one considers polygonal pieces that consist of a finite number of squares. Then the plane region of $\mathfrak{P}_{m+1}$ is formed by excluding from $\mathfrak{P}_m$ a finite number of smaller pieces of the plane.

You convince yourself that also in this way, that what you assumed about the possibility of $\mathfrak{P}_{m+1}$, is excluded.

The second point is similar. Surely one *can* to a given P and $\mathfrak{Q}'$, $\mathfrak{Q}''$, possibly choose the region H'_μ such that it joins to H'; it is impossible to do that for all regions H'_μ, as follows from my last proof. Two polygons $\mathfrak{Q}'$ and $\mathfrak{Q}''$ as I considered them, determine always a region in which not *each* of the available H'_μ can be joined to H'. Maybe I will explain this in more detail in my MS.

Now I readily agree, also today, that I could have inserted all of that in the article.

It would be useful to clear this up this in a coherent context. I intend therefore to let one my students work this out in detail, for of a dissertation. Some of this can be transferred immediately to 3-dimensional space, for example the theorem that a polygon is a *simple curve* is easy to prove as in Chapter V § 3.—

Your own proof for the paths I still don't understand. I refer you to the earlier mentioned example. Take a point set that looks like the curve $y = \sin \frac{1}{x}$.

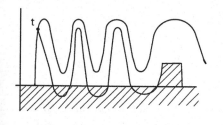
As far as I understand your notation [.?.] that part of $\mathfrak{P}_\nu$ whose distance from l is not smaller than $\frac{3}{2}\varepsilon_\nu$, assigns to the path l its *exterior* rather than its interior, and the maximal change will than necessarily be larger than $2\varepsilon_\nu$. For sufficiently large ν it is of course different, but from your argumentation that can't follow, because that refers to *every* ν.

Would you please be so kind as to write to me as soon as possible; then I will send you as soon as possible my MS.

Sincerely

Affectionately yours [92]
A. Schoenflies

[Signed autograph – in Brouwer]

1909-11-09a

To C.S. Adama van Scheltema [93] — 9.XI.1909[a]   Blaricum

Dear Carel, [Beste Carel]

Yesterday I heard from Bertha, [94] that you are still in Florence at your piano III. Why didn't you drop me a note about that? I send you now my inaugural address on the off chance. [95] Did you, by any chance, not get my last card? Your postal system is hopeless, I still remember that from the last time. I wish that we could go round together. Lucas van Leyden's Christ in the Tribuna, the self-portraits of Bellini and Michelangelo in the Uffizi, the drawings of Da Vinci, and the Spanish Chapel, the glory of dogmatism! Are you still not intrigued by those?

This summer the world's foremost mathematician [96] was in Scheveningen; through my work I had already been in contact with him, now I have again and again walked with him, and talked as a young apostle with a prophet. He was 46, [97] but youthful in soul and body, a strong swimmer who eagerly climbed over walls and barbed wire fences.

It was a beautiful new ray of light through my life. Now do write me a few words right away.

Your Bertus

[Signed autograph, postcard – in Scheltema]

---

[92] *Hochachtungsvoll – Ihr ergebener.*   [93] To 'Sig. C.S. A. v. Scheltema (all indirizzo della Sign.a. Rossi) Lugarno Vespucci 32 (piano III) Firenze.'   [94] Scheltema's sister. [95] [Brouwer 1909c].   [96] David Hilbert.   [97] 47, in fact.

909-12-15

From H.A. Lorentz — 15.XII.1909   Leiden

Dear Mr. Brouwer, [Zeer geachte Heer Brouwer]

I still have to thank to you for so kindly sending me the 'Public Lecture' [98] with which you began your activity as a privaatdocent. [99] Insofar as my ignorance of several of the subjects discussed did not prevent me from following you, I have read your lecture with much interest and pleasure, and enjoyed the wealth of ideas by which it distinguishes itself.

Of course I have paid special attention to the part referring to questions related to the principle of relativity. I do understand the point of view you embrace, but it still takes me some effort to share it. In fact, I still have the idea (or like to fancy that I do) that space and time can be completely separated, and in relation to that, that there is just *one* time, and that hence the proposition that two facts occurring at different places are simultaneous, can have only *one* meaning; it remains of course a question whether one can succeed in observing that simultaneity. Under this point of view one arrives at considering *one* system x, y, z, t as the 'true' one; the introduction of another x', y', z', t' is nothing but a mathematical transformation. Even when I would, by way of precaution, not base myself on this idea, and so do not assign a special meaning to *one* system x, y, z, t with respect to all the other just mentioned systems, I think it would be advisable, for clarity's sake to agree once and for all that by coordinates x, y, z we will mean those that are measured with measuring rods which are at rest with respect to the ether, and that t will mean the time measured by clocks that are in the same situation and that are synchronized by means of light signals. In that case we can call this *the* time, and (x, y, z) *the* space; again x', y', z', t' will be mathematical basic quantities.

This point of view presupposes that one assigns the ether a kind of substantiality, in so far that one can speak about being at rest or being in motion with respect to the ether, where again it doesn't matter whether we can *determine* whether a system moves with respect to the ether. In favor of this 'substantiality' of the ether is the circumstance that it can be the carrier of electromagnetic states and the accompanying energy.

If one frees oneself completely from the old conception of the ether as a 'substance', then we loose the tool that we just used to assign a special

---

[98] I.e. the inaugural address of a privaatdocent, called – *'Openbare les'*.
[99] [Brouwer 1909c], 12.X.1909.

meaning to *one* system (x, y, z, t) out of many others, and one is forced to suppose that one system (x, y, z, t) is as good as the other (x', y', z', t'), and we hardly have another way out than to drop the sharp distinction between space and time. This point of view has the advantage that we can see immediately that the phenomena in a system can't change by a translation imparted to the system (so we bring the relativity principle to the foreground); in the old representation it always remains a somewhat 'arbitrary' result that a translation never has any influence on observable phenomena.

Please, don't hold this exposition against me, and allow me to wish you all the best for your future activity.

With kind greetings, sincerely

Obediently yours $^{\langle 100 \rangle}$
H.A. Lorentz

[Signed autograph – in Brouwer]

1909-12-19

From A. Schoenflies — 19.XII.1909   **Königsberg**
                                                                 Haarbrückerstr. 12

Partially changed $^{\langle 101 \rangle}$

Dear doctor, [Sehr geehrter Herr Dr]

If you think that you have 'more or less a right to request the deletion of the words you objected to' and if you furthermore suggest that I would have violated our agreement, then I quite firmly must resist that. I cannot acknowledge this in any way.

First I remind you that I expressly reserved the right of a free hand in relation to Chapter IV § 12. Second, our agreement referred to the revisions, and that only means that changes that the one thinks necessary will be submitted to the other for inspection and for information and for comment, before the imprimatur is given. That one can consider such changes necessary, seems self-evident to me.

$^{\langle 100 \rangle}$ *Met vriendelijken groet hoogachtend – Uw dienstwillige.* $^{\langle 101 \rangle}$ Note on top of page.

At least it happens to me, and as far as I can see, to many others, that only after reading the printed version of the manuscript one can reach a correct judgement whether one has found the right formulation for what one wanted to say.

Precisely this I have left open for myself by our agreement. This doesn't mean at all that you are obliged to endorse my changes. But I must definitely reject the implicit reproach in your words. I think I acted in fact almost with exceeding correctness by pointing out to you in my letter everything that I changed or added. So much about things in general.

On page 1, line 7 I will be happy to replace the words 'für die ersten Paragraphen' [102] by the words 'den allgemeinen Aufbau', [103] in line with your wishes. Instead of 'in ihnen' [104] I say then 'in den ersten Paragraphen' [105]. Moreover, I enclose the corrected page.

About the end of page 2, allow me the following remark. To decide which facts are to be presented in my article, is up to my own judgement, and not a stranger's. That is after all what this is all about. When I consider it necessary that the actual changes that must be made to theorems XIII and XV should be mentioned in more detail in § 1 and tentatively in the introduction, I have every right to do so; and I will give you the motivation, though I'm not obliged to. I have done my best to give a positive turn to my article, by showing that my line of thought in Chapter V can be retained, namely by 1) stressing the reachability and the simple curves and 2) replacing the erroneous Theorem XIII in Chapter IV by a more restricted one, which however suffices for the proof. The latter I could have been inserted already in the second paragraph of page 2 on the place marked ⊗; that it is at the end of page 2 instead, is indeed not important. You yourself acknowledged this interpretation of the exterior boundary in the letter I quoted. I referred to these words only, to marshal your own words against the opinion expressed in your letter of 8/12, in which you label the changes in Theorem XV formal and trivial. I already had for various reasons the plan to treat the outer boundary [106] in more detail in § 1 and in the introduction, especially also in connection with the letters exchanged earlier with you. You will have to admit that this concerns new facts, and that the question where and how I insert them into my article only depends on me. You are entitled to expect that I will refrain from any polemic or restrictive remarks against your article — and that is what I did. To let this come out even more clearly, I have made yet another change which you can see on the

---

[102] for the first sections.   [103] the general structure.   [104] in these.   [105] in the first sections.   [106] *Aussenrand.*

enclosed page. However, I cannot forego an explicit mention of the contents
of the end of page 2.

I hope and wish that this concludes our exchange of letters on these matters. Anyway, it is not possible for me to adopt further material changes.

    Sincerely yours [107]
    A. Schoenflies

    Anyway, I don't believe that anyone could think that I want
    to belittle your article or lessen my errors. In my opinion the
    opposite is the case.   D.O. [108]

    Because you have sent the revision already to Mr. Blumenthal,
    I have sent the printer, in order to speed up things, a page with
    the same additional corrections as the enclosed page. You will
    receive then the final revision from me. [109]

[Signed autograph – in Brouwer]

1909-12-24a

From J. Hadamard — before 24.XII.1909[a] [110]

Dear colleague, [Mon cher collègue]

Thank you very much for your two articles. [111]

The following is a very simple proof of the invariance of a point under a univocal [112] transformation of the sphere. It suffices to apply the
proposition contained in your first note [113] as applied to the vector distribution, [114] taking for the vector at any point M the tangent to the great
circle arc connecting this point to its corresponding point M'.

Such a vector must become indeterminate for (at least) one position of
M: this means that for such a position either M' will coincide with M, or
M' will be diametrically opposite to M.

---

[107] *Hochachtungsvoll und ganz ergebenst.*   [108] *der Obige* = the above.   [109] Note by
Schoenflies in the margin of the last page.   [110] Undated. As *Brouwer to Hadamard,
24.XII.1909* is a reply to this letter (see also *Brouwer to Korteweg, 24.XII.1909[b]*),
the 'before' is evident.   [111] Most likely [Brouwer 1909a] and [Brouwer 1909d].
[112] single-valued.   [113] *brochure.* See Freudenthal's comments in Brouwer, Collected
Works II, p. 428.   [114] vector field.

The choice between these two hypotheses doesn't need the *indicatrix*. To obtain that, it is convenient to phrase your proposition on vector distributions not in the form given by you, but in that given by Poincaré in his investigations of curves defined by differential equations (Journal de Mathématiques, 1881, first Memoir, I think). From the results of Poincaré (or, if you wish, from Euler's theorem) it follows that at least one of the singularities of the vector distribution (say M_0) must be such that when a point M describes a small circle around M_0, the corresponding vector will make one turn (in total) in the same sense as M. For this to be the case the point M_0' corresponding to M_0 has to coincide with M_0 if the transformation preserves the indicatrix, and is just opposite it in the other case.

Sincerely yours, [115]

J. Hadamard

[Signed autograph – in Brouwer]

909-12-24b

To D.J. Korteweg — 24.XII.1909[b]   **Amsterdam**

Professor

I have become convinced that the proof of Hadamard cannot be salvaged, not even by further developing his train of thought. But the treatise of Poincaré, which gave him the idea, has in a different way suggested to me a good proof, which is quite a bit simpler than the one I gave last year. However, the result appears by surprise, while in my original proof I gradually build up the transformation, and I see myself gradually being forced to admit the invariant point.

Enclosed the copy of the letter to Hadamard, which I dispatch at the same time as this one. [116]

Sincerely yours

L.E.J. Brouwer

[Signed autograph – in Korteweg]

---

[115] *Croyez, je vous prie, à mes meilleurs sentiments.*   [116] Cf. [Johnson 1981] p. 139 ff.

1909-12-24c

To J. Hadamard — 24.XII.1909^c Amsterdam $^{\langle 117 \rangle}$

Dear Sir, [Cher Monsieur]

Thank you very much for having pointed me to out Mr. Poincaré's memoirs on algebraic vector distributions. But concerning your proof, could it be that you are mistaken? You say that in a univocal $^{\langle 118 \rangle}$ transformation that preserves the indicatrix (to begin with, for a univocal but not biunivocal $^{\langle 119 \rangle}$ this expression, 'preserving the indicatrix', does not always have a meaning, so let us replace 'univocal' by 'biunivocal') the vector of a point M that describes a small circle around M_0 will rotate in the same sense if M_0' coincides with M_0, and in the opposite sense if M_0' is diametrically opposed to M_0. But for a general biunivocal and continuous transformation neither of these properties exists. As far as I can see, these tangents to arcs of great circles, which I have tried as vectors myself (cf. p. 8 of my note $^{\langle 120 \rangle}$), do not succeed in attaining our objective. It is only in the case of the elliptic plane that they are sufficient (cf. p. 9 of my note).

Reading the memoirs of Mr. Poincaré quoted by you, I have had another idea. First we remark that if we adapt the concept of an 'index' (quoted from the first memoir, p. 400) to general continuous vector distributions, corollary I of page 405 becomes the following: '*If the singular points are finite in number, each of them has a finite index and the algebraic sum of all the indices is equal to 2.*'

Let us now assume a biunivocal and continuous transformation of the sphere into it itself. Let A be a point that is not invariant, B its image, and C the image of B. Then let M be a variable point of the sphere and M' its image. To define the vector at the point M, we draw a small circle through M, M' and B, and as vector in M we choose the tangent to the arc MM' of that circle that doesn't contain B. The vector distribution has then as singular points 1°. the invariant points of the transformation, 2°. the points A and B. If the former don't exist, we will only have the points A and B,

---

$^{\langle 117 \rangle}$This letter is an reply to the undated letter *Hadamard to Brouwer before 24.XII,1909^a*). See also the draft *Brouwer to Hadamard, 4.I.1910*, (Paris), also in *CWII*, pp. 426–427, [Y17]; Freudenthal's comments p. 422 ff. Place of dispatch – from *Brouwer to Korteweg 24.XII.1909*. $^{\langle 118 \rangle}$single-valued. $^{\langle 119 \rangle}$bijective. $^{\langle 120 \rangle}$[Brouwer 1909a].

and because the index of A equals -1, and the one of B is $+1$, the sum of the indices will be zero, which is impossible.

Sincerely yours,

L.E.J. Brouwer.

[Signed autograph, draft – in Brouwer; copy in Korteweg]

———————————

Chapter 3

1910 – 1919

To D. Hilbert [1] — **1.I.1910**   **Paris**

Dear Geheimrat [Mein lieber Herr Geheimrat]

Warmest wishes to you and to your dear spouse for the new year, for your health and for your scientific work.

I am staying here during the winter holidays with my brother, the geologist. Unfortunately my wife couldn't accompany me. In the middle of January my lectures start again, and I will return.

The good relations with Mr. Schoenflies have been restored, I am certain, mostly through your intervention. I enclose his last two letters, to which I have answered that I am satisfied with his last version and that I consider the matter settled.

May I add a few remarks about the univocal [2] (not necessarily (1-1) [3]) continuous mapping of a sphere κ onto a sphere λ? If one imposes the condition that it is *both ways* continuous, then it is a (1-1) continuous image of a rational function of the complex variable. By the condition of continuity both ways, I mean that a closed Jordan curve around a point L of λ, that converges to L, for each point K of κ that has L as image will correspond to a closed Jordan curve around K that converges to K.

If we now have *two* of these maps satisfying these conditions from a sphere (or a more general closed surface) K to a sphere L and to a sphere

---

[1] No addressee; from the text it follows that the letter was addressed to Hilbert; see also Freudenthal's remark in *CW II*, p. 425.   [2] single-valued; in letter *eindeutig*.   [3] in letter *ein-eindeutig*.

M, then the question arises which additional conditions must be satisfied in order to conclude that the correspondence between L and M is a complex algebraic one in the sense of Analysis Situs. Returning to the general one way univocal and continuous correspondence between two spheres, for each of those a finite number n as its *degree* can be given, in such a way that all relations of the same degree, and only these, can be transformed continuously into each other. In particular, all correspondences of the nth degree can be transformed continuously into rational functions of the nth degree in the complex variable.

To define this degree we introduce homogeneous coordinates x, y, z on κ, and homogeneous coordinates ξ, η, ζ on λ and then we consider the injective mapping that is domain-wise determined by a correspondence

$$\xi : \eta : \zeta = f_1(x, y, z) : f_2(x, y, z) : f_3(x, y, z),$$

where f_1, f_2, f_3 are polynomials.

Next we assume a positive orientation on both spheres, and we choose in each point of κ this positive orientation, then each point in general position of κ occurs p times with positive orientation and q times with negative orientation. Then one can show that for each point of λ in general position $p - q$ is a constant, which we will call the *degree* of the mapping.

If the correspondence between x, y, z and ξ, η, ζ is not determined by polynomials, then one can approximate it by such polynomial correspondences, and it is easy to show that these approximating correspondences have a constant degree, which we can also assign to the limit correspondence. This degree is always a finite positive or negative number. In particular, a (1-1) [4] continuous transformation of the sphere into itself will have degree $+1$ if it doesn't change the orientation, and -1 otherwise.

Now you know my theorem that each (1-1) continuous transformation of the sphere into itself that does *not* change the orientation will always have at least one fixed point. This theorem can be extended in the following manner, namely that each *univocal* continuous transformation of the sphere into itself whose degree is *not* -1, will always have at least one fixed point.

And I have succeeded to extend the theorem in this form to the n-dimensional sphere. There it reads as follows: Each univocal continuous transformation of the n-dimensional sphere into itself has at least one fixed point. The exception is for odd n the transformations of degree $+1$, and for even n the transformations of degree -1.

---

[4] Brouwer in the margin: 'stricter formulation'.

(1-1) transformations therefore have necessarily a fixed point, either for odd n and reverse orientation, or for even n and unchanged orientation. Even more general is the result for univocal continuous transformations of the interior of the n-dimensional sphere into itself, for these have anyway a fixed point.

Once more best wishes and greetings for you both

Your ever revering [5]
L.E.J. Brouwer

[Autograph draft – in Brouwer; also in *CW II*, p. 420 ff.]

910-01-04

To J. Hadamard — 4.I.1910   Paris [6]
6 Rue de l'Abbé de l'Epée

Dear Sir [Cher Monsieur]

I can at present communicate to you several extensions of the fixed point theorem for (1-1) continuous transformation of the sphere. They concern univocal [7] continuous transformations of the sphere. To such a transformation one can assign a finite number n as its *degree*. Starting from a transformation of degree n, one can obtain by means of continuous variations each other transformation of degree n, but no others. In particular one can always obtain a rational transformation of degree n of the complex sphere.

To determine this *degree*, let us introduce homogeneous coordinates (in the double sense). Write x, y, z for the original sphere and ξ, η, ζ for the image, divide the sphere into a finite number of regions and first consider transformations defined by the relations:

$$\xi : \eta : \zeta = f_1(x, y, z) : f_2(x, y, z) : f_3(x, y, z),$$

where f_1, f_2, f_3 are polynomials, which might well be different for different regions of the sphere. Let us call this transformation a polynomial transformation. Let us define an indicatrix on the sphere: then every point P

---

[5] *Ihres immer verehrenden.*   [6] Draft without addressee. In view of earlier correspondence, and topic, Hadamard is clearly the recipient; letter also in *CWII*, pp. 426–427, with Freudenthal's comments.   [7] single-valued.

of the image *in general position* occurs a number of r_p times with positive indicatrix, and a number of s_p times with negative indicatrix. One shows that $r_p - s_p$ is a constant: that is the degree of the polynomial transformation.

Let us return to the general univocal and continuous transformation. It can be approximated by a series of polynomial transformations; one shows that the latter all have the same degree. This is furthermore the degree of the limit transformation.

The degree is always a finite positive or negative integer. The degree of the (1-1) transformation is $+1$ if the indicator stays the same, and -1 if it is reversed.

Now the generalized fixed point theorem becomes the following: Each univocal and continuous transformation of the sphere into itself for which the degree is not -1 has at least one invariant point.

Moreover, I have extended this theorem to spheres of m dimensions. It is then stated in the following manner: Each univocal and continuous transformation of the m-dimensional sphere into itself contains at least one fixed point, except when a) m is odd and the degree n equals $+1$, b) when m is even and the degree n equals -1.

In particular when the transformation is (1-1) [8] there exists at least one fixed point a) if m is odd and the indicatrix is reversed and b) is m is even and the indicatrix is invariant.

For the volume of an m-dimensional sphere in the space of $m + 1$ dimensions (if we mean the sphere itself by it) I have lately succeeded in establishing a still more general result, to wit: every continuous univocal (not necessarily biunivocal) transformation of the volume of an m-dimensional sphere into itself possesses at least one fixed point.

On the general vector distributions of the sphere soon two more articles [9] by my hand will appear, where I study several questions connected with the principle of Dirichlet and with the decomposition of a field in a part that is 'source free' [10] and a part that is 'rotation free'. [11] For this I first establish the most general form that tangent curves (or characteristics, after Poincaré) can assume. As the main result of the first article one must consider the property that every characteristic that does not approach a singular point is a spiral whose two limit cycles are characteristics themselves. The property that the existence of at least one singular point is necessary, is basically nothing but an extra corollary, on which I would not have insisted,

---

[8] Brouwer's remark in text: 'stricter formulation.'   [9] [Brouwer 1910d, Brouwer 1910e].
[10] '*quellenfrei*'.   [11] '*wirbelfrei*'.

were it not that it was the first result that could be simply formulated, and
also because to me there seems to be an close relation between this theorem
and the one about the fixed point of the sphere, a relation that is nowhere
clarified, except in your correspondence. In the second article I have inserted
your beautiful direct and more complete proof of the existence of at least
one singular point.

My address will be in Paris until January 15. Maybe there is an opportunity to meet you?

Sincerely yours, [12]

L.E.J. Brouwer

[Signed draft/copy – in Brouwer]

To D. Hilbert — 18.III.1910   Amsterdam

Dear Geheimrat, [Sehr geehrter Herr Geheimrat]

A few months ago I cited in an article that appeared in the Amsterdam
Proceedings, [13] my small Annals note on 'Transformations of surfaces into
themselves' [14] as to be found in volume 68. Meanwhile part 3 of this volume
is already appearing now, so that probably my note will not get its turn in
volume 68.

Could you perhaps arrange it so, without disturbing the regular course,
that my note, which in print is only 4 to 5 pages, finds a place in part 4
of volume 68? Because of the above mentioned citation this please me very
much, and I would be greatly indebted to you.

Just in case that you are away on holiday and this letter doesn't reach
you in time, I am writing to Mr. Blumenthal in the same vein.

In July I hope to have the opportunity to come to Göttingen for a while.

Apart from a new group theoretic communication, [15] I am preparing
an article to be submitted to the editors of the Annalen. In this article I
solve the problem of invariance of dimension insofar that I prove that in any

---

[12] *Agréez, monsieur, mes salutations distinguées.*   [13] *KNAW, Proceedings.*   [14] *Über
Transformationen von Flächen in sich* [Brouwer 1910g].   [15] [Brouwer 1910c].

case spaces of even and odd dimension cannot be continuously and one-one mapped onto each other.

In the Amsterdam communications [16] I continue my work on transformations of surfaces into themselves and on continuous vector fields.

Recently I studied your article about Dirichlet's principle and am very much interested in the sequel that you mentioned to me last year. [17]

Also on behalf of my wife I wish you all the best for the coming Easter, and a cordial 'auf Wiedersehen'

Ever your revering [18]
L.E.J. Brouwer.

[Signed autograph – in Hilbert; fragment 'Apart from ... each other' also in *CWII*, p. 429]

———————

1910-09-00

To D.J. Korteweg — late summer 1910 [19]

Now that the course is long enough behind me, and calm thinking back and complete consideration of the matter has been possible, I have come to the unshakable conviction that lecturing without having been invested with authority, as I have done in the past course (of which I have a wretched recollection). is tantamount to throwing my energy into a pit and I cannot and may not prolong that, and extending my duties, as implied by the proposal discussed recently with de Vries, is a fortiori completely excluded. Already before we started this experiment last year, I had, as you know, these negative expectations about it, but I thought that given your opinion, I had to suspend mine. However, now that the result has not vindicated you, I again carry the full responsibility for my position. This position is determined by the fact that I was only prepared to accept an appointment as privaatdocent if the point of departure was that the interest of the university demanded an expansion of mathematical teaching, and that I would undertake that task for free as long as the authorities wouldn't consent to

———

[16] *KNAW, proceedings.*   [17] The discussions in October 1909 between Brouwer and Hilbert in Scheveningen, see [Van Dalen 1999], p. 128.   [18] *Ihres immer verehrenden.*
[19] Handwritten draft of a letter by Brouwer; the addressee has to be Korteweg; undated – sometime during, or after, the summer of 1910 is a fair guess; Brouwer had been a privaatdocent for one year.

the creation of a new post. In that case, however, there would be in the granting of this task a hint to the authorities, which would be fittingly emphasized by a compulsory examination [20] (which really would not violate the law any more than a compulsory examination by a professor), and this would be the more fitting because almost all other disciplines of our faculty have seen a substantial personnel expansion in the last years, even without there being a vacancy for the purpose. [21] I think that neither Schoorl, nor Cohen, nor de Meyere, nor Zeeman, at their first appearance as lecturer in Amsterdam, were appointed in an existing vacancy.

Thus I feel compelled to tighten up my attitude as follows: I am only *then* prepared to continue my activities at the university if the authorities are seriously urged to create a lecturer position for me, if need be with meagre pay, which automatically would carry with it compulsory examinations; and moreover if, in case of proven refusal of the authorities in this matter, you would make an examination with me compulsory. If you cannot cooperate in this, then *sans rancune*, but then I will end the sad enterprise that has disrupted the harmony of my life for a year.

[Handwritten draft – in Brouwer.]

Editorial supplement

[The following is a draft of a letter possibly written after a discussion with Korteweg following the above letter. It would be dated also sometime in October - November 1910.]

A couple of the points, touched upon this afternoon, make me once more pick up the pen. First the remark that I don't learn anything from my lectures — on the contrary —, and that consequently the time spent on them is wasted as long as I have no guarantees that at least the students learn something from it. A remark of yours this afternoon forces me to reconsider the present situation from another point of view.

---

[20] The central courses had examinations at the end of term. These examinations were called *tentamens*, they did not have the same legal status the final exams had. A course without such a tentamen was not taken serious by students and staff. Hence Brouwer's insistence on this mark of recognition.   [21] Scratched out by Brouwer: 'When Cohen as assistant of Bakhuis Roozeboom, Schoorl as assistant of de Bruyn, de Meyere as assistant of Sluyter was appointed; also when Zeeman was appointed as lecturer, there wasn't a vacancy, I believe.'

One of the reasons that has induced me to initiate the matter, was the hope that eventually the reinforcement of the mathematical teaching staff in our university will become possible and that an extraordinary professorship will come to me, and in your work; though it has become fairly clear to me, being passed over in Delft, that there is no prospect for me there if I do not show first what I can do as teacher. In that case I seem, in the present manner, to be putting the cart before the horse. For I can't expect that you come to hear and evaluate my lectures; so you will have to get your impressions of the quality of my teaching from the members of the audience, who because of their 'don't-work-for-it' attitude must think it abnormally difficult; if I understood you correctly this afternoon, such information has reached you already, and you seem to have attached some value to it.

In contrast to this, I would argue that those who have attended my lectures at my request, people with experience of lectures in mathematics, and who certainly would have held up to me the unvarnished truth if they had cause to, have stated that they never attended such clear lectures. But whatever is the case, 'show what I might be as teacher' I certainly cannot do in the present circumstances, I can only damage my reputation as such in an undesirable and wrong way.

This point of view could only be eliminated, if either you or de Vries or any other person who was completely competent in my eyes would come to my lectures under some kind of pretext to hear and judge.

For the rest, I'm still convinced that when a person of any scientific value is teaching badly, it is always because of indifference, never because of incompetence.

your
L.E.J. Brouwer

I enclose 2 reprints.

[Signed autograph draft – in Brouwer]

––––––––––––

911-00-00

To O. Blumenthal — 1911 [22]

Let us now imagine T_n to be a ring surface in three dimensional space, ϑ to be a contractible closed Jordan curve on T_n, A the simply connected domain determined by ϑ on T_n, B the complement in T_n. Then by the definition of linked varieties, [23] Γ will certainly intersect A (resp. A'), because ϑ can be contracted to a point inside A. Inside B however ϑ *cannot* be contracted to a single point. The existence of intersection points of Γ with B is consequently *not certain*. If then furthermore the invariance of dimension is *not* certain, Γ could be completely contained in A, and would then constitute only *one single domain* α. This domain α would possess *in that case no boundary* that one could approximate by *'point pairs'* T_p *that are linked with* T_n.

For the mentioned simple case one can by the way also easily prove the existence of intersection points of Γ with B (resp. B'). To achieve the same thing for arbitrary n and p and arbitrary T_n, one must prove that the closed polyhedral manifolds Γ always have an even number of intersections with $T_n = A' + B'$. But it seems to me that to carry out the proof of this 'evident' fact is extremely laborious. Actually a similar difficulty occurs already in the justification of the *definition* of the linked varieties for arbitrary p and n.

Hence my position about the second part of the note of Lebesgue is that he has quite correctly proved a very beautiful theorem for three dimensional space, but that for higher dimensional spaces he only has stated 'evident' extensions without proving anything. However, one precisely needs the invariance proof for the higher dimensional spaces; so in my opinion the mentioned part of Lebesgue's note doesn't contain anything at all pertinent to the invariance.

Best greeting

Your
L.E.J. Brouwer

[Autograph draft – in Brouwer; also reproduced as [Y5] in Brouwer *CW II*, p. 452]

---

[22] Last page of a draft letter, part of the correspondence concerning the Lebesgue affair. Clearly O. Blumenthal is the addressee. In view of the available correspondence and the publications on the dimension invariance, the draft dates back to 1911.   [23] In text everywhere: *variétés enlacées*.

1911-03-00a

L.E.J. Brouwer - *note on Lebesgue's proof* — **III.1911^a Amsterdam** [24]

[Handwritten remark in the margin:] 'Accepted, Hilbert'.

REMARK ON THE INVARIANCE PROOF OF MR. LEBESGUE
BY L.E.J. BROUWER AT AMSTERDAM

The proof of invariance of the dimension number, given by Mr. Lebesgue on page 166–168 of this volume, contains a gap on page 187, lines 6–13. Namely, from the property, that I_{h-1} extends from any manifold $X_i = X_i^0$ to any manifold $X_i = X_i^0 + 2l$, $(i = h, h+1, \ldots, n)$, one cannot immediately conclude that I_h extends from each manifold $X_i = X_i^o$ to each manifold $X_i = X_i^0 + 2l$, $(i = h+1, h+2, \ldots, n)$. Hence the existence of all I_p is not certain. In any case, considerable further considerations are required here.

Concerning the arguments of Mr. Baire, which Mr. Lebesgue used, the unproved theorems to which the problems are reduced there, lie deeper than the problem itself.

[Typescript – in Brouwer]

Editorial supplements

O. Blumenthal to D. Hilbert — *27.X.1910*   *Aachen* [25]

In the vacation we have made a very nice trip to Paris; however, I have unfortunately not seen mathematicians, they were all still on holiday. I did nonetheless get acquainted with Lebesgue, who happened to be in Paris. He is a very interesting man, and he told me that he is already for a long time in the possession of not one, but of several proofs of the invariance of the dimension number, which Brouwer has proved now in the Annalen. [26] He has sent me one of these proofs

---

[24] On this sheet a handwritten letter by H. Lebesgue, see *Lebesgue to Blumenthal III.1911.* See also the remark at the end of *Blumenthal to Brouwer, 28.III.1911.*
[25] Transcription, only of the part of the letter with relevance to Brouwer and Lebesgue is reproduced here.   [26] Brouwer submitted the paper in June 1910, he lectured on the theorem in a meeting of the Dutch mathematics society in October 1910; the paper was published in 1911.

for the Annalen, which looks very clever. I have not scrutinized it in detail for the correctness of the proof, but only for the correctness of the idea. For the details one can trust such a sharp-witted man. But if you want to carry out a detailed check, the article is at your disposal.

[Signed typescript – in Hilbert]

O. Blumenthal to D. Hilbert — 14.III.1911   *Aachen* [27]

I find the matter Brouwer-Lebesgue highly unpleasant, and in fact I am completely on Lebesgue's side. That means the following: Lebesgue says explicitly that he accepts certain theorems as proven; these theorems refer to certain linear equations and inequalities, and these can certainly be proved; in other words, the problem seems to me not to lie in these equations, and the whole set-up of the proof of Lebesgue is in my opinion an altogether passable and beautiful road to reach the dimension proof. But anyone reading Brouwer's note will not get that idea at all; the note is in my opinion phrased in an unfriendly and unpleasant manner. Therefore I had planned to ask Brouwer to withdraw the note for the time being, particularly because volume 70 will be completed only in the middle of May, so there is no hurry at all. Moreover, I am personally (just like Lebesgue (according to an earlier communication)) not able to understand Brouwer's proof, and I consider it very well possible that there are several gaps there as well.

For these reasons I would think it best not to accept Brouwer's note for the time being, but to ask him to wait in any case till the end of the volume, and then also to phrase the note in a completely different tone. In the case of an emergency, i.e. if he doesn't accept this, the editors could add an objective note, just like Noether has done very successfully in the Sannia-Zinder conflict.

Please tell me your views. I am willing to negotiate with Brouwer, however I believe that an intervention from your side would carry more weight.

[Signed typescript – in Hilbert]

----

[27] Transcription, only of the part of the letter related to Brouwer and Lebesgue.

1911-03-00b

H. Lebesgue to Blumenthal — III.1911^b Paris [28]

If I understand the remark of Mr. Brouwer correctly, it amounts to this: I have announced that I was going to accept facts that I qualified as quite evident, and that doesn't replace a proof of these facts.

On this point I agree with Mr. Brouwer, I merely add that if I haven't written out my proof completely, it is only because I have promised already for some time an article on this topic to the Secretary of the Société Mathématique de France.

I willingly admit that my phrasing is quite poor, because Mr. Brouwer has been able to believe that I didn't see the necessity of proving all of it, and that he now thinks it useful to point out this necessity to other readers.

H. Lebesgue

[Signed autograph – in Brouwer]

1911-03-25

From O. Blumenthal — 25.III.1911   Aachen

Dear Mr. Brouwer! [Sehr geehrter Herr Brouwer]

Allow me a few words in the matter of your dispute with Lebesgue. To begin with, that I have informed Mr. Hilbert of the considerations that I will put to you, that he agrees with me and that he has asked me to negotiate with you in the name of the editors of the Annalen.

First, I can inform you that the last issue of the present volume of the Annalen will be published only towards the end of May. Hence there is no hurry for you to submit your note against Lebesgue, but you can take your

---

[28] No addressee. The letter is written on a typed document '*Bemerkung zu dem Invarianzbeweis des Herrn Lebesgue*', dated March 1911, which had been submitted to Blumenthal. The latter must have passed it on to Lebesgue for comments. Lebesgue returned the document with his evasive comments. It is reasonable to date it March 1911. See also the remark at the end of the letter *Blumenthal to Brouwer 28.III.1911*.

time for that until further correspondence with Lebesgue has cleared up the matter.

Indeed, your haste in publishing this remark raises the suspicion that you don't expect anything from further discussion with Lebesgue. This assumption would however be unjustified, for on the one hand it appears from the letter of Lebesgue to you that he had a clear conception of the proof of the provisionally assumed theorems, on the other hand he himself writes to me literally: 'Writing the proof in detail does not take very long and I am about do so, but really, it seems impossible to make my results come out piecemeal in this fashion, and I think that your readers, more generous than Mr. Brouwer, would be willing to give me credit until my definitive memoir appears.' [29]

To me it seems to follow from this statement that it wouldn't be right to publish your note before you have ascertained that not only Lebesgue's short note in the Annalen, but really his *entire method of proof* is deficient. I am convinced that Lebesgue will make his manuscript, which he has finished, available to you for checking. If necessary I would be willing to mediate in this sense. I would in fact like to point out to you — and now I come to the core of my proposal — that *your note is phrased in a very rude form*, and that everybody will necessarily interpret it as saying that the gaps emphasized by you cannot be filled, which means that you consider the proof of Lebesgue *false*, because false and incomplete are in this case the same. In my opinion, however, you can only make this reproach to a man of Lebesgue's importance if you are entirely certain of your case.

So I would like to ask you again urgently to reconsider the matter concerning your remark once more. If you insist to publish it, I will of course do so, but then I will of course ask Lebesgue as well to send me his new manuscript, which I will then publish as soon as possible.

Finally, let me point out to you again something also stressed by Lebesgue, that *nobody doubts or contests your priority for this fundamental proof*. The priority belongs no doubt to the one who publishes first. But that your note was already there in print, when Lebesgue wrote his one, is clear from his text. So Lebesgue is in his own opinion and in that of the world not your rival, but your follower. So I think you should by all means leave him the time to present his proof in extenso.

By the way, Lebesgue has withdrawn the reply that he sent to you for the Annalen, and he has submitted to me another one, which I enclose. Please return it to me. I find this second text just as incomprehensible for

---

[29] Lebesgue is quoted in French in the letter.

the uninitiated as the first one, and I would ask Lebesgue for a different formulation if it is comes to that.

Summarizing once more my view, it is as follows: in the way you have written your remark, it will generally be understood that you consider Lebesgue's proof irreparably false, more precisely that you believe that the theorems assumed to be true by Lebesgue constitute essentially the core of the whole proof. Publishing such a remark seems to me appropriate only then, if you are positively convinced that Lebesgue does not possess the missing completions. So I advise you to publish nothing about the matter for the time being. When, however, you have acquired that conviction, and when you can prove it, then the Annalen will of course be gladly at your disposition. Because then a warning for Lebesgue is [30] in the general interest.

I hope I have formulated this letter correctly in all its parts, and that it doesn't lead to misunderstandings, which can easily happen in a complicated situation. My aim is merely of a conciliatory nature, I don't favor one party, at the very least I want to guarantee factual correctness; but, if possible, I would like to spare the Annalen an unnecessary polemic, so I advise you to strike only when you are certain of the deficiency of Lebesgue's proof.

The correspondence that you have sent to Hilbert, I return hereby. Also the one copy of your note with the answer of Lebesgue. [31]

Sincerely yours
O. Blumenthal

[Signed typescript – in Brouwer]

1911-03-27

To O. Blumenthal —27.III.1911   **Blaricum**

Dear Professor [Sehr geehrter Herr Professor]

Immediately after Mr. Lebesgue had informed me that he had prepared a complete version of his proof, I have informed him and Mr. Hilbert that I would withdraw my submitted remark; I have at the same time just asked

---

[30] Blumenthal wrote first 'is' and then 'would be' but didn't cross out one of the two
[31] See note *Brouwer, III.1911* and note *Lebesgue III.1911*.

Mr. Lebesgue to be so kind send me his elaborated version for my information.

In my opinion it would be most desirable that Mr. Lebesgue would
publish the additions to his proof in the 70th volume of the Annalen, because
every reader will be baffled at the place in question, and it seems to me that
the deficiency of the argument is not sufficiently stressed by the footnote.

I very much appreciate the explanation in your letter; for the rest please
believe me that the submission of my remarks was determined exclusively
by scientific reasons, and not by any self-seeking motives. Also because the
readers of the Annalen could have read something different in the text, I am
glad to consider myself relieved from this disagreeable duty.

Sincerely yours,

L.E.J. Brouwer

[Signed autograph, copy – in Brouwer]

911-03-31

To D. Hilbert — 31.III.1911   Blaricum

Göttingen dated 1.4.1911
To Blumenthal for [his] information,
with the request to be so kind to send
the letters back to Brouwer by registered mail. With best greetings Hilbert.

Dear Geheimrat, [Lieber Herr Geheimrat]

Enclosed I send you for your kind information the continuation of my
exchange of letters with Mr. Lebesgue, in the course of which I have withdrawn my submitted remark.

This withdrawal pleased me very much, because the letters of Lebesgue
(as well as those of Blumenthal later) showed me that my remark was interpreted by Lebesgue as a priority charge, which wasn't at all what I intended.

To me it remains inexplicable why Lebesgue doesn't want to bring the
elaboration contained in his last letter (the contents of which, by the way,
remained obscure to me after a first glance) to the notice of the readers of
the Annalen.

Yesterday I conversed very pleasantly a couple of hours with Weyl; maybe I see him again today.

With the best greetings.

Your
L.E.J. Brouwer.

[Signed autograph – in Hilbert (signed draft in Brouwer)]

1911-05-09

To O. Blumenthal — 9.V.1911   **Blaricum** [32]

Dear Professor [Sehr geehrter Herr Professor]

The situation with Mr. Lebesgue is now as follows:

1) It is impossible for me to discuss things further with him, because it is the second time now he has lost sight of politeness towards me.

2) In his latest letter he takes back his earlier statement that he had worked out a complete version of his proof in the Annalen. [33] The elaborations he announced to me earlier as such and which he sent to me, he now calls a 'hasty formulation' [34], for which he rejects every responsibility.

3) After the abandonment of this statement of Lebesgue, which was the ground for the withdrawal of my Annalen "Remark", I consider it my duty to resubmit my note.

4) The considerations that Lebesgue earlier called the 'complete version' are riddled with false conclusions, and they are beyond repair.

5) Recently Lebesgue has published in the Comptes Rendus a second proof which is likewise irreparably wrong. He still clings to the correctness of that, notwithstanding what I communicated with him (compare the paragraph of his letter that is marked in pencil).

6) I consider the submission of the enclosed "Remark", as I already remarked, as my duty, but as a disagreeable duty. When the editors would

---

[32] The handwriting is not Brouwer's. As Brouwer mentions a possible trip to Limburg (close to Aachen), the recipient must be Blumenthal. The 'Sehr geehrter Herr Professor' indicates that Brouwer observed a measure of formality, called for by the content of the letter. Moreover, the letter would probably be passed on to Hilbert.   [33] [Lebesgue 1911a].
[34] rédaction hâtine.

consider the publication not in the general interest, I would derive from this judgement the liberty to withdraw it.

7) My "remark" has been cast in a strictly objective formulation, and it remains silent about our exchange of letters; I would ask you to demand, in the possible answer from Lebesgue, the same from him.

I enclose the final part of the exchange of letters with Lebesgue.

I will probably come to Limburg within a few days; maybe I can see you? In that case I might learn from you what you think of my submitted article, better than in writing.

Sincerely yours,
(was signed) L.E.J. Brouwer

[Signed autograph, copy – in Brouwer]

Editorial supplement

[The following document is most likely the Bemerkung that the above letter refers to. It is dated May 1911. The text is an expansion of the note of March 1911.]

BEMERKUNG ZU DEN INVARIANZBEWEISEN DES HERRN LEBESGUE [35]

VON L.E.J. BROUWER IN AMSTERDAM

In the derivation of the invariance of dimension, which Mr. Lebesgue communicated in Vol. 70 of the Mathematische Annalen (p. 166–168), certain facts are assumed as 'quite evident'. [36] I have to remark that the justification of these 'evident' properties constitute the kernel of the whole proof.

In the Comptes Rendus (vol. 152, p. 841, March 27, 1911) the same author has developed a second method, where he uses the following lemma: 'For every regular closed manifold T_n that lies in a R_{n+p+1}, there exists an arbitrary small T_p that is linked with T_n.' From this it is inferred that T_n is not everywhere dense in R_{n+p+1}.

---

[35] A comment on the proofs of invariance of Mr. Lebesgue. [36] *bien évidents.*

Now there are two cases possible with respect to this argument.

Either R_{n+p+1} designates a Cartesian space; but then it is clear without any lemma, that T_n cannot fill R_{n+p+1} everywhere dense, because T_n is a closed and R_{n+p+1} an open manifold; consequently this trivial property is of no significance for the solution of the invariance problem. Or R_{n+p+1} designates a regular space in the general sense, and we must understand the lemma as follows: 'In every regular closed manifold T_n that lies in a R_{n+p+1} there exists an arbitrary small manifold T_p, which is linked with T_n in a certain neighborhood.' But then the proof of Lebesgue of this lemma implicitly presupposes the invariance of dimension, because in the case that T_n fills R_{n+p+1} everywhere dense, there may not be such open domains α in Γ, whose boundaries are used.

Finally, at the end of Lebesgue's proof in the Annalen some elaborations of Mr. Baire are adduced, but contrary to what Mr. Lebesgue states, these do not essentially solve the problem, but merely elucidate its connection with deeper theorems.

[Autograph manuscript – in Brouwer]

1911-06-11

To O. Blumenthal [37] **— 11.VI.1911**   **Amsterdam**

Dear Sir, [Cher Monsieur]

You have made clear to me that in the proof of invariance of dimension of a space I might have made the work for the reader lighter by prefacing the reasoning with a succinct explanation of the main ideas.

My proof is the rigorous elaboration of the following principles:

Let K be a q-dimensional cube, lying in a q-dimensional space E_q, with a center denoted by M, and a boundary denoted by F. If there would be a continuous (1-1) correspondence between E_q and a $(q+h)$-dimensional space E_{q+h}, such that K corresponds to a set k, and M to a point m of

---

[37] The document does not show the addressee, but from the content it is clear that it was written to Blumenthal. The letter was written in French, so that Blumenthal could forward it to Lebesgue.

the space E_{q+h}, each polyhedral set p of dimension q, resulting from a small continuous deformation of k, will be nowhere dense in E_{q+h}. Hence there exists in the space E_q a set P corresponding to p, which is nowhere dense in E_q, and which results from a small continuous deformation of K.

Next I show that each set π resulting from a small continuous deformation of K is everywhere dense in the neighborhood of M. First I show this for polyhedral sets π of dimension q; hence the same property ensues for more general sets π to which P belongs.

Preserving all details of this proof one can modify it slightly, considering the images f and F of F that belong to p and to π respectively, and then prove by that method that M is separated by F from infinity, whereas evidently m is not separated from infinity by f.

It seems to me that this is the modified proof that Mr. Lebesgue had in mind in the first part of his Note in the Comptes Rendus of March 27, 1911.

As for two other proofs published by Mr. Lebesgue, they hardly, in my opinion, merit that name.

In the one of the Mathematische Annalen, which you have published following mine, Mr. Lebesgue bases himself on certain facts said to be 'quite evident', suggesting that they are very simple properties whose proofs can be left to the reader. Mr. Lebesgue has as a matter of fact affirmed to me that this was indeed his idea, adding that it would suffice to project every I_p on the manifold $(x_2 = x_2^0, x_3 = x_3^0, \ldots, x_{p+1} = x_{p+1}^0)$, to deduce the existence of I_{p+1} from that of I_p.

I believe that Mr. Lebesgue is mistaken; that the proof of these facts constitutes a separate problem, which is more difficult than that of the invariance.

With respect to the second proof of the Comptes Rendus of March 27,1911, [38] it contains a vicious circle, in it the invariance is tacitly assumed as proven. Actually, if one isn't certain of invariance, it could happen that T_n fills Γ and that there exists no boundary at all of the domain a at a finite distance of T_n.

If E_{n+p+1} would refer exclusively to Cartesian spaces one could remedy this mistake and choose the manifold Γ in a special way. Indeed, if E_{n+p+1} is a Cartesian space, the reasoning that deduces from the lemma about the linked manifolds the theorem '*that T_n doesn't fill E_{n+p+1}*', and from there the invariance, doesn't make sense, because T_n is a closed manifold and E_{n+p+1} an open manifold, from which it follows immediately that T_n doesn't fill E_{n+p+1}, which is trivial and without importance for the invariance proof.

---

[38] [Lebesgue 1911b].

Hence one must assume that E_{n+p+1} can be a closed manifold, but then the vicious circle cannot be cured.

I cannot finish this letter without expressing my regret that the correspondence with Mr. Lebesgue, as a result of his article in the Mathematische Annalen, could not get us to agree. But I don't see that I was in any way wrong in this matter. This is what happened: Mr. Lebesgue publishes a proof of invariance where he assumes certain facts as 'quite evident'; I find that I don't see the evidence, and I turn to the author, who—having written it—must be assumed to understand it, and who has the scientific duty to explain himself to the first reader who asks for it. However, when Mr. Lebesgue answered me, not only did he not say anything precise about the question, but he also admits that he never worked out the requested argument. So Mr. Lebesgue didn't have the right to use the term 'quite evident', and in the volume of the Mathematische Annalen containing his article, a rectification would be necessary indeed. To that purpose I proposed to Mr. Lebesgue a remark by me, by the way leaving him the choice if he preferred another form. Mr. Lebesgue, getting angry, informs me that he has no objection at all against my remark.

A little later Mr. Lebesgue starts corresponding again and declares formally that he now possesses a complete version of the proof of the properties in question. That changed everything completely, I withdraw my remark from the Mathematische Annalen, and I ask Mr. Lebesgue to send me this complete version. Mr. Lebesgue complies with my request, and I study his argument several times, but it remains obscure to me, and moreover contains nothing that anybody can't see right away. I ask for new explications: Mr. Lebesgue excuses himself by rejecting any responsibility for the version he sent me, qualifies it as a 'premature version', in short retracts his former statement that made me withdraw my remark for the Mathematische Annalen. Then I have found it impossible to continue the correspondence. Where in this whole story have I done something to reproach myself? I'm not conscious of such a thing.

Sincerely yours, [39]

L.E.J. Brouwer

[Signed autograph copy – in Brouwer]

---

[39] *Agréez, monsieur, l'expression de mes sentiments cordiaux.*

From O. Blumenthal — 14.VI.1911   Aachen

Dear Mr. Brouwer! [Lieber Herr Brouwer!]

Thank you very much for your letter about Lebesgue. I believe that it is in principle quite felicitous, though you might also understand Lebesgue's wish for moderation of some terms. I have now, after your letter [arrived], studied once more the note of Lebesgue and I did it as precisely as I could; for my orientation I would like to ask you a few more questions.

First the article in the Annalen. The difficulty here is already in the construction of I_1, isn't it? At least it seems to me now, after I have grown suspicious, that the boundary of e_1 is already so complicated, that I don't see how Lebesgue works with that. As far as I remember, you thought that the problem only started at I_2. I'd like to ask you for information on this.

Now the Comptes Rendus. That you have recognized in the main proof a modification of your proof pleases me very much. Whether Lebesgue thinks that he gives another proof of the invariance of dimension in the note, or that he views the theorem about the linked manifolds as a further result, is not clear from the text. The title suggests that he considers the theorem as a result by itself, not as a lemma for another proof of the dimension theorem. It could also be possible that the final statement, that with these methods one can prove the dimension theorem in three different ways, was slipped in afterwards. In this respect I wouldn't be too hard on him. Now I would like to know: suppose that the dimension problem is solved, is then the theorem about the linked manifolds correct and rigorously proved? I don't dare to make a decision: it would be almost too beautiful if the theorem were true. I am especially suspicious because of the next-to-last theorem: A T_n does not fill $E_{n+p+1}(p \geq 0)$ and divides E_{n+p+1} into domains for $p = 0$ and only in this case. [40] That would be the reverse of the Jordan theorem for dimension n, and then, as far as I can see without the condition of reachability, hence a quite impossible result. Am I wrong there? Or did Lebesgue really goof so badly? That would surprise me very much.

Likewise, I would like to be informed about the theorem that Lebesgue uses as lemma in his Annalen proof, namely that $n+1$ domains always have a point in common. Can it be proved by your methods? Indeed, to me it seems that the theorem by itself is nice and important.

---

[40] Blumenthal quotes this sentence in French.

Finally I have a question that doesn't relate to Lebesgue. I talked with you here about the generalization of the Jordan theorem to space. Afterwards it occurred to me that you misunderstood me, and that you thought of a more difficult theorem than I did. The theorem I imagined, and which I hoped would be provable, runs more or less as follows: Let G be a closed point set in R_3 that remains completely finite and let p, r_1 and r_2 be fixed numbers. It must be possible to isolate around every point P of G a subset g of G, which contains P containing only such points that have a distance $\leq p$ to P, and that can be mapped continuously and one-to-one onto a plane domain that completely contains a circle of radius r_1 and on the other hand lies completely in the interior of a circle of radius r_2. Then G divides the space into two parts.

I mean, such a thing should be provable by the invariance of dimension. Maybe it is this theorem that Lebesgue had in mind.

I will forward your letter to Lebesgue only when I have your answer, because I need the explications which I ask from you, for my accompanying letter, in which I will have to discuss matters quite in depth.

Your visit to us in Aachen is also for my wife and me a very pleasant memory. I hope very much we will soon meet again. That should be possible somehow. Please give my best greetings to Mrs. Brouwer. My wife's health is still the same. She greets you most cordially.

Your
O. Blumenthal

[Signed typescript with handwritten insertions. – in Brouwer]

1911-06-16a

From O. Blumenthal — 16.VI.1911[a]   **Aachen**

Dear Mr. Brouwer! [Lieber Herr Brouwer!]

Thank you very much for your letter. Your first explanations about I_1 I cannot understand. You must have expressed yourself not clearly. I understand you as follows: in the case of dimension 2, the totality of all polygons that intersect a line $x = 0$ must be bounded inside the cube by a line that contains the entire interval $0 < y < 2l$. It is clear that this can't be the theorem, but that the length of this interval must be distributed over several parallels $x = \lambda$. See the figure at the end. The interval I_1 would then be the totality of the dashed lines in my figure, in other words a discontinuous structure. The existence proof for such an I_1 can of course only be given by your indirect method. Now what is the problem with I_2? Is it that I_1 already consists of separated manifolds? Would you please explain that to me once more, and in French too, so I can send it to Lebesgue and afterwards incorporate it in the article for the Annalen.

I have thoroughly examined the theorem about the linked manifolds, also for three-dimensional space, and I found no errors. I am glad that it seems correct to you too. My objections regarding the deduction of the Jordan theorem are indeed resolved by the observation that Lebesgue only speaks about the easy part of the Jordan theorem.

I didn't know, by the way, that this part is easy to prove, I myself had only occasionally and superficially thought about it, and nothing came out of it. When I think that it would be valuable to prove the Jordan theorem for space, I mean of course the whole theorem, and not for the special purpose I have in mind, but for the general interest. The conditions I wrote down, should be merely the conditions that a point set is a one-to-one continuous image of a sphere. It is clear they are necessary but I wasn't clear about the sufficiency. You have put me at ease about this point.

Would you be so kind as to give me the explanation about I_1 and I_2 once more in extenso and in French, and especially the reason why the existence of I_2 can't be inferred with the same indirect reasoning as in the I_1 case. Please explain this very clearly, so the reader of the Annalen, and I too, really understand it, because if Lebesgue missed your objections and still doesn't understand them, you must assume that they are not right away evident.

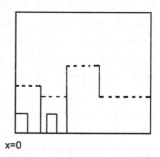

x=0

With best greetings

Your
O. Blumenthal

[Signed typescript – in Brouwer]

1911-06-19a

To O. Blumenthal — 19.VI.1911[a]   Amsterdam

Dear professor, [Lieber Herr Professor]

You are asking for a more detailed exposition not only for yourself, but
also for Lebesgue and for the readers of the Annalen. However, I would like
to explain this first to you personally, if you don't mind. Because I have
explained all of this to Mr. Lebesgue in letters, so extensively and repeatedly
that nothing new can be added. Just because of that, an explanation for
his attitude gradually forced itself upon me at last, namely that he saw his
error right after my first letter, but that he was too vain to admit it, and
that his further conduct was determined by the hope that he would perhaps
later find a proof of the presupposed theorems and by the necessity to win
time for that.

As far as the reader of the Annalen is concerned, in the sequel to the
public French discussion they will naturally find the necessary explanations.
For, I envision this sequel could be as follows: you urge Mr. Lebesgue in
your accompanying letter to produce now the complete proof he announced
both to me and to you he had ready. Then there are three cases possible: [41]
first: he produces again the same so-called complete proof, of which he
already sent a copy to me. This first case will occur of course if and only if
Lebesgue has been honest until now. Then I send you a new French letter
for the Annalen, [42] in which I reveal the irremediable errors of this proof
and add a proof of my own.

second: he corrects the errors of the article of the Annalen and produces a
completely new and correct proof, which he may have found in the meantime.
Then he is obliged to apologize for his behavior, but for the rest the matter
can be considered as finished.

third: he doesn't give any proof, and tries, as before, to back out of the
public discussion. In this case as well I publish in the Annalen my own
proof, starting with an explanation of the problems of Lebesgue's proof; I
must add to this explanation that Lebesgue doesn't know how to solve these
problems, which is clear from a correspondence with him! Thus the reader
of the Annalen will in all cases be completely informed, if you agree to this
plan, and he will get a complete and rigorous proof.

---

[41] For a similar list, see *Brouwer to Baire, 5.XI.1911* (draft).   [42] See *Brouwer to
Blumenthal 11.VI.1911.*

I would like to ask you, by the way, not to tell Mr. Lebesgue for the moment about the existence of my proof. I myself have on purpose kept silent about my own considerations, in my correspondence with Lebesgue, as in my French letter, which I submitted to you. For I am of the opinion that first Lebesgue must have stated clearly and plainly his views on his own subjects, before my proof or its existence can come into play.

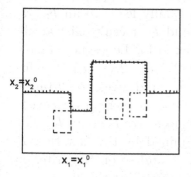

Now to business, and to begin with your plane figure.

In the case of two dimensions there is in the boundary of e_1, only *one single* (the line that dotted with small crossbars in the figure) one-dimensional space, which stretches from $x_2 = x_2^0$ to $x_2 = x_2^0 + 2l$; this we choose as I_1; the other one-dimensional spaces in the boundary of e_1, drawn as dashed lines, cannot extend from $x_2 = x_2^0$ until $x_2 = x_2^0 + 2l$.

Because I_1 is connected and stretches from $x_2 = x_2^0$ to l $x_2 = x_2^0 + 2l$, two subsets of I_1, of which the first contains the subinterval that borders $x_2 = x_2^0$, and the second one contains the subinterval that borders $x_2 = x_2^0 + 2l$, must have at least a point in common, so there exists certainly a point I_2, *hence for two dimensions there is no problem.*

For three dimensions it gets worse. For example, if we partition the edges of the main cube into 8 equal parts, then the cube is partitioned into 8^3 smaller cubes, and parallel to an arbitrary plane it can be split up in 8 layers of $[8^2]$ [43] small cubes each. I now assume that (as is possible) e_1 is composed of *in the first place* the whole first layer parallel to the plane $x_1 = x_1^0$, *secondly* of the second layer parallel to $x_1 = x_1^0$ the small cubes that are shaded in the figure here on the side; *in the third place* the entire third layer parallel to the plane $x_1 = x_1^0$, while e_2 and e_3 are composed of the two first layers parallel to respectively $x_2 = x_2^0$ and $x_3 = x_3^0$.

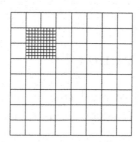

---

[43] 8^2 inserted by Freudenthal.

The boundary of e_1 inside the main cube is then composed of two connected two-dimensional spaces (one simply connected space $_\alpha I_1$ and one space $_\beta I_1$ that has the connectivity of a cylindric surface, which stretches both from $x_2 = x_2^0$ until $x_2 = x_2^0 + 2l$ and from $x_3 = x_3^0$ to $x_3 = x_3^0 + 2l$. *We must choose* one and only one of these spaces as I_1; keeping both together doesn't work, because all further conclusions rest on the *connectedness* of I_1; *which one* we must choose, Lebesgue doesn't say. Two spaces are considered here, but only one of them, namely $_\alpha I_1$ finally leads to an I_3; $_\beta I_1$ leads, as one easily sees, to a disconnected I_2, and I_3 doesn't exist at all. Still, $_\beta I_1$ possesses the property that is fundamental for Lebesgue, namely that it stretches from $x_\alpha = x_\alpha^0$ until $x_\alpha = x_\alpha^0 + 2l$ $(a = 2, \dots, n)$. So this property is completely worthless for the argument, and gives no certainty for the n-dimensional space that among the different $_\nu I_1$ there always exists one that finally leads by a suitable choice of the successive I_α to an I_n. The choice of I_1 from the $_\nu I_1$ will have to be determined by the fact that I_1 represents the *outer boundary* of e_1, but for I_2, I_3 and so on the criterion fails, so that I don't believe that one can achieve something this way.

As I said already above, these explanations do not contain anything new for Mr. Lebesgue. I hope very much that I have expressed myself completely comprehensibly now, and finally, I would, once more, like now to hear from you which theorem of the Analysis Situs you mentioned to me as absolutely necessary for the continuity proof [44] of the existence of polymorphic functions on Riemann surfaces? From your last letter it seems that I must conclude that it is not the Jordan theorem.

With best greetings

[Autograph draft – in Brouwer; partly as in *CWII* pp. 446–447 (from 'Now to business' on), with Freudenthal's comments]

1911-06-20

To O. Blumenthal — 20.VI.1911 [45]

For a further elucidation of my letter of yesterday I add that if in my three-dimensional example $_\beta I_1$ is chosen as I_1, then we obtain as 'boundary

------

[44] Cf. [Van Dalen 1999] section 5.3.   [45] No addressee – continuation of *Brouwer to Blumenthal 19.VI.1911*; see also Freudenthal's note in *CWII* p. 448.

of that part of I_1, which is contained in the elements of e_2 that do not completely belong to e_1' the following point set that lies in the plane $x_2 = x_2^0 + \frac{1}{4}l$:

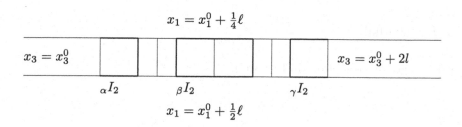

$$x_1 = x_1^0 + \frac{1}{4}\ell$$

$x_3 = x_3^0$ ⟶ $x_3 = x_3^0 + 2l$

$_\alpha I_2 \qquad _\beta I_2 \qquad _\gamma I_2$

$$x_1 = x_1^0 + \frac{1}{2}\ell$$

Of the three connected subsets of this point set $_\alpha I_2$ is completely contained in 'the elements of e_3 that do not entirely belong to e_1 or e_2'; $_\beta I_1$ and $_\gamma I_2$ lie however completely outside 'the elements of e_3 that do not entirely belong to e_1 or e_2'.

One should have to choose therefore as I_3 the boundary between $_\alpha I_2$ and $_\beta I_2 +_\gamma I_2$. But this boundary doesn't exist, and there is no I_3.

[Autograph draft/copy – in Brouwer; also in CWII, p. 448]

011-06-22

From O. Blumenthal — 22.VI.1911 Aachen [46]
Rütscherstrasse 48

Dear Mr. Brouwer! [Lieber Herr Brouwer!]

Please send me *immediately* your second proof copy. In March a batch of proofs that I sent to Teubner was lost, which I found out already in a different way. Probably I had put your proofs in the same package. So I ask you to send me quickly your second copy. — The proof that a closed surface divides the space in at least two parts is indeed very simple. I have figured that out already. For the purpose of function theory one needs *certainly* the division into only **2** subspaces. Whether one has to prove also reachability, I don't know yet, it is possible that one can do without. But

---

[46] Date and place - postmark.

even more important would be the reverse statement under the assumption of all possible reachability.

Many greetings,

your
O. Blumenthal

[Signed autograph, postcard – in Brouwer]

1911-07-02

To O. Blumenthal [47] **— 2.VII.1911**   **Amsterdam**

Dear Sir, [Cher Monsieur]

The following is a proof of the 'evident' theorem on which Mr. Lebesgue based a proof of the invariance of dimension.

We start from the concepts of *n-dimensional element* and *two-sided n-dimensional manifold (open or closed)* that were introduced in my memoir 'Ueber Abbildung von Mannigfaltigkeiten'[1], and we mean by *two-sided n-dimensional system* a finite set of n-dimensional elements belonging tot to one or more two-sided n-dimensional manifolds.

By the *boundary* of such a system we will mean the $(n-1)$-dimensional sides belonging to a single element of the system. The points of the system not situated on the boundary form the *interior* of the system.

The boundary of the system is composed of a finite number of closed two-sided $(n-1)$-dimensional manifolds. It is true that several p-dimensional $(p < n-1)$ sides belonging to a single or several of these manifolds can overlap or coincide, but in the following we will abstract from this circumstance.

From the indices of the elements of the system we derive the indices of the elements of its boundary by a familiar method.[2]

Coming back to the article of Mr. Lebesgue, we designate by C_h the side of the interval I determined by the equations $x_p = x_p^0, (p = 1, 2, \ldots, h)$.

---

[1]Mathematische Annalen, vol. 71, p. 1, 2, 5 of the proofs.   [2]ibid., vol. 71, p. 12 of the proofs.

[47]Cf. *Blumenthal to Brouwer 16.VI.1911*: '[...] in extenso and in French [...]'

Let I_h be a two-sided $(n-h)$-dimensional system, represented 'simpli-cially'[3] on C_h in such a way that the interior of I_h is represented on the interior of C_h and that the boundary of I_h is represented on the boundary of C_h. Then this representation possesses a certain *degree*,[4] which is an entire number that we will suppose to be equal to $+1$. It follows that the boundary of I_h is represented on the boundary of C_h with the same degree $+1$.

Those elements of the boundary of I_h whose interior is represented on the interior of C_{h+1} form a two-sided $(n-h-1)$-dimensional system S_{h+1}, whose interior is represented on the interior of C_{h+1}, and whose boundary is represented on the boundary of C_{h+1}. The degree of these representations is still equal to $+1$.

Let us destroy in I_h an element q_1, of which all image points have a coordinate x_{h+1} less than $x_{h+1}^0 + l$; we are left with a two-sided $(n-h)$-dimensional system I_h'. Those elements of the boundary of I_h' whose interior is represented on the interior of C_{h+1} *or* on the interior of C_h form a two-sided $(n-h-1)$-dimensional system S_{h+1}', whose image projected onto C_{h+1} gives a simplicial representation by virtue of which the interior of S_{h+1}' is represented on the interior of C_{h+1} and the boundary of S_{h+1}' is represented on the boundary of C_{h+1}. The degree of these representations still is $+1$, because there exist regions in the inside of C_{h+1} where the image of S_{h+1}' for the new representation is identical to the one of S_{h+1} for the original representation.

Let us now destroy in I_h a series of elements $q_1, q_2, q_3, \ldots, q_m$, one by one, who all have only image points with coordinates x_{h+1} less than $x_{h+1}^0 + l$, and among whom are all the elements I_h whose image touches C_{h+1}; we are left with a two-sided $(n-1)$-dimensional system $I_h^{(m)}$. Those elements of the boundary of $I_h^{(m)}$ that possess for the original representation images whose interior lies in the interior of C_h form a two-sided $(n-h-1)$-dimensional system of $S_{h+1}^{(m)}$. Repeating for each q_α the reasoning applied to q_1 we will find that the image of $S_{h+1}^{(m)}$ for the original representation projected onto C_{h+1} gives a simplicial representation by virtue of which the interior of $S_{h+1}^{(m)}$ is represented with degree $+1$ on the interior of C_{h+1} and the boundary of $S_{h+1}^{(m)}$ is represented with a degree $+1$ on the boundary of C_{h+1}.

Let us denote $S_{h+1}^{(m)}$ by I_{h+1}; operating on I_{h+1} like on I_h, and so forth, we will not stop until we reach I_n, in other words to a system of points

---

[3]ibid., vol. 70, p. 162.   [4]ibid., vol. 71, p. 7 of the proofs.

represented with degree $+1$ on the point C_n, hence consisting of at least one point. [5]

You see that the ideas and methods used in my proof of invariance reappear, all more complicated than the proof of the 'evident' theorem of Mr. Lebesgue, so that the Note of Mr. Lebesgue (and likewise the one of Mr. Baire about the same problem) does not have any other merit for invariance than reducing it to a more difficult problem.

Cordially yours

L.E.J. Brouwer

[Signed autograph draft/copy – in Brouwer]

1911-07-08

To O. Blumenthal — 8.VII.1911   **Blaricum**

Dear Professor [Lieber Herr Professor]

As a matter of fact I noticed yet another gap in the so-called third proof of Lebesgue. It is in the words: 'Let us reduce α in size in such a way that it is bounded by a finite set of polygonal manifolds T_p'. [48] But is, even for a domain in Cartesian space such a 'small reduction' possible? In three-dimensional space it is always possible, because there the boundary of a domain determined by a finite number of planes (as such the 'reduction' can of course always be constructed) is composed of a finite number of two-dimensional manifolds. This property vanishes already in four-dimensional space, as seen from the following example:

At the point O we place four mutually orthogonal three-dimensional co-ordinate spaces. They partition the neighborhood of O into 16 parts, which can be distinguished by the signs of the coordinates. From these 16 domains

---

[5] One might slightly modify the preceding reasoning and consider instead of the degree of the representation of I_h on C_h the parity of the number of points of the intersection of I_h with a plane manifold $x_p = x_p^0 + b, (p = h + 1, h + 2, \ldots, n; 0 < b < 2l)$. But this modification (analogous to the one that is contained in the Note of the Comptes Rendus of March 27, 1911, of Mr. Lesbesgue about my invariance proof) doesn't affect the basis of the argument.

[48] quote in French.

we select eight which consecutively have the following coordinate signs.

$$+ + ++; + + + -; + + - -; + - - -; - - - - -; - - - - +; - - ++; - + ++$$

The domain G composed of these eight subdomains determines a ring shaped domain on a sphere K around O, and this domain is bounded by a torus ρ; the boundary g of G will be found by projecting ρ from G. This point set g is *not* a three-dimensional manifold, and can also not be composed from a finite number of three-dimensional manifolds: one only has to consider a neighborhood of O in g. Perhaps one can prove that the boundary of a 'small reduction' of a such domain like G not only in this simple case, but also in complete generality, can be assembled from closed manifolds. But in any case that is a problem on its own. Maybe it is not hard, but I doubt that Lebesgue has been aware of this problem.

[crossed out part:] This new difficulty is of a quite different sort than the one that occurs in the basic definition of 'linked manifolds', and it would justify a criticism of the third proof in a form which differs form the original one. Or should one interpret the entire second part of the Comptes Rendus Note, that I can't give any definite interpretation in more than three dimensions, purely as a communication of an idea without any pretense of rigor?

As far as the Annalen article is concerned, such an interpretation is impossible.

One cannot subject a Note in the Comptes Rendus to the same requirements as an article in the Annalen.

Criticism on: 'it follows' etc.

[Handwritten draft – in Brouwer]

Editorial supplement

[Remarks by Brouwer in Dutch and German — notes jotted on top of the above letter. The middle one in German, the other two in Dutch.]

In a multiply connected space E_{n+p+1} not every T_n can be contracted to a point. What happens then with the definition of 'linked manifolds'? (interpreting E_{n+p+1} exclusively as Cartesian won't do, for an earlier mentioned reason). And can we take each T_n as *some* boundary of a space of dimension $(n+1)$ in E_{n+p+1}?

Precisely because in this part of the Comptes Rendus-Note one cannot think of anything definite in more than three dimensions, it is so difficult to find in it a starting point for a constructive criticism, after I had to give up on the 'vicious circle' criticism.

For an even number of intersections of two closed spaces the proof is easy if one of the spaces is a line; because then on such a broken line we make each time a jump at one bend.

And we must check two kind of crossings, namely for a line interval with an $(n-2)$-edge, and of moving points with an $(n-1)$-edge.

But in general there are more kinds and much more difficult crossings to check.

1911-07-14b

To D. Hilbert — 14.VII.1911^b   **Amsterdam**

Dear Mr. Geheimrat, [Lieber Herr Geheimrat]

I will spend a few weeks in the Harz, and will travel via of Göttingen, I will stay there a few days. I am very much looking forward getting acquainted with people and things there, and more in particular to see you and Mrs. Geheimrat again. I hope to travel on coming Monday or Tuesday from here to Göttingen.

Enclosed you find the tragic end of the correspondence with Lebesgue.

Together with this letter I send to you the proof of invariance of an n-dimensional domain, for publication in the Annalen. [49] Immediately after my return I will prepare the proof of the Jordan theorem for space for publication in the Annalen. [50]

My wife regrets very much that this time she can't come with me, because of the pharmacy, and greets you most cordially.

Your
L.E.J. Brouwer.

[Signed autograph – in Hilbert]

---

[49] [Brouwer 1911c].   [50] [Brouwer 1911d].

911-08-19a

To O. Blumenthal — 19.VIII.1911^a Amsterdam

Dear professor [Lieber Herr Professor]

As per our agreement I inform you that in the end of June or the beginning of July I received the proofs of the figures (without text) of my article 'Beweis des ebenen Translationssatzes', and, as indicated in these, I have sent them immediately back to Teubner, and since that time no proofs of either text or figures have arrived here.

I have thought more about the difficulties in the second part of the Note by Lebesgue in the Comptes Rendus, and I am now convinced that the justification of the (by the way undoubtedly correct) definition of *linked* manifolds T_n and T_p of Lebesgue is a very deep problem. I did succeed in determining a clarification for a more restricted concept, namely for *linked*, manifolds T_n and T_p, *measured in a certain way*; and because I restricted myself to this more narrow concept, I could reconstruct the course of Lebesgue's proof. The scope of the theorem of Lebesgue is then considerably restricted for arbitrary n and p, but for $p = 0$ it also says in the narrower version, that *in R_{n+1} a one-to-one continuous image of the n-dimensional sphere determines at least two domains*, i.e. the first part of the Jordan theorem in arbitrary dimensions.

I venture to communicate to you also my second proof of the theorem used by Lebesgue in his article in the Annalen. Neither the concept of degree of a mapping nor the sequence of the I_p are used. In the e_p I disregard the points that belong to the boundary of I. I denote by f_p the $(n - p)$-dimensional interval contained in the boundary of I that satisfies the equations $x_1 = x_1^0, x_2 = x_2^0, x_3 = x_3^0, \ldots, x_p = x_p^0$; I denote by g_p the $(n - 1)$-dimensional interval contained in the boundary of I which satisfies the equation $x_p = x_p^0$; and h_p denotes the point set consisting of the $(n - 1)$-dimensional intervals $x_{p+1} = x_{p+1}^0; x_{p+1} = x_{p+1}^0 + 2l; \ldots; x_n = x_n^0; x_n = x_n^0 + 2l$ contained in the boundary of I.

Then the boundary of e_1 is composed of a finite number of closed $(n-1)$-dimensional manifolds; we denote the one among these that contains f_1 by μ_1; μ_1 consists of f_1, of parts contained in h_1, and of parts contained in the interior of I.

The part of μ_1 contained in $g_2 + e_2$ is bounded by a finite number of closed $(n-2)$-dimensional manifolds; we denote the one among these that contains f_2 by μ_2; μ_2 is composed of f_2, of parts contained in h_2 and of parts contained in the interior of I.

The part of μ_2 that is contained in $g_3 + e_3$ is bounded by a finite number of closed $(n-3)$-dimensional manifolds; we denote the one among these that contains f_3 by μ_3; μ_3 is composed of f_3, of parts contained in h_3 and of parts contained in the interior of I.

Proceeding in this fashion we finally arrive at a point pair μ_n containing the point f_n, and as there is no h_n, the second point lies inside I. This point belongs as well to $e_1, e_2, \dots, e_n$, as to an E_i that is not contained in any e_p, and this proves the Lebesgue theorem.

My complete Jordan proof [51] is now also finished for n-dimensional space, and I hope to finish writing it up this month. Can I send the article then to Aachen [52]?

Will you get the extended memoir of Lebesgue for the Annalen?

Many greetings and goodbye for now!

Your

L.E.J. Brouwer

[Signed handwritten draft – in Brouwer]

1911-08-19b

To C.S. Adama van Scheltema [53] — 19.VIII.1911[b]   Amsterdam

Dear Carel [Beste Carel]

Coming home after protracted wanderings I find the Faust, [54] which must have been here already for six weeks or so. I am glad it has finally appeared, and I believe that with this translation you have achieved the achievable; but what a mass of diligence, concentration, and dedication you have sacrificed on the altar of piety for your great predecessor!

---

[51] [Brouwer 1911d].   [52] Blumenthal's hometown.   [53] Addressed - 'Bergmann-strasse 62[4], München'.   [54] Scheltema's Faust translation into Dutch, [Adama van Scheltema 1911].

I didn't know you had taken your task so conscientiously. From what sentiment do you draw the strength for it?

Your chum
Bertus

[Signed autograph, postcard – in Scheltema]

11-08-26

From O. Blumenthal — 26.VIII.1911   Aachen

Dear Mr. Brouwer! [Lieber Herr Brouwer!]

Many thanks for the manuscript and the accompanying letter. [55] Apparently the proof that you have held out in prospect to me on the way back in Aachen, was the Lebesgue proof.

It is very simple and clear. We will see what Lebesgue does. I still think that he also has such a simple proof, which he has condensed so strongly in his manuscript that it is not possible anymore to read his true intentions in it. For what we at the time discussed in the Linzenshäuschen, was definitely not even the beginning of a proof. I have written now to Lebesgue and asked him to give me his 'Mémoire étendu' [56] for the Annalen, but as yet I have no answer.

I would like now to deal quickly with the automorphic functions. At the moment I'm not really up to date on the topic. Fricke is certainly much more competent than I. Altogether, it would surprise me if one could manage it with the simple Jordan theorem including reachability, but without any converse. At least with elliptic functions I always do the proof by using the reverse of the Jordan theorem, but it is possible that this is not necessary. But I believe that it would be easy for you to get information by yourself about this problem. In Klein's article in Mathematische Annalen 21 [57] the problem is completely and clearly formulated from a set theoretic point of view, even though the answer given there doesn't satisfy the standard for rigor. I strongly advise you to go through the matter there, and *not* in the fat Fricke and Klein, [58] where one trips again and again over details that obscure the general idea.

---

[55] Probably   [Brouwer 1911d].   [56] Extended   memoir   [57] [Klein 1882].
[58] [Fricke 1897, Fricke 1912].

Now as to your manuscript. I believe I have understood the proof itself in broad outline: before I get to a *real* understanding of things in the Analysis Situs always takes some time. But I have to say that I would have wished the exposition to be somewhat different. It is about a fundamentally important theorem, which will be used and read by many people. So I find it really awkward that for the reading of your paper there is so much cross reference to other papers, and not only to yours but also to the Comptes Rendus Note [59] of Lebesgue. I would thus strongly advise you to make a more complete version of this article, and explain in it as far as somehow possible all of the various concepts occurring in it, such as pseudo-manifold, net, fragment, and thus only for the *theorems* about these matters refer to your earlier papers; but in the very first place give the proof that the Jordan surface divides space into at least two parts in the article itself. This is all the easier, because the proof is so brief anyway. At least I had prepared once a proof that seemed perfect me to be and that could be given in a few lines. The citation of Lebesgue can of course remain, but then it will be a pleasant and courteous extra, and not an essential ingredient.

Please understand me correctly: if you don't want to make any changes, I will of course accept your work, also in its present form, but I believe that you would do yourself and your readers a great favor, if you complete your article in the way I indicated. If you are willing to make the change, I will send you back the manuscript.

Best greetings to you and Mrs. Brouwer.

Your
O. Blumenthal

[Signed typescript – in Brouwer]

––––––––––

1911-09-14

To D.J. Korteweg — 14.IX.1911   Blaricum

Professor,

I will tell my informants about the arrangement concerning the Proceedings. [60]

––––––––––

[59] [Lebesgue 1911b].   [60] *KNAW, Verslagen, Proceedings.*

Already about a week ago I had asked the Library to order Klein-Fricke for me from Delft. In the meantime I have as yet not received it; I have now written that I prefer the library to send me the copy of the Society. [61]

As far as the lectures in projective geometry are concerned, I am not convinced by you. By the 'authorities' I didn't mean the board [62] of the university (about whom I can indeed be assumed to know nothing) but the Mayor and Aldermen themselves, [63] who are not bound by the advice of the governors, and who bear full responsibility when they wish to perpetuate a shortage of teaching staff that has been pointed out to them; they who are so liberal with respect to other subjects of our faculty.

In the case of physics, chemistry, botany, and zoology they even arranged for a far more ample staffing than at other universities; for mathematics plus mechanics plus astronomy they still think two professors are sufficient. In Groningen and Utrecht those two are available for mathematics exclusively, which makes a big difference.

The rejection of your request by the Mayor and Aldermen was in my view an insult, not only for me and you, but even more so for our science, which should not be an appendix, but the crown of the faculty.

You say that mayor and aldermen have granted me the position of 'privaatdocent'. But although it is true that in other cases they really grant something substantial, namely an opportunity to get some visibility, or a means to collect fees from the students, in my case Mayor and Aldermen know after your request as well as you and I do, that it is I who grants the Mayor and Aldermen something, namely my assistance in teaching, and by their refusal they have qualified me as maybe the only municipal employee that isn't worth a wage.

Persons in governing positions are usually very far removed from our science, and as a consequence they are more or less insensitive to its needs. But to bow my head without protest for this insensitivity, doesn't seem to be my responsibility, even though the term 'insipid' in my previous letter was perhaps not quite well chosen.

With cordial greetings

Your

L.E.J. Brouwer

---

[61] The Dutch Mathematical Society   [62] curators.   [63] The University of Amsterdam was a municipal university. The governors (members of the board) were called 'Curatoren', and the Mayor of Amsterdam was qualitate qua president of the board.

I already dropped by you, but this summer the Kostverloren Vaart [64] is more poisonous to me than ever, so I come to Amsterdam only for the most urgent administrative matters.

[Signed autograph – in Korteweg]

1911-10-08

From O. Blumenthal — 8.X.1911   **Aachen**

Dear Mr. Brouwer! [Lieber Herr Brouwer!]

I have the bad luck that in my function theory business I stumble each time upon geometric problems, which I don't trust myself to handle in full rigor, even when I have an outline of proof that seems reliable to me if done correctly. Usually I trust that these questions must seem very easy and childish to you. I need the following theorem for four-dimensional space. In my opinion the difficulty is the same for each dimension > 2.

Let a continuous closed manifold M be given that lies completely in a finite part of R_4. Now I consider the totality of all planes, i.e. three-dimensional linear manifolds, that have points in common with M. The totality of all these common points apparently is composed of continuous manifolds and single points, which form a nowhere dense set in the plane.

Theorem. There are always planes that have only nowhere dense points, but no continuous manifolds in common with M.

It would be even nicer when there were planes that only have a finite number of points, or even one point in common with M. But that is maybe hard to prove, and maybe not even correct at all.

Can you prove the theorem? Until now I don't quite see how one should do that. For the case of a three-dimensional analytic M in R_4 one can work out a simple proof with the indicatrix, but that is apparently a detour, because the only thing that matters is that the manifold is closed and lies in a finite part.

I would be very grateful for a speedy answer.

I was very sorry that I missed you in Karlsruhe. Bernstein told me that you argued with Koebe about the continuity proof. It is no pleasure to have

----

[64] A canal in Amsterdam.

a discussion with him; but until now he has never made a mistake, therefore I am inclined to give him credit in this case,[6] especially because I believe that I can just about see what he can do with the deformation theorem [(65)]. However, Bernstein ascribed opinions to him that would be very disputable. But I think this is a misunderstanding of Bernstein.

Many greetings to you and your wife from us both.

Your
O. Blumenthal

I have received your proofs. And I also thank you very much for your reprints.

[Signed typescript – in Brouwer]

)11-10-12

From O. Blumenthal — 12.X.1911   **Aachen** [(66)]
Rütscherstrasse 48

Dear Mr. Brouwer! [Lieber Herr Brouwer!]

I must correct my latest letter insofar that I don't need the indicated theorem anymore, I'm pleased to say. My results could be obtained in a simpler way. Nonetheless I keep thinking the thing is right and also interesting by itself. I can't find a more or less clear proof. All the same you must be able to do it, because you know how to 'add dimensions'. If in three-dimensional space each support plane of arbitrary direction had a continuous structure in common with the continuous closed manifold, then one would have on the manifold at least ∞^1, ∞^2, ∞^3 points. That is of course no proof, because first of all one has difficulties with such lines that are common to a whole sheaf of support planes. That is however so far the way I got closest to approaching a proof. Another approach, wherein I wanted to prove that a closed surface with the property demanded in my theorem, must necessarily

---

[6][handwritten remark] just like earlier Lebesgue, so maybe also unjustifiably.

[(65)] *Verzerrungssatz.*   [(66)] Date and place - postmark.

have a corner, was too hard for me to think through. I hope anyway that
you will publish your uniformization of the automorphic functions in the
Annalen?

Many greetings

Your
O. Blumenthal

[Signed autograph, postcard – in Brouwer]

1911-10-28

From R. Baire — 28.X.1911   **Dijon**
                                                       Université de Dijon,
                                                        Faculté des Sciences

Dear sir and colleague,   [Monsieur et cher collègue]

I thank you cordially for sending me your publications, and I congratulate you very much with the progress you have made in the field of modern
Analysis Situs.

As for me, I feel obliged for several reasons to postpone developing the
methods that I had indicated in my publications of 1907. At that time I was
too much occupied with working out my ‘*Leçons sur les théories générales
de l’Analyse*’, [67] and after completion of that work I have unfortunately
fallen ill, and for some time I had to leave aside my research.

Dear colleague, renewing my thanks to you, I hope you accept my best
wishes for a beautiful scientific career.

René Baire

[Signed autograph – in Brouwer]

----

[67] Lectures on the general theories of Analysis

)11-11-02

From R. Baire — 2.XI.1911   **Dijon**
                                                        Université de Dijon,
                                                        Faculté des Sciences

Dear colleague, [Mon cher Collègue]

For the reasons related to my health that I mentioned to you, I am at
this moment not able to pay sufficiently sustained attention to the study
of the questions that are raised in your letter. If I am not too indiscrete,
permit me to ask you who are the authors about whom you complain? One
is no doubt Lebesgue. It so happens that since several years I haven't had
personal relations with him, for reasons that have nothing to do with pure
science.

I haven't studied his proof in the Mathematische Annalen in depth, and
his exposition was anyway too condensed.

As far as my method is concerned, there still was some work required to
make it valid for n dimensions (I speak about the definition of the outside
and the inside of a surface; the method was indicated by a phrase in the
middle of my Note in the Comptes Rendus of 1907). I am convinced that
there is in principle no difficulty, and no doubt that is what Lebesgue wanted
to say.

I had hoped to improve these methods, and to give more complete theorems, but I didn't get around to it right away I had left that work provisionally aside, and since then I have been taken by surprise by the illness

I apologize that my great weakness in German and my ignorance of
English don't allow me to follow your publications quickly enough. That is
a deficiency of us French, that we have poor knowledge of other languages.

Cordially yours, dear colleague [68]

René Baire
24 rue Andra

[Signed autograph – in Brouwer]

––––––––––––––

[68] *Recevez, mon cher collègue, l'expression de mes sentiments les plus cordiaux.*

1911-11-05

To R. Baire — 5.XI.1911   **Amsterdam**
                                                     Overtoom 565

Dear colleague, [Mon cher Collègue]

I vividly enjoy continuing our correspondence.

The mathematicians that I think I have to complain about are Zoretti [69] and Lebesgue.

About your studies of 1907, they aim at the proof of the *invariance of the n-dimensional domain,* in other words the theorem that in the space E_n the (1-1) continuous image of a set without boundary points forms a set which itself also has no boundary points.

However, the equivalence of the invariance of the number of dimensions is the following theorem, which is much more restricted:

'In the (1-1) continuous image of a set that doesn't contain anything but non-boundary points, the non-boundary points form an everywhere dense set.' (See for this subject my note in volume 70 of the Mathematische Annalen, and the article of Mr. Fréchet in volume 68 of the same journal).

In my eyes the great merit of your studies of 1907 is that they show that the invariance of the n-dimensional domain can be deduced from the following theorem:

'In E_n the (1-1) continuous image of a closed manifold of $n-1$ dimensions determines *at least* two domains.'

This remark was a step forward in the solution of the extremely important problem of invariance of the n-dimensional domain, because its solution allows to use the continuity method in a perfectly rigorous manner for the uniformization of algebraic functions (see Poincaré, Acta Mathematica 4, p. 276–278).

Now, for the invariance of the number of dimensions, the theorem where you stopped, didn't lead to any progress, because the theorem is – in my view – much more difficult than the invariance of dimension. As I see it, the outline you give in the Comptes Rendus leaves the main difficulty untouched. For a long time I have been searching for a proof; for $n = 3$ it is easy, for arbitrary n I only have found it this summer, by means of a reasoning which

---

[69] Zoretti had in his review of Schoenflies' 1908 *Bericht* mentioned Baire, Lebesgue and Brouwer (in this order) as having made a decisive step forward in the matter of the dimension invariance.

I then rediscovered in the second part of the Note of Lebesgue (Comptes Rendus, March 27, 1911) [70], where by the way it is in a form that is almost completely incomprehensible, and inexact if one reads it literally.

This proof is complicated in a way differing completely from that of the one about the invariance of dimension, and it seems to me that one will not be able to simplify it.

Earlier I had succeeded in proving the invariance of the n-dimensional domain by means of the following lemma.

'In E_n the (1-1) continuous image of a closed part of closed manifold of $n - 1$ dimensions determines only one domain.'

And afterwards I have completed the result of Lebesgue by proving that in E_n the (1-1) image of a closed $n - 1$-dimensional manifold determines *precisely* two domains.

Concerning the Note of Lebesgue on pages 166–168 of volume 70 of the Mathematische Annalen, the characterization of the sequence $I_1, I_2, \ldots, I_n$ is very unsatisfactory, because it can happen that it already stops at I_3. And with this characterization the whole proof collapses.

This is what Lebesgue recognized immediately, when I pointed it out to him, and he has answered me by trying to complete the characterization of the I_p. Now, these additions turned out to be still insufficient; later Lebesgue has given a new proof of his lemma, in which the I_p did not play a rôle anymore. Neither I, nor Mr. Blumenthal (Editor of the Mathematische Annalen) have been able to understand this proof (taken literally it was wrong, but that was maybe because of an awkward formulation); well, Mr. Lebesgue refuses not only to give us new explanations, but he also doesn't want to come back to the subject in the Mathematische Annalen and correct the reasonings that he already has recognized to be wrong.

I myself have composed a proof of the lemma of Lebesgue, a few days after its publication, but I think I shouldn't publish it and leave Mr. Lebesgue the opportunity to acquit himself of his duty.

Sincerely yours, dear colleague

L.E.J. Brouwer

[Signed autograph – in Baire]

---

[70] [Lebesgue 1911b]

Editorial supplement

[Private note in Dutch by Brouwer concerning the above letter] [71]

Letter to Baire 5.11.11

Explication of the 'lie deeper' than invariance. The continuity method of Poincaré.

Last summer I found the proof of the lemma of Baire, but then I recognized the proof in the 2nd part of the Comptes Rendus Note [72] of Lesbesgue. *Before that* I had found my other 'Proof of the invariance of the n-dimensional domain'. *Later* the proof for the complete Jordan theorem. — *About the Annalen piece of Lebesgue.* [73] I point out to Lebesgue the insufficiency of the characterization of the I_p. Lebesgue tries in his first letter to tidy up that characterization. For me they still are insufficient. Then Lebesgue tries to get there without the I_p. This last proof incomprehensible, for me and Blumenthal. However Lebesgue refuses $1°$. further information $2°$. to correct his error in the Annalen. I myself had proved the Lebesgue Annalen theorem a few days after it appeared, but I don't publish this proof, because I must give Lebesgue the opportunity to fulfill his duties.

The priority of my 'Invariance of domain' is not upset:

a. *publicly*, because with Baire-Lebesgue not yet everything has been published, but from Lebesgue one may expect supplementary arguments, as he *promised* them so emphatically, and hence doesn't seem to think they are trivial.

b. *privately*, because, when Lebesgue informed me that he could get half of Baire right with his 2nd Comptes Rendus, I already possessed my 'invariance of domain'.

c. *publicly and privately*, because Lebesgue didn't formulate 'invariance of domain', neither in Comptes Rendus, nor in his letters, and even Baire only mentioned 'invariance of domain sets'.

---

[71] Cf. also *Brouwer to Blumenthal 19.VI.1911.*   [72] [Lebesgue 1911b].
[73] [Lebesgue 1911b].

In the next letter to Baire point out that Lebesgue wrote in his remark in the Annalen 'because of Baire, whom he knew to be very neurasthenic', and that Lebesgue should have mentioned me when he corrected his Annalen article in the Comptes Rendus.

That I don't speak in my first of the 3 Annalen articles about the Baire-Lebesgue proof, is just because I understood this *formally erroneous* proof only after I had found it myself all over again; but then my first Annalen article was already submitted.

[Autograph draft – in Brouwer. English translation in *CWII* p. 441–442, with Freudenthal's comments.]

●11-11-07a

To C.S. Adama van Scheltema — 7.XI.1911[a]   Amsterdam

Dear Carel, [Beste Carel]

Your report that I have not only foregone Annie's sauce, but also a fair with roller coaster, has of course aggravated my regret and remorse not a little.

But fortunately, this winter we can walk across the moor together and we can contrast our lonely lives: self-confidence, faith and creative power against universal denial, passive contemplation and a little vandalism.

Although I am nowadays rather fertile, and gradually have acquired some international fame and envy, don't get too serious an impression of my work. For I still harbor the intimate certainty that mathematical talent is analogous to an abnormal development of the nail of the big toe.

On congresses I perform for the popes of science the rôle of enthusiastic ensign, but when I sketch in spirited conversation 'mit flammender Begeisterung' [74] the perspectives that are the soul of my work, my apparently absorbed gaze lavishes itself on the monomania of their expressions, and sees desolately trapped heroes in some, in others poison brewing goblins, and in the latter the anonymous torturers of the former. And while I am

---

[74] With blazing enthusiasm.

physically imbued with the feeling of being in hell, my eyes radiate the sadistic lust of sympathy.

My productivity will never bring forth a grand creation, because it is only fertilised by the derisive analysis of existing things.

None of the colleagues, however, will ever fathom this, though some of them in the long run are feeling uneasy in my presence, and those then make the rounds calumniating.

Every now and then I talk to Bertha [75] in the train; she told me that you are going to publish your Italian diaries, [76] and she asked what it actually was that made Faust beautiful. She had asked others, but without result. It seems I have been somewhat successful, because the next time I saw her, she was engrossed in Faust.

Thanks for your letter, and your handshake across the German tankards — which always cheers me up, — and warm greetings to you and Annie, also from Lize,

Your friend Bertus

[Signed autograph – in Scheltema]

1911-11-21

To O. Blumenthal — 21.XI.1911   Amsterdam

Dear Professor [Lieber Herr Professor]

Can the enclosed Correction and Addendum [77] still be included in the last issue of Vol. 71? I would be most grateful for that.

The contents of the Addendum I had sent last week to the printer, to be included in a footnote. Unfortunately it was already too late, which I have regretted very much.

I owe you more information about the publication of my uniformization. Koebe claimed in Karlsruhe that he was already for a long time in possession of all arguments lectured about by me, except of the invariance of domain; and had partly stated these already in his articles, but he could not right

---

[75] Scheltema's sister.   [76] [Adama van Scheltema 1914].   [77] In text *Berichtigung* and *Nachtrag*; published as *Berichtigung* and *Bemerkung* in [Brouwer 1911b, Brouwer 1911a].

away name places. Therefore I can't make up my mind about publication; the invariance of domain appears now by itself; the remainder does not seem very profound to me, and I can very well imagine that it is completely trivial for the automorphic professional. In any case Koebe will probably refer to this in future publications as something completely self-evident; and that could then rob the, anyway not all too great, importance of my possibly available publication.

I have now sent my Karlsruhe talk [78] to Fricke, and I am very curious about his — indeed the most competent — opinion.

In the Bulletin des Sciences Mathématiques of October 1911 (S. 287) Baire (sic! [79]), Lebesgue and I, in this order, are quoted as founders of the invariance of dimension. This matches exactly with the opinion which I thought I could read between the lines of Lebesgue's Note in the Annalen, [80] as I recently wrote to you. You can see from this how much to the point my critical footnote is, in more than one respect. Indeed, didn't in fact Lebesgue officially throw down the gauntlet for me with his Annalen Note? [81]

Best greetings!

Your
L.E.J. Brouwer

[Signed autograph draft – in Brouwer]

11-12-05

From R. Baire — 5.XII.1911   **Dijon**
                                                      Université de Dijon,
                                                      Faculté des Sciences

My dear colleague, [Mon cher Collègue]

I want to thank you immediately for sending your articles, though I can't promise you that I will study them right away.

---

[78] [Brouwer 1912b]. The Letter to Fricke, [Brouwer 1912d], was dated 22.XII.1911. [79] inserted by Brouwer. [80] [Lebesgue 1911b]. [81] followed by the crossed out 'and this I must publicly take up.'

Formerly I have been a close friend of Lebesgue, my comrade at the École Normale. Our separation has come about by an act of his, as a consequence of malicious procedures he used against me, in matters of career, not of a scientific nature. That today he tries to give himself a beautiful rôle by praising my works out of 'pity' (?!), it's one more malicious procedure, especially in a letter addressed to a third party and a foreigner.

I don't ask him to advertise for me. I think that you will be the first to recognize that my very pronounced neurasthenia didn't stop me to push for clarity in my work at least as far as Lebesgue.

To return to the scientific question, I regret that unfortunate circumstances have prevented me from keeping the promise made in my notes of 1907. I still believe that by following the method that I indicated very *succinctly* in the Comptes Rendus, one can prove *without fundamental difficulties, but maybe requiring a longish exposition,* the propositions that I need. But on the other hand these propositions form a less complete set than your statements 1, 2, 3 of p. 314. [82]

Cordially yours, [83]

René Baire

[Signed autograph – in Brouwer]

1911-12-10a

To H. Poincaré — before 10.XII.1911 [84]

I take the liberty to send you with this letter three small articles that recently appeared in the Mathematische Annalen, [85] as well as the unpublished text of a talk given by me at the German Congress in Karlsruhe on September 27, 1911. I can, however, not decide to publish this communication without asking you. [86]

---

[82] See [Brouwer 1911d].   [83] *Avec mes meilleurs sentiments de cordialité.*   [84] As Brouwer questions the analyticity of the correspondence mentioned below, and Poincaré showed in his letter of 10.XII.1911 surprise at this, it is not too far fetched to date this letter before December 10. Furthermore, in view of the reprints Brouwer enclosed, the letter may be dated after November 16.   [85] Probably [Brouwer 1911c, Brouwer 1911d, Brouwer 1911e], which appeared 16.XI.1911.   [86] An insertion is missing here.

My 'Beweis der Invarianz des n-dimensionalen Gebiets' [87] has been inspired by the reading your 'Méthode de Continuité' in volume 4 of the Acta Mathematica. [88] It was in the course of this reading that I had the impression that on the one hand one didn't know in the general case if the one-one and continuous correspondence between the two $6p - 6 + 2n$-dimensional varieties concerned, is analytic, and on the other hand that in order to be able to apply the method of continuity, one has to start by proving the absence of singular points in the variety of modules of Riemann surfaces of genus p; this last demonstration, incidentally, turns out to be fairly easy. Now after having read somewhere in an article by your hand (I believe about the equation $\Delta u = e^u$ in the Journal de Liouville) [89] that you considered your exposition of the method of continuity as perfectly rigorous and complete, I started to fear that I had poorly understood your memoirs in the Acta, and I have published my article 'Beweis der Invarianz des n-dimensionalen Gebiets' without indicating there the application to the method of continuity, restricting myself to an oral communication on this subject on September 27, 1911 at the Congress of the German mathematicians in Karlsruhe, of which communication I join the text to this letter. At the occasion of this talk Mr. Fricke has expressed to me his doubts to me that at the start I had formulated exactly the result of your arguments of pages 250–276 of the *Acta*. Meanwhile I continue to believe that I have interpreted you exactly.

In fact, if the conditions of this statement, in which the word 'uniformly' (uniformément) is the key word, are satisfied, the *reduced polygons of the sequence of groups converge also uniformly to the boundary of the $(2n + 6p - 6)$-dimensional cube*, and because of your arguments there exists at least a *reduced limit polygon* that only has parabolic angles on the fundamental circle, corresponding for that reason to a limit Riemann surface, for which either the genus is decreased, or the singular points have become coincident.

Would I ask too much of your benevolence and your precious time, asking you to be so kind as to convey briefly to me your opinion about the disputed points, to wit *1°* whether I have formulated the result of pages 250–276 of the Acta correctly and *2°* whether I was wrong saying on the first page of the attached communication that pages 276–278 of the *Acta* tacitly assume 'Theorem 1' and 'Theorem 2'?

---

[87] Proof of invariance of the n-dimensional domain. [88] [Poincaré 1887]; Poincaré uses here the 'method of continuity' to solve the equation $\Delta u = e^{2u}$. [89] [Poincaré 1895].

I would be extremely obliged if you could thereby deliver me from my doubt on these points.

Yours deeply revering [90]

L.E.J. Brouwer

[Autograph draft – in Brouwer]

1911-12-10b

From H. Poincaré — 10.XII.1911[b] [91]

Dear Colleague, [Mon cher Collègue]

Thank you very much for your letter; I don't see why you doubt that the correspondence between the two manifolds would be analytic; the modules of the Riemann surfaces can be analytically expressed as functions of the constants of Fuchsian groups; it is true that certain variables only can have real values, but the functions of those real variables preserve nonetheless the analytic character.

Now in your eyes the difficulty arises from the fact that one of these manifolds doesn't depend on the constants of the group but does depend on the invariants. If I recall correctly, I considered a manifold depending on the constants of the fundamental substitutions of the group; so to a group there will correspond a discrete infinity of points of this manifold; next I subdivided this manifold in partial manifolds, in such a fashion that to a group corresponds a single point of each partial manifold (in the same way as one decomposes the plane in parallelograms of the periods, or the fundamental circle in Fuchsian polygons). The analytic character of the correspondence doesn't seem to be altered to me.

With regard to the manifold of the Riemann surfaces one can get into problems if one considers those surfaces as Riemann did; one may for example wonder if the totality of these surfaces doesn't form *two* separate manifolds. The difficulty vanishes if one views these surfaces *from Mr. Klein's point of view*; the continuity, the absence of singularities, the possibility to go from one surface to the other in a continuous way become then almost intuitive truths.

---

[90] *Agréez, monsieur, l'expression de ma profonde vénération.*   [91] Postmark as mentioned by Mrs. C. Jongejan.

I apologize for the disjointed fashion and the disorder of my explications; I have no hope they are satisfactory to you, because I have presented them very poorly to you; but I think they will lead you to make the points that bother you more precise, so I can subsequently give you complete satisfaction. I am happy to have this opportunity to be in contact with a man of your merit.

Your devoted colleague, [92]
Poincaré

[Signed autograph – in Brouwer; also in [Alexandrov 1972]. See also [Zorin 1972].]

11-12-21

To A. Schoenflies — 21.XII.1911   **Amsterdam**
                                                                                                   Overtoom 565

Dear Professor [Sehr geehrter Herr Professor]

When I was last summer with Mr. Fricke in Harzburg, [93] the conversation turned to the new edition of your Bericht [94], and we thought that you might not be averse to a little help in correction of the proofs, thereupon I said that I personally would be happy to collaborate in this way.

Just now I hear from Fricke than he has conveyed my offer to you and that you are in favor of it, So I have to the honor to inform you most obediently that I am at your service. I am glad to be able to express in this way how much I feel obliged to your Bericht. With cordial greetings

Your
L.E.J. Brouwer.

[Signed autograph – in Brouwer]

---

[92] *Votre bien dévoué collègue.*   [93] Brouwer regularly stayed in Harzburg, see [Van Dalen 1999], p. 306.   [94] [Schoenflies 1900, Schoenflies 1908]; most of the corrections and revisions concerned the second part, which contained the basics of topology.

1911-12-22

To R. Fricke — 22.XII.1911   Amsterdam

Dear Geheimrat, [Hochgeehrter Herr Geheimrat]

With reference to our last conversation I inform you about some remarks related to the topological difficulties of the continuity proof, which I have presented at the meeting of Naturforscher in Karlsruhe.

Let κ be a class of discontinuous linear groups of genus p with n singular points and with a certain characteristic signature; for this class the *fundamental theorem of Klein* holds, if to every Riemann surface of genus p that is canonically cut and marked with n points there belongs *one and only one* canonical system of fundamental substitutions of a group of class κ.

In the continuity method, which Klein uses to deduce his fundamental theorem, the following six theorems are applied.

1. The class κ contains for every canonical system of fundamental substitutions that belongs to it *without exception* a neighborhood that can be represented one to one and continuously by $6p-6+2n$ real parameters.

2. During continuous change of the fundamental substitutions within the class κ the corresponding canonically cut Riemann surface [95] likewise changes continuously.

3. Two different canonical systems of fundamental substitutions of the class κ cannot correspond to the same cut Riemann surface. [96]

4. When a sequence α of canonically cut Riemann surfaces with n designated points and genus p converges to a canonically cut Riemann surface with n designated points and genus p, and when each surface in the sequence α corresponds to a canonical system of fundamental substitutions of the class κ, then the limit surface likewise corresponds to a canonical system of fundamental substitutions of the class κ.

5. The manifold of cut Riemann surfaces contains for every surface belonging to it *without exception* a neighborhood that can be one to one and continuously represented by $6p - 6 + 2n$ parameters. [97]

---

[95] Crossed out footnote of Brouwer's draft: 'For the sake of brevity I write 'covered Riemann surface' rather than 'a covering surface constructed with signature σ of a Riemann surface of genus p with n designated points.' [96] Crossed out footnote of Brouwer's draft: 'Two covered Riemann surfaces are considered identical, if and only if the corresponding uncovered surfaces can be mapped conformally onto each other such that corresponding return cuts and stigmata behave identically with respect to the formation of the covering surfaces.' [97] In the draft followed by a footnote number identical to that of the preceding footnote.

6. In the $(6p-6+2n)$-dimensional space the one to one continuous image of a $(6p-6+2n)$-dimensional domain is also a domain.

I am ignoring here theorems 1, 2, 3, 4. For the case of the boundary circle they have already been completely treated by Poincaré in Vol. 4 of the Acta Mathematica; for the most general case only theorems 3 and 4 await an exhaustive proof; in this matter also this gap will be filled in by Mr. Koebe in papers that are to appear soon.

Theorems 5 and 6 are those which constitute the topological difficulties of the continuity proof that are emphasized in your book about automorphic functions.[7] However, of these theorem 6 is settled by my recently published article 'Beweis der Invarianz des n-dimensionalen Gebiets',[8] whereas the application of Theorem 5 can be avoided by carrying out the continuity proof in the following modified form:

We choose $m > 2p - 2$[9] and consider on the one hand the set M_g of automorphic functions belonging to the class κ that only have simple branching points and with m simple poles in the fundamental domain,[10] and on the other hand the set M_f of Riemann surfaces covering the surface, of genus p, with n signed points, and with m numbered leaves and with $2m + 2p - 2$ numbered simple branching points not at infinity, for which the sequential order of the leaves and the branching points correspond to the canonical relations in the sense of Lüroth-Clebsch.[98]

The set M_f constitutes a continuum, and possesses for each of its [99] corresponding surfaces *without exception* a neighborhood which is one to one and continuously representable by $4p - 8 + 2n + 4m$ real parameters.

For an arbitrary automorphic function φ belonging to M_g there exists in M_g a neighborhood u_φ which can determined by $4p - 8 + 2n + 4m$ real parameters; these parameters are the m complex places of the poles in the fundamental domain, the $m - p - 1$ complex behaviors of the $m - p$ arbitrary pole residues, and the $6p - 6 + 2n$ parameters of the canonical systems of fundamental substitutions. The value domain of the parameters belonging to u_φ constitutes a $(4p - 8 + 2n + 4m)$-dimensional domain w_φ.

[7]Cf. Vol. 2, p. 412, 413. [i.e. *Fricke-Klein, Theorie der automorphen Funktionen, Vol. 2, p. 413.*] [8]Mathematische Annalen 71, p. 305–313. Cf. also the articles of Baire and Lebesgue, quoted in the same volume p. 314. [9]In order to be more specific, we henceforth suppose $p > 1$. [10]Automorphic functions that only differ by an additive and multiplicative constant we consider as identical.

[98]Crossed out footnote of Brouwer: 'Two of these surfaces are considered identical if and only if the corresponding not-covered surfaces can be mapped so much similarly onto each other that corresponding return cuts and stigmata behave the same with respect to the construction of the covering surface.'. [99]Here Brouwer corrects a grammatical mistake related to the gender of the German word for 'set'.

With the function φ there corresponds a finite number of surfaces belonging to M_φ. Furthermore we conclude from the theorems 1, 2, 3 and the remark that possible birational transformations into itself not only for the single Riemann surface, but also for the totality of Riemann surfaces belonging to u_φ, cannot become arbitrarily small, [11] that with a sufficiently small w_φ in M_f there corresponds a finite number of one to one and continuous images, and hence because of Theorem 6 *a domain set*. However, *then the total set M_g in M_f corresponds with a domain set G_f too.*

Now we formulate Theorem 4 in the following form:

'When a sequence of canonically cut surfaces of M_f converges to a canonically cut surface of M_f and when each surface of the sequence corresponds to a canonical system of fundamental substitutions of the class κ, then the limit surface also corresponds to a canonical system of fundamental substitutions of the class κ.' [12]

This property immediately entails that the domain set G_f cannot be bounded in M_f, and *hence it must fill the whole manifold M_f.* This proves the fundamental theorem for every Riemann surface of genus p on which there exist algebraic functions with more than $2p - 2$ simple poles and with exclusively simple branching points, i.e. just for any Riemann surface of genus p.

Sincerely yours,
L.E.J. Brouwer

[Autograph draft – in Brouwer]

[Brouwer's copy of the proofs of 22 February (returned 26 February) carries a few comments in Dutch:]

this note goes further than the one of the Jahresbericht 1) because of the completely different use of Theorem 4. 2) by virtue of the completely different way in which *here* is abstracted from the continuity of both 'sets' that are compared.

---

[11] According to the treatise of Hurwitz in Vols. 32 and 41 of the Mathematische Annalen both the ordering of the periodicity and the number of fixed points must for these birational transformations remain below a certain finite bound, and hence the periodicity of a sufficiently small birational transformation must be transferred to the simply connected covering surface that winds aperiodically around its fixed points. But then this covering surface would admit a periodical conformal transformation with fixed points into itself, which is a contradiction. The property used in the text can, by the way, probably also be understood in a much more direct way. [12] As Mr. Koebe has expounded on the Naturforscherversammlung in Karlsruhe, this theorem can be most elegantly concluded from his deformation theorem (*Verzerrungssatz*).

For 'set' I use here everywhere the word 'Menge'; 'manifold' I use only where connectedness is implicitly 'alluded to'.

Editorial comment

There are several manuscripts of the 'letter to Fricke'. The final version of the letter, up to corrections and additions to the proofs, is printed as 'Über die topologischen Schwierigkeiten des Kontinuitätsbeweises der Existenztheoreme eindeutig umkehrbarer polymorpher Funktionen auf Riemannschen Flächen, (Auszug aus einem Brief an R. Fricke). *Nachrichten von der koeniglichen Gesellschaft der Wissenschaften zu Göttingen*, (1912), pp. 603–606. The proofs of February 8, 1912 bear the different title 'Über den Kontinuitätsbeweis der . . .'. A typescript version of the letter in the Brouwer archive has the simple title 'Über die topologischen Schwierigkeiten des Kontinuitätsbeweises.' Freudenthal, in his notes to *CWII* pp. 577–580, gives a history of the various handwritten and typewritten manuscripts (*CWII* pp. 581–583). In particular the footnotes underwent drastic changes. The draft (with numerous corrections and insertions) is dated '22. Dezember 1911'. Brouwer's handwritten copy carries no date. The final version was dispatched to Fricke on 30.XII.1911.

11-12-30b

To A. Hurwitz — 30.XII.1911[b]   **Amsterdam**
                                                        Overtoom 565

Dear Professor [Hochgeehrter Herr Professor]

Please excuse me for taking the liberty to turn to you with a question. The fact is that I need the following theorem:

'A birational transformation τ of a Riemann surface of genus $p > 1$ into itself can never transform a canonical system s of cuts (consisting of p pairs of return cuts that are connected in a point C) into an *equivalent* canonical system s' of cuts.' (s and s' are called equivalent when they can be transformed into each other by a continuous motion of the surface.)

I have convinced myself of the correctness of this theorem in the following way:

'Let the transformation τ have n fixed points. We construct an extended canonical system of cuts S, consisting of s and n cuts from C to the fixed points, and which is taken by τ into S'. If now s' would be equivalent to s, then *by the periodicity of* τ S' would be equivalent to S. (S and S' are called equivalent, when they can be transformed into each other by a continuous motion of the surface *without moving the n fixed points*).'

Now we construct for this Riemann surface the simply connected covering surface, as is customary in the theory of automorphic functions, which winds aperiodically around the n fixed points, and which can be mapped conformally onto the interior of a circle, so that the n fixed points and their reproductions are moved to the circumference of the circle. The transformation τ then corresponds to a conformal transformation *without fixed points* of this circle interior into itself, which must be *periodic* because of the equivalence of S and S', which is a contradiction, because a periodic conformal transformation of the interior of a circle into itself always has a fixed point.

It seems to me very probable that the theorem in question can be grasped much more simply from the standpoint of the combinatorial construction of 'regular' Riemann surfaces, and even that it is an immediate consequence of your earlier investigations in this field.

Am I correct in conjecturing this? And has the theorem already been stated somewhere? I would be very grateful for your kind communication about this.

Sincerely yours

Your most obedient
L.E.J. Brouwer

[Signed autograph – in Hurwitz; also in *CW II* p. 616–617 (with Freudenthal's comments)]

1912-01-04a

From H. Poincaré — before 4.I.1912[a]

My dear colleague, [Mon cher Collègue]

Thank you very much for your successive letters; I will study the matter in detail as soon as I will have time. I still believe that the simplest way to prove the absence of a singular point would be to not use the Riemann surfaces in the form given by Riemann, in other words with stacked flat leaves and cuts, but in the form given by Klein; an arbitrary surface with a convenient connection and some law (with a representation that is or isn't conformal) for the correspondence of the points of this surface with the imaginary points of the curve $f(x, y) = 0$.

Already many years ago I have expounded my ideas about this point during a session of the Société Mathématique de France; but I didn't publish them, because Mr. Burkhardt, who was present at that session, told me then that Mr. Klein had already published them in his *autographically prepared* lecture notes [100]; maybe you can avail yourself of these.

It all amounts to this. Let $f(x, y) = 0$ be a curve of genus p; to this curve I let correspond a Riemann-Klein surface S and a law L of correspondence between the real points of this surface and the complex points of the curve $f(x, y) = 0$. Next I consider surfaces S' and laws L' that differ infinitesimally [101] little from S and L. On first must prove that there are ∞^{6p-6} such surfaces S' (which are not considered distinct if they can be transformed into each other by birational transformations); and then one can always pass from an arbitrary S', L' to another arbitrary S', L', without moving too far from S, L and without passing **by** S, L.

Your devoted colleague,
Poincaré

[Signed autograph – in Brouwer]

––––––––––––––

12-01-13

To F. Klein — before 13.I.1912^a [102]

I was very sorry to hear how you have completely overworked yourself by your indefatigable and unselfish efforts in the interest of science and the common good. Would that you take a bit more care of yourself in the future: we need you for a long time as our leader and master.

––––––––––––––––––––

[100] 'autographiées' in text; [Klein 1892]   [101] In text: infin.^{t}.   [102] Reproduced as [Y6] in *CW II* p. 584; with Freudenthal's comments on the dating and the addressee.

In correcting the galley proofs [103] I will be happy to take your remarks into account, and I will designate the relevant theorem as your *'general fundamental theorem'*, and I will point out that my theorem 6 will be superfluous as soon as the analytical relation between the two manifolds has been shown [104] (unfortunately I have until now not succeeded in giving this proof). About your proof as presented by Poincaré in Acta Mathematica 7, of the finiteness of the birational transformations for $p > 1$, I would like to permit myself the remark that this proof presupposes the uniformization, so it cannot be applied in my line of thought. Moreover I don't need merely the finiteness of the transformations of a certain surface, but of *all* surfaces of genus p. (It would be a priori possible that indeed the smallest birational transformation of a given surface possesses a finite maximal deformation F, but that this F can become arbitrarily small if the surface is varied. Noether recently informed me that also he didn't know a proof of the last theorem, independent of Hurwitz's). I would be very happy to compose a comprehensive version of the article for the Mathematische Annalen.

With best wishes for your speedy and complete recovery and with cordial greeting

your admiring [105]

[Autograph draft/copy – in Brouwer]

1912-01-16

From O. Blumenthal — 16.I.1912   Aachen [106]

Dear Mr. Brouwer! [Lieber Herr Brouwer!]

I would like very much to have the absolutely shortest and best proof of invariance of dimension in the Annalen. Therefore I promise you publication in the next issue, together with your translation theorem on the condition that the new proof doesn't exceed 3 pages. But I would like you to write it sufficiently elaborately so everybody can understand it. That will, I guess,

---

[103] of [Brouwer 1912d].   [104] Page 2 footnote 4.   [105] *Ihr verehrender.*   [106] Date and place - postmark.

fit into three pages, as you think yourself that only one page is necessary. The issue appears in the beginning of March.

Thank you very much for your kind condolences at the death of my parents. I hope that everything is well at your end. Best greetings.

Your
O. Blumenthal

[Signed typescript, postcard – in Brouwer]

12-01-21

To F. Engel — 21.I.1912   **Amsterdam**
                                                                 Overtoom 565

Dear Professor [Sehr geehrter Herr Professor]

In your review on p. 194 of vol. 40 of 'Fortschritte der Mathematik'[107] you raise two kinds of objections against my paper. First, you think that I have imposed overly strong restrictions on the problem, and second, you find that even accepting these restrictions my exposition is not completely watertight.

With respect to the first point I would be very grateful, if you would be so kind as to inform me precisely which more comprehensive problems you envisage, as I have not succeeded to get a completely clear picture from your indications.

The phrasing of your review would roughly indicates that you wish the following assumptions:

'Suppose that an n-dimensional manifold μ carries a continuous *parametrizable set* of transformations containing the identity, which has in a neighborhood of the identity in the first place the group property, and which in the second place can be one-to-one and continuously represented by p real parameters.'

These conditions would certainly not be sufficient, because for every p-dimensional group g one can construct in many different ways a p-parameter

---

[107] F. Engel, review of [Brouwer 1909b] in *Jahrbuch über die Fortschritte der Mathematik* 40, p. 194.

set of transformations, which is identical with g in a certain neighborhood of the identity, but which is outside this neighborhood neither identical with g nor possesses the group properties at all. For in the set theoretic version of the problem, there is no possibility to infer under the assumption of the analytic dependence on the p parameters of the transformations, properties arbitrarily far away from the identity from properties close to the identity. Indeed, in the set theoretic version of the problem one doesn't have the possibility to conclude from properties close to the identity to properties arbitrarily far away because of analytic dependency of the p parameters of the transformations.

Hence in any case the conditions must be phrased in the following more restricted form:

'Assume that an n-dimensional manifold μ carries a continuous group [108] of transformations containing the identity, [109] which in a neighborhood of the identity can be one-to-one continuously represented by p real parameters.'

From these assumptions it follows immediately that the whole group can be mapped one-to-one and continuously onto a p-dimensional 'parameter manifold', and that the group consists of pairwise inverse transformations. This last property in its turn implies that on the n-dimensional manifold μ the transformations are *everywhere one-to-one*, so we have recovered all the conditions of my Annalen article.

But maybe I have in the above, completely misinterpreted the ideas of your review? For example, the meaning of your words 'unnecessary restriction that the group must be closed' has remained totally obscure to me.

For the second point of your criticism, namely the incompleteness of my exposition, you give two examples:

1) Certain obscurities in the formulation of the conditions in § 1. I would really like to know which obscurities or uncertainties you have found here, because I believed that I have satisfied every demand of set theoretic precision and exactness.

2) The 'lies certainly' on p. 255, l. 20. This is quite self evident indeed. For when a point x moves continuously from 0 to a, then the point $2x$ moves in the same direction from 0 to $2a$. The moment the point $2x$ reaches a, the point x is at the required point b.

---

[108] In the older literature the identity is not always included in the group definition.
[109] Note that the terminology in group theory had not been generally agreed on.

I would attach great importance to reach agreement on the above with a group theorist of your authority.

Sincerely yours [110]

L.E.J. Brouwer

[Signed autograph, draft – in Brouwer; also in *CWII* p. 141–142, with Freudenthal's comments]

●12-01-28

From F. Engel — 28.I.1912 [111]   **Greifswald**
                                                   Arndtstr. 11 [112]

Dear Doctor! [Sehr geehrter Herr Dr.!]

I am very glad that you have directly contacted me with your letter of the 21st. Only, it's a pity that all these kinds of things are so difficult to discuss in writing. Verbally it would be much easier to come to an understanding.

To begin with, I would regret it if you would have read in my review any kind of disapproving judgement about your work. I didn't mean anything like that, but judged 'mit Bewunderung zweifelnd, mit Zweifel bewundernd', [113] as one should according to Lessing's scala judge when confronted with a Master.

With that I don't actually want to recognize you straight away as master, but just express my general esteem for your articles, although on the other hand I acknowledge you unconditionally as my master in set theory, the application of which is not my line at all.

I must maintain that the conditions indicated on p. 247 leave very much to be desired in clarity of formulation. Maybe they appear to be clear to a died in the wool set theorist [114] but I must say: 'the expressions of the system sound dark to uncircumcised ears.' [115]

---

[110] *Mit ausgezeichneter Hochachtung bin ich – Ihr ergebenster.*   [111] Letter in 3 parts: dated 28.I.1912, 30.I.1912, 31.I.1912; postmark 31.I.1912.   [112] Envelope.   [113] 'Admiring with doubt or doubting with admiration' – well-known quote of Gottfried Lessing about the correct attitude of critics with respect to masters.   [114] in German '*eingefleischter Mengentheoretiker*'.   [115] Allusion to Jeremiah 6:10.

On p. 247, l. 3–11 you have packed so much into one sentence, that already for that reason it is impossible to be clear, at least for me. Also l. 17–19 is much too succinctly worded to be clear.

As regards the other points, I cannot befriend myself with them, [116] just as before.

[From Engel] — **30.I.1912**   [Continuation of 28.I.1912]

that you assume from the beginning that the transformations are one-to-one and invertible throughout the whole manifold. I cannot see that the case where one has a group of analytic multi-valued transformations, can be reduced to the one-to-one case downright.

When I proposed to assume to begin with the one-to-one property of the transformations only in the neighborhood of a point, and likewise the group property only in the neighborhood of the identity transformation, then I meant of course: I have a parametrized family of transformations that are uniquely invertible in the neighborhood of a point, this family contains the identity transformation and two transformations of the family that lie in a certain neighborhood of the identity transformation produce again a transformation of the family. [13] Of course the transformations of the family must generate a group when they are arbitrarily often performed successively, and we only have to assume that this group doesn't contain other transformations than those in the original family.

In this way not only the original family of transformations will be extended, but at the same time one gets the original transformations defined outside of the range on which they were originally defined. So one gets an extension of the transformations also when one can't make use of the analytic continuation. Therein the tremendous power of the group concept manifests itself.

Now whether from these assumptions the ones that you made follow, I cannot judge, and I don't trust myself to say something about it. Should it be the case and should you already have checked that yourself, I would, to be honest, disapprove of your not saying so in your article. Then you would have omitted the reduction to the assumptions you made to the most simple and natural possible ones, or you would not have fathomed the true meaning of the group concept, namely that it by itself produces a principle

---

[13]I would for the sake of for simplicity also assume pairwise inverse transformations.

[116] *ich kann mich nicht damit befreunden.*

for analytic continuation.

I hope that it is now clear to you what I meant when I said that one shouldn't assume at the outset that the group be closed; one shouldn't even think of the whole group but only of a piece of it. In all applications one finds oneself anyway in this case.

That the 'lies' on p. 255 is so self evident, I still can't see. I don't see that without further assumptions it amounts to the theorem that a continuous function that takes two values also takes every intermediate value. In whatever way one proves this theorem, it seems that a proof here is necessary, or the reduction to an earlier proven theorem. Maybe I am mistaken, but I have the feeling that a proof is necessary.

For the rest, I can't get around admitting that you have, by your article and your letter, perhaps unintentionally, brought me a new insight. For it now no longer appears practical to me, also when one considers analytic groups, to extend the equations that arise by analytic continuation in the usual way, but one should [14] only consider those analytic continuations that arise from the application of the group concept. In this manner one will in some cases not exhaust the entire domain to which the functions can be analytically continued, but instead of that one will not leave the domain in which the group transformations are all uniquely invertible.

[From Engel] — **31.I.1912**   [Continuation of 30.I.1912]

This letter is written in several installments, because aside from the fact that I am nowadays fairly busy as a dean, I have also been selected by lottery for jury duty and that makes even more demands on my time, which is already so restricted.

Finally I would like to mention that, to be quite honest, I don't think that the profit of this kind of research does not quite match the effort spent and the necessary investment in acumen. You may consequently still think me a heretic. I would all the more be glad, if you now also would work on group theory itself, for there still is a lot to be done.

Sincerely yours

F. Engel

[Signed (on 31.I.1912) autograph – in Brouwer; also in *CWII* p. 144–146]

-----

[14]at least in the case of transitive groups

1912-02-03

From O. Blumenthal — 3.II.1912 [117]   Aachen

Dear Mr. Brouwer! [Lieber Herr Brouwer!]

May I ask you for a service for the Annalen? I send you herewith an article by Lennes, 'Curves and Surfaces in non-metrical space', [118] which seems to me closely related to your articles. More specifically I see a part of the Jordan theorem formulated there, and actually just what you have proved in detail in the Annalen. Hence I request your opinion on the article: whether it is correct, and in which relation it stands to yours, and whether you think it deserves to be published in the Annalen.

Many thanks in advance and many greetings!

Your
O. Blumenthal

[Signed typescript, postcard – in Brouwer; also in *CWII* p. 487]

1912-02-04b

From F. Engel — 4.II.1912[b]   Greifswald
                                                        Arndtstr. 11 [119]

Answer to my short question, whether
according to Mr. E. the one-to-oneness
isn't a consequence of the pairwise
inverseness. [120]

Dear Dr.! [Sehr geehrter Herr Dr.!]

I think that an example will the best thing to explain to you what I mean.

---

[117] Date and place - postmark, Amsterdam postmark 4.II.1912.   [118] [Lennes 1911]. The paper was rejected by the *Mathematische Annalen*. The paper played a role in the later dimension discussion, where it served to show that Brouwer knew in 1911 the modern definition of connectedness, as it occurred in this paper of Lennes. Brouwer had independently formulated the notion in [Brouwer 1911c]; see Freudenthal's historical comments, *CWII* p. 486.   [119] From envelope.   [120] Brouwer's note.

I consider a one-parameter group, generated by a multivalued infinite transformation. For example if we have the infinite transformation

$$x' = x + \frac{1}{2}\sqrt{x}\delta t$$

then we obtain the one-parameter group of two-valued transformations

$$\sqrt{x'} = \sqrt{x} + t$$

or:

$$x' = x + 2t\sqrt{x} + t^2$$

Here every transformation is two-valued, but coincides with its inverse transformation, or perhaps better: it changes by analytic continuation into its inverse.

Now let $x_0 \neq 0$ and $\sqrt{x_0}$ be one of the two root values. We take our departure from the one-parameter set [121] of transformations:

$$x' = x + 2t \cdot \sqrt{x_0}\left(1 + \frac{x - x_0}{x_0}\right)^{\frac{1}{2}} + t^2$$

where $(1 + \alpha)^{\frac{1}{2}} = 1 + \frac{1}{2}\alpha + \dots$. This set is defined for every t and für $|x - x_0| < |x_0|$. Its transformations have the form:

$$x' - (t + \sqrt{x_0})^2 = (x - x_0)\left(1 + \frac{t}{\sqrt{x_0}}\right) + \dots$$

where the omitted terms are of second and higher order in $x - x_0$.

Let us call these transformations of this set S_t, then the transformation $S_t^{-1} \cdot S_{\tau+t}$ defines the manner how the point into which x_0 is transformed by S_t, is transformed by S_t.

I choose $\tau = -2\sqrt{x_0}$, then:

$$S_{-2\sqrt{x_0}} : x' - x_0 = -(x - x_0) + \dots,$$

hence this transformation takes the point x_0 back to x_0 again. But:

$$(S_{-2\sqrt{x_0}})^{-1} : x' - x_0 = -(x - x_0) + \dots$$

$$S_{t-2\sqrt{x_0}} : x'' - (t - \sqrt{x_0}) = (x' - x_0)\left(\frac{t}{\sqrt{x_0}} - 1\right) + \dots$$

---

[121] 'Schar'.

so it follows:

$$(S_{-2\sqrt{x_0}})^{-1} \cdot S_{t-2\sqrt{x_0}} :$$

$$x'' - (t - \sqrt{x_0})^2 = (x' - x_0)\left(1 - \frac{t}{\sqrt{x_0}}\right) + \cdots$$

i.e. this transformation is not S_t but S_{-t}. However, one obtains by group theoretic continuation from S_t the inverse transformation S_{-t}, just like when one makes an analytic continuation of S_t and goes once around the zero point.

This seems to prove clearly that the assumption of univaluedness of the transformations in the neighborhood of a point and the assumption that the group for each of these transformations also contains its inverse in the neighborhood of this point does not imply at all the univaluedness, let alone the unique invertibility in the domain which this point reaches under the transformations of the group.

Also when the infinite transformation is one-valued, the transformations of the one-parameter group can be multi-valued, e.g. $x' = x + e^{-x}\delta t$ gives

$$x' = lg(e^x + t)$$

But enough of this now. With best greetings

Yours truly
F. Engel

To my consternation I see just now, that I dispatched only half of my letter. [122]

[Signed autograph – in Brouwer; also in *CWII* p. 147–148]

1912-02-12a

From O. Blumenthal — 12.II.1912[a]   Aachen [123]

Dear Mr. Brouwer! [Lieber Herr Brouwer!]

I have received Lennes' paper and your criticism. Thank you very much. I thank you especially for having justified your evaluation so thoroughly and

---

[122] Written upside down on top of first page.   [123] Date and place - postmark.

precisely. Consequently I have of course returned the paper to Mr. Lennes, with your comments enclosed, and I have moreover referred to your paper on the Jordan theorem, which he doesn't seem to know.

Again best thanks and many greetings

Your
O. Blumenthal

[Signed autograph, postcard – in Brouwer; also in *CWII*, p. 487, with Freudenthal's comments]

)12-02-12b

From P. Koebe — 12.II.1912[b]   **Leipzig**

Dear Mr. Brouwer! [Geehrter Herr Brouwer!]

Ad A. Mention is lacking of the important indispensable proof item [124] that two canonically cut Riemann surfaces of genus p can always be continuously transformed into each other. (One can do without that only for the boundary circle theorem, but not for the other fundamental theorems). Klein's proof, Annalen, 21, [125] for this is not possible, the way it is done there, in an exact manner, because the regularity of the analytic boundary correspondence is interrupted in the corners. Therefore I map the surfaces that are cut open with p separated return cuts (as in Annalen 69) [126] onto a normal [127] region with altogether regular boundary correspondence. These normal regions form *one* continuum, from which it follows that not only the surfaces of genus p form *one* continuum, but that also the ones with p return cuts do so. From this it further follows t hat surfaces that are cut open in *any* fashion also are continua.

Ad B. Poincaré had certainly not planned to prove Theorem 4. Poincaré rather represents the interpretation of *closed* continua by adding *limits of polygons,* a viewpoint which you too, still constantly emphasized in Karlsruhe. Poincaré recently informed me in conversation that the continuity method cannot be used *at all* if one wants to prove the no-boundary-circle theorem, because these manifolds are *not closed.* Your way of presenting it

---

[124] *Beweispunkt.*   [125] [Klein 1882].   [126] [Koebe 1910].   [127] *schlicht.*

is therefore only an interpretation of Poincaré's views that afterwards was constructed under the impression of my communications in Karlsruhe, and in its kind a very original achievement. Also *Fricke-Klein's* 'Vorlesungen über automorphe Funktionen' has adopted extensively the view of Poincaré about *closedness*.

Ad C. More to the point and further taking into account the meaning of the achievement roughly as follows:

> while for the most general case in particular the theorems 3, 4 and A still lack an exact justification, which however according to his provisional communications in the Göttinger Nachrichten (see more in particular also the most recent communication 'Begründung der Kontinuitätsmethode im Gebiete der konformen Abbildung und Uniformisierung' (1912)) [128] has been achieved completely by Mr. Koebe, and which will be soon published in full extent in the Mathematische Annalen. The proofs found by Mr. Koebe extend to the case of boundary circle uniformization, the only oneconsidered by Poincaré, and imply a *life giving* advance, because of the liberation from the thoughts introduced by Poincaré and copied by Klein-Fricke about polygonal limits and closed continua, an advance which is at the same time a return to Klein's old standpoint of non-closed continua which was vigorously attacked by Poincaré. By the way, Koebe's continuity method represents also in relation to Klein a remarkable fundamenta advance because Koebe actually doesn't use Theorem 4, although this theorem can very well be proved, as Mr. Koebe told me in connection with his proof method, by enlisting the help of the 'choice-convergence theorem'.

NB: B and C can be best put in the form of a *footnote*, because it is not the text of a letter. [129]

Ad D. This goes indeed much better with the normal surface with $6p - 6$ *parameters*, which is the Abelian integral of the first kind with $(p-1)$ modulo 6 and which gives a $2\pi i$, as I told you already in Karlsruhe.

---

[128] Foundation of the continuity method in the domain of conformal mapping and uniformization, [Koebe 1912]   [129] Note that Brouwer is requested to endorse a text that Koebe is withholding from Brouwer. For a discussion of the Brouwer-Koebe conflict see Freudenthal's comments in *CW II*, p. 572 ff. Furthermore [Van Dalen 1999] section 5.3, p. 189 ff. See also *Brouwer to Hilbert 9.III.1912*

Ad E. This footnote can be cancelled after the explanations ad C.

Ad F. Here A must be now considered.

I will soon send you my corrected page proofs. Please send me yours again.

With best greetings

P. Koebe

[Signed autograph – in Brouwer]

1912-02-14

To P. Koebe — 14.II.1912   **Amsterdam**
                                                                        Overtoom 565

Copy

Dear Mr. Koebe [Geehrter Herr Koebe]

Fortunately I am still in possession of the abridged text of my Karlsruhe talk, which I enclose, so that you can no longer maintain that I used in Karlsruhe in the talk or in discussions the *'closed'* manifolds of Poincaré!

That you could make such a statement only proves that modern set theory must be absolutely unfamiliar to you. For, the elaborations of Poincaré who works with the so-called 'closed manifolds' are pure balderdash, and can only be excused by the fact that at the time of their formulation there was not yet any set theory.

That the proof of the "Weierstrass Theorem" (in Klein's terminology) and therefore the continuity proof for the case of the boundary circle can nonetheless be carried out on the basis of the other elaborations of Poincaré, was precisely the content of my lecture in Karlsruhe.

Through your communications I have acquired the further insight that by means of your deformation theorem my method can be carried over to the most general fundamental theorem.

What you recall from my lecture or from our conversation about the 'closed manifolds' used by me, refers to the following: I consider in the enclosed text automorphic functions as identical when they only differ by additive or multiplicative constants, and thereby I achieve that there corresponds to every internal point of the cube a *closed* manifold of functions with m poles. Only this justifies the word '*Alsdann*'[130] on p. 3, l.19 of the enclosed text, since only because of this closedness one can obtain certainty that a point sequence in M_π always will have a limit point that belongs to M_π, if the corresponding point sequence of the cube has a limit point inside the cube.

Please return the enclosed text to me after a few days.

With best greetings

L.E.J. Brouwer

Why don't you send me a copy of your manuscript, as I did, and as you promised me?

[Typescript copy – in Brouwer; also in *CW II* p. 585, with Freudenthal's comments; a signed autograph copy was attached to *Brouwer to Hilbert, 24.II.1912*]

1912-02-24

To D. Hilbert — 24.II.1912   **Amsterdam**

Dear Geheimrat, [Lieber Herr Geheimrat!]

I request your help and protection in a very disagreeable matter.[131] On January 2 I sent Koebe a copy of my letter to Mr. Fricke, which was sent in December and presented to the Göttinger Gesellschaft der Wissenschaften on January 13, and about one week later I received the enclosed postcard. This card was followed on February 14, *not* by the promised manuscript, but by the letter that is enclosed here together with my answer; in it I have marked in blue pencil the statement that my refutation refers to (all the rest is nonsense).[132]

Koebe can however not really mean the statement concerned, just as little as anyone who has heard my talk in Karlsruhe. Hence I sense in Koebe's statement merely his intention to give in his next note the matter the appearance that my letter to Fricke contains certain thoughts that I have learned

---

[130] Consequently   [131] i.e. the Koebe affair.   See [Van Dalen 1999] section 5.3.
[132] Letter *Brouwer to Koebe, 14.II.1912*.

in conversations with Koebe, while the true state of affairs in Karlsruhe was that I contributed to those conversations the complete continuity proof of the boundary circle case, whereas Koebe only contributed some inkling that his deformation theorem [133] could be used somehow in the continuity method. In fact he said in the session of September 27 at the end of my talk the following: 'Because on the basis of my deformation theorem *nothing can happen* during continuous change of modules, the achievements of Brouwer are in my train of thought dispensable.' To this I emphatically answered: 'The deformation theorem can only extend the boundary circle result obtained by Poincaré, and thereby also at the same time extend my continuity proof to the most general case; in this extension my contributions remain as before necessary in their full strength.' Then Koebe spoke the nonsensical words: 'What Mr. Brouwer has shown, I do with Poincaré sequences.' and then Klein closed the discussion.

Only after long private discussions, in which also Bieberbach, Bernstein and Rosenthal took part, Koebe learned subsequently from me between September 27 and 29 which partial result (incidentally formulated already by Klein in Annalen 21, and at that time called "Weierstrass Theorem" by me) can be obtained by means of his deformation theorem, and which remaining part remains to be treated by my contributions. And in those conversations I have, as the just mentioned gentlemen must know exactly, brought up *all* details of my present note.

However, several warning voices told me already at that time: 'All that you now are explaining to Koebe, you will only with the greatest difficulty be able to claim as your property, as soon as he will have understood it', and indeed certain symptoms in Koebe were visible that seemed to prove these voices right, so when I had returned home I wanted, in order to avoid an unpleasant fight with Koebe, to abandon any publication about this matter which is anyhow rather far removed from my interests and with which I had only occupied myself in passing on Klein's request. Only after Blumenthal had urged me and I moreover had heard that Klein would like to see a publication by me, it came to this note of January 13.

My request would now be the following: Just as I didn't receive Koebe's earlier promised manuscript, he will not, I believe, send me the now promised page proofs *before* my note is declared ready for printing, so that I will not for instance be able to adjust in time the text to refute Koebe's claims in advance. May I ask you now to arrange that I get the Koebe page proofs directly from the printer? And in case I then would find that they contain

---

[133] *Verzerrungssatz.*

the above mentioned or other falsehoods, to make him extirpate them so as to avoid unpleasant polemics?

I would be most grateful to you for that.

With best greetings

your

L.E.J. Brouwer.

On my enclosed letter of 2/14 I haven't heard anything more from Koebe. The text of my talk he hasn't returned to me either.

[Signed autograph – in Hilbert]

1912-02-27

To D. Hilbert — 27.II.1912   **Blaricum**

Dear Geheimrat [Lieber Herr Geheimrat]

For your better information I am sending also to you an abridged text of my lecture in Karlsruhe. I hope that you will be able to recall that, so to speak, every word of this text was also spoken in the lecture. In any case you must be able to call to mind that in my lecture I applied the continuity method neither to the Klein polygon continuum nor to the allegedly 'closed' group continuum of Poincaré (as Koebe claims) but to *the continuum of automorphic functions with m poles*, and that this indeed constituted the core of the matter.

Will the Wolfskehl Symposium about the foundations of mathematics go through, which you planned last summer for the Easter vacation?

Many greetings to both of you, also from my wife

Your
L.E.J. Brouwer.

[Signed autograph – in Hilbert]

)12-03-06a

To F. Engel — 6.III.1912[a]   **Amsterdam**
                                                                    Overtoom 565

Dear Professor, [Sehr geehrter Herr Professor!]

Absorbed by many activities, I was until today not able to reply to both
of your extensive letters, which made your ideas now very clear to me. Now
I address once more all points of my first letter.

The word 'obscurity' which you used in your first review must, according
to your more detailed explanations, be interpreted as 'difficult to understand
for the uninitiated'. About such a *subjective* view one can of course not
argue, but many a reader will, contrary to your intention, have received
from your words the impression that I have not defined my fundamental
concepts with sufficient precision, which would be a very *objective* error,
which I most emphatically must reject (the only purpose of the 'rectification'
in the beginning of the second Correction in Bd. 69 [134] was not to exclude
from the outset certain singular connectivity situations that do not occur in
finite continuous groups).

The 'certainly lies' on p. 255 amounts, as you will now certainly see
yourself, without further ado to the theorem that every continuous function
('continuous monotone function' would even suffice for this case) which takes
two values, also takes every intermediate value; the reader of the Annalen
hardly needs to be reminded of such a trivial theorem. Yet also here, as I
believe, many a reader of your review will have got the impression, that my
article contains *several objective gaps*, of which you pointed out the one just
mentioned one merely as an example.

Therefore you would do me a great pleasure if you could decide to insert
in a possibly forthcoming review of my second communication a remark to
rehabilitate me.

As far as the inner foundation of my general assumptions is concerned, I
believe that I can clarify in a most complete way by means of the following
reflections.

Let us call a point set m *concatenated*, [135] if according to some law
certain infinite point sequences f that belong to m are assigned certain
points p_f that likewise belong to m and that are characterized as *limit points*

---

[134] [Brouwer 1910a].   [135] *verkettet* in German, which is translated as 'bound together',
'connected', or 'concatenated'. We have opted for the last term.

of f in such a manner that for every point p_f there is always a *subsequence* of f that only has this single point as limit point, and that the limit points of a subsequence of f constitute a subset of the limit points of f, while finally the following property holds: if α_μ is the only limit point of the sequence $\alpha_{\mu\nu}$, and α the only limit point of the sequence α_μ, then every $\alpha_{\mu\nu}$ contains such a final segment $\alpha_{\mu\pi}$, that the sequence of these final segments *only* has α as limit point.

A point set that is not concatenated, i.e. a point set without limit points, is called *discrete*.

By a *neighborhood* of a point p that belongs to m we mean a subset of m that contains infinitely many points of every point sequence in m that has p as limit point.

(Let us now construct around every point p of m a neighborhood u_p, and let us choose two arbitrary points p_1 and p_2 in m. If, independently of the choice of u_p and of p_1 and p_2 it is possible to put a *finite* point sequence in m such that two consecutive points of this sequence belong to one and the same u_p, then m is called a *connected* point set. [136])

Let us call a map from one point set m to a point set m' *continuous*, when every limit point of a point sequence of m corresponds to a limit point of the corresponding point sequence of m'.

Let us call the point set m *homogeneous*, when for every neighborhood u_p of an arbitrary point p of m each other point of m has a neighborhood that can be mapped one to one and continuously onto u_p.

Now let an arbitrary point set m carry an arbitrary group γ containing the identity and pairwise (single- or many-valued) inverse elements. We then cover m with a point set μ such that any two coinciding points of m will also be considered identical in μ if and only if each transformation of γ will take them into two coinciding points of m. Furthermore in μ the point π will be considered to be a limit point of the sequence F if in the first place the point p in m corresponding to π is the limit point of the sequence f that corresponds to F in m, and in the second place this relation between p and f will be preserved by an arbitrary transformation of the group γ. Finally the transformation τ in γ will be considered as limit element of the infinite sequence φ of transformations, when every arbitrary point of μ is taken into such a point π by τ and by φ in such a point sequence F, that π is a limit point of F.

Consequently both the 'transformation manifold' μ and the 'parameter manifold' γ are *homogeneous point sets*, and in reference to μ the transforma-

---

[136] Here Brouwer defines the notion of 'connectedness'. In [Brouwer 1911c] he introduces the definition that is now universally accepted. For a discussion of the history of 'connected' see Freudenthal in *CW II* p. 486.

tions of γ are not only pairwise inverse, but also *one to one and continuous,* while the point set m now appears as a *folding* (i.e. as a single valued — not one to one — and continuous image) of μ.

Hence every group of (single- or many-valued) pairwise inverse transformations of an arbitrary point set *from a homogeneous group of one to one and continuous pairwise inverse transformations of a homogeneous transformation manifold* (a group of this latter kind will be called a *canonical group,* when both the transformation manifold ⟨137⟩ and the parameter manifold ⟨138⟩ are closed [15]) *be obtained by folding of the transformation manifold.*

Now only the following types of closed homogeneous point sets are until now known (and probably no others exist):

a) discrete point sets.

b) finite dimensional manifolds R_n according to my definition.

c) countably infinite dimensional manifolds R_ω (compare the relevant articles of Fréchet).

d) point sets of order type ζ of disconnected, nowhere dense, perfect point sets of R_n.

e) '*product sets*' constructed from sets of the four previous kinds (e.g. a discrete set of order types ζ of three-dimensional spaces).

And the most general canonical group, for which both the transformation manifold and the parameter manifold belong to type e), can be composed in a simple way from canonical groups for which both the transformation manifold and the parameter manifold belong to one of the types a), b), c), d), which therefore can be called *prime groups.*

Examples of prime groups are the finite substitution groups (parameter manifold and transformation manifold of type a)), the Fuchsian and Kleinian groups (transformation manifold of type b), parameter manifold of type a)), the finite continuous groups according to the definition of my Annalen article (parameter manifold and transformation manifold of type b)), the infinite continuous groups (transformation manifold of type b), parameter manifold of type c)), the ζ-groups to which I called attention in the Amsterdam Proceedings of April 1910 ⟨139⟩ (parameter manifold and transformation manifold of type d)).

In complete agreement with the above we get the example of your last letter (the group $\sqrt{x'} = \sqrt{x} + t$) from the translation group of the plane,

---

[15]We call a point set *closed* when there exists for every point a neighborhood in which every fundamental sequence has a limit point which also belongs to the point set.

⟨137⟩In text: T.M.   ⟨138⟩in text: P.M.   ⟨139⟩[Brouwer 1910f].

i.e. from a finite continuous group in my definition, *through folding* (in this special case by a two-to-one mapping) *of the transformation manifold.*

The nature of this folding is, by the way, not subject to any limitation; because it is completely arbitrary, one can make it in specific cases also so complicated that the group cannot be expressed by analytic formulas. The related canonical group will however not be influenced by that; it remains a finite continuous group in my definition.

I would be very glad if the above has clarified why in my view the restriction of the problem formulated in my Annalen article, namely 'determine all finite continuous groups', is a completely natural one, and does not in the least entail an artificial restriction.

With best greetings [140]

Sincerely yours
L.E.J. Brouwer

[Signed autograph, draft – in Brouwer; also in *CWII* p. 149–152 (with Freudenthal's comments)]

1912-03-06b

From P. Koebe — 6.III.1912[b] [141]

Dear Mr. Brouwer! [Geehrter Herr Brouwer!]

I am looking forward with interest to the publication of your talk in Karlsruhe. However, I cannot agree to the publication of parts of your letter to Fricke [142] because you have no right to the, so to speak arbitrational, presentation given there, and because the achievements of Poincaré and me appear there in an unworthy and incorrect light. Also, in view of the publication of your talk in the Jahresbericht the publication of the letter is anyway superfluous.

Yours truly [143]
P. Koebe

[Signed autograph – in Hilbert]

---

[140] *Ihr ganz ergebener.*   [141] Date postmark.   [142] [Brouwer 1912d].   [143] *Ergebenst.*

)12-03-07

To D. Hilbert — 7.III.1912   Blaricum [144]

Dear Geheimrat [Lieber Herr Geheimrat]

Let me add to my preceding letter that Koebe is in my opinion *obliged* to
send me his proof sheets, and this for the following reason: When I believed
in November I had to conclude from a letter from Fricke that Koebe was
almost ready with a note for the Göttinger Nachrichten on the continuity
proof, I proposed to Koebe to edit our notes in mutual agreement, and *only
after Koebe had accepted this proposal,* I have sent him first my manuscript
and then my proofs. When he now for his part sends me neither the one
nor the other, he is guilty of the most outrageous faithlessness. Such a thing
one does not have to accept! Moreover he stubbornly refuses to send back
to me a manuscript of my lecture in Karlsruhe which three weeks ago I lent
him for a few days. All of this is so mysterious to me! Or does Koebe's note
perhaps not yet exist, and does he behave in this way only to gain time? In
that case I would like ask you not to wait any longer for him, and to get my
note now printed. Please, write me a line!
Best greetings! [145]

Your
Brouwer.

[Signed autograph, postcard – in Hilbert]

●12-03-09b

To D. Hilbert — 9.III.1912[b]   Blaricum

Dear Geheimrat [Lieber Herr Geheimrat]

After mailing my last letter to you I received the enclosed card from
Koebe. It brings neither the recantation of his false statements about my
talk in Karlsruhe that I desired, nor the promised page proofs of his note

---

[144] Postmark Blaricum. The address of the sender is given in handwriting: 'Overtoom
565, Amsterdam'.   [145] *Schöne Grüsse.*

that he owes me. I'll now have to give up the hope that he will return to reason, and therefore I ask you to get my note for the Göttinger Nachrichten printed now. [16] In the meantime I set great store by rebutting here to you, the objections to my note that Koebe has raised in his letter [146] and on the enclosed postcard. [147]

ad A) and E) of the letter. Koebe apparently doesn't know Fricke's cube theorem [148], otherwise he would understand that the premise that the cut up surfaces constitute a *single* continuum does not play a role in my proofs.

ad B) of the letter. The correctness of my quotation concerning Poincaré will be substantiated by the publication of my Karlsruhe talk.

ad C) of the letter. Here Koebe moves in a vicious circle, because on the one hand he demands from me that I extensively praise his paper which hasn't appeared yet, on the other hand he tries to prevent me from seeing this article.

I emphasize again that I don't know anything about Koebe's achievement except the vague idea he formulated in Karlsruhe, namely to use the deformation theorem [149] for the continuity method, and that I nonetheless quote therefore Koebe only in a very specific way, because I have justified for myself, in all detail, that Theorem 4 follows completely and generally from the deformation theorem.

Ad D) of the letter. Koebe apparently doesn't understand that a not one-to-one but continuous specification of a set by r real parameters doesn't guarantee at all that this set is an r-dimensional manifold without singularities.

To the statement on the card that of both publications in the Jahresbericht and in the Göttinger Nachrichten [150] *one makes the other superfluous.* The similarity of both notes is a purely superficial one; in their contents they supplement each other, and the role of the article in the Jahresbericht amounts to the justification of both footnotes 1) (p. 2) and 1) (p. 4) of the Göttingen note.

That the planned note of Koebe doesn't contain any falsehoods or insinuations concerning me, is, by the way, 1 more in Koebe's interest than in

---

[16] At the same time I send my second page proof to the printer, which contains a small subsequent change (insertion of the word 'recently' [*neulich*, ed.] on p. 2 l. 6 from below).

[146] *Koebe to Brouwer 12.II.1912.*   [147] *Koebe to Brouwer 6.III.1912.*   [148] *Würfelsatz.*
[149] *Verzerrungssatz.*   [150] [Brouwer 1912d, Brouwer 1912c].

mine, because in my eventual refutation I will probably not be able to avoid to disgrace him irreparably.

    With best greetings

    your
    L.E.J. Brouwer

[Signed autograph – in Hilbert]

12-03-26

From F. Engel — 26.III.1912   **Greifswald**

    Dear Dr.! [Sehr geehrter Herr Dr.!]

    Thank you very much for your letter of the 6th of this month. I would have liked to answer a long time ago, but notwithstanding the vacation, I was all the time hampered.

    In the review of your second communication, which I am just now preparing, I give an explanation of the sort you wish. I hope that you will be satisfied by it.

    I agree fully with the considerations in your letter, and I freely admit that in this manner your restriction of the problem seems completely natural. But I miss in both of your articles any indication of the fact that thereby also a much more general problem is dealt with, and such a hint seems to me quite necessary, for which reader will figure that out by himself?

    On the other hand I as yet lack the comprehensive view to see that thus now also the case that only a piece of the group is given, in the neighborhood of the identity transformation, and in the neighborhood of a point, is completely settled. Because in the groups one really meets, one actually always knows in advance only such a piece.

    Furthermore even a group that one knows in its complete extension, can be given in such a form that, so it seems to me, difficulties arise.

    For example, if one writes the general projective group homogeneously and with canonical parameters, [17] then the coefficients are everywhere convergent power series of the parameters, but ∞ many parameter systems give the same transformation. One should really always be on guard for some-

---

[17] I would be very grateful to you if you wouldn't use the word 'canonical' in yet another new meaning. That can cause confusion.

thing like that. Can you now always replace such a parameter manifold by one in which the relation between the points and the transformation is one to one? I shudder for the generality of such considerations and I cannot arm myself against the fear that it isn't feasible to exhaust all possibilities.

Anyway I wish very much to be able to speak to you in person at some time.

With best greetings

yours truly
F. Engel

[Signed autograph – in Brouwer; also in *CWII* p. 153]

1912-03-29

To F. Engel — 29.III.1912   **Blaricum**

Dear professor [Sehr gelehrter Herr Professor]

For your promise in your letter about your review [151] of my second communication I thank you most kindly.

As far as your example of a p-dimensional parameter manifold γ is concerned, in which infinitely many points correspond with the same transformation, it follows from my previous letter that this parameter manifold γ is from my point of view not the true parameter manifold, but that it changes into the true and likewise p-dimensional parameter manifold γ' only by identifying all points that correspond to the same transformation into a single point, so that γ' contains for every transformation only a single point. Naturally γ' will in general be quite differently connected than γ, more in particular, if γ has the simple connectivity of the p-dimensional *number space*, then this property will be generally lost for γ', so for purposes of calculation one will be often obliged to return to γ.

Now I come to the case mentioned by you, that initially only an n-dimensional piece of space τ in the neighborhood of a point P is given, which carries a p-dimensional set π, that lies in the neighborhood of the identity and that also contains it, and that consists of one-one and continuous and pairwise inverse transformations.

---

[151] [Engel 1913].

If we can speak of a group generated by this system (τ, π), then of course also a procedure F must be given by which the transformations of π also are meaningful for all points into which the points of τ are transformed by arbitrary repetitions of π; in the case of transformations given by power-series such a procedure F will naturally consist of analytic continuation.

Assuming this, the point set m which is generated from τ by arbitrary repetition of π is acted upon by a *group* γ consisting of these repetitions of π. This group of pairwise inverse (one- or many-valued) transformations is a so-called 'prime group' (cf. my previous letter) for the *homogenous* point set μ formed by the 'unfolding' of m, *and we will say that 'the system (τ, π) determines a finite continuous group' if and only if μ coincides with τ in a certain neighborhood of P and γ with π in a certain neighborhood of the identity.* (Indeed, if τ and π are assumed totally arbitrary, then one will generally find that μ and γ are manifolds of a *higher* dimension number than τ and π, mostly even of countably infinite dimension.)

At the same time it is clear that if one knows that the system (τ, μ) determines a finite continuous group, the procedure F is completely determined, for it must necessarily consist of the 'group theoretical continuation'.

With the best greetings, and also hoping for my part that I can meet you soon in person,

Your most truly [152]
L.E.J. Brouwer

[Signed autograph, draft – in Brouwer; also in *CWII* p. 154–155, with Freudenthal's comments]

─────────

•12-05-16

H. Weyl to F. Klein— 16.V.1912   Göttingen

Dear Geheimrat, [Sehr geehrter Herr Geheimrat]

About the *factual* differences between Koebe and Brouwer I am only very insufficiently informed. In carrying out the continuity proof three things play a role:

─────────────────

[152] *Ihr ergebenster.*

1) the group continuum,

2) the continuum of Riemann surfaces of genus p,

3) the mapping of both of these onto each other.

In 1) *Brouwer* relies throughout on earlier investigations (Klein, Fricke, Poincaré), that prove that one is dealing with a single connected continuum. *Koebe* takes up this part again and simplifies it substantially by using the deformation theorem [153] which relieves him from the investigation of all degeneracies (boundary parts of the continuum) and at the same time offers the possibility to expand the continuity theorem to all further cases (*Brouwer* only considers the boundary circle case). However I am not certain whether I assess the role of the Koebe deformation theorem correctly, because the course of the proof is completely unknown to me.

Ad 2): Here it seems that the tool of extension of the dimension number is necessary for Brouwer, and also a precise formulation of the circumstances under which two Riemann surfaces of the same genus can be held to be 'little different from each other' (precise formulation of the continuity concept in the manifold of Riemann surfaces). *Koebe* thinks that the extension of the dimension number is something of very secondary importance in the whole proof and claims (which *Brouwer* has disputed) that the theory of functions and integrals on the surface would yield $3p - 3$ modules that correspond in the strict sense *one to one invertible* and continuous with the Riemann surfaces; e.g. one would obtain them with Riemann, as one maps the given surface by means of a suitable normed integral of the first kind.

Ad 3) That the theorem proved by Brouwer about the invariance of the n-dimensional domain is here the decisive argument is admitted without restrictions by *Koebe* too.

Koebe seems to present the matter to be that this, but also only this, is Brouwer's merit, namely that he has ascertained by this theorem the foundation of all continuity proofs, whereas he claims for himself: to have developed in a 'drastic' and 'plain' way those tools that in the specific case of the uniformization problem make the realization of the continuity proof possible. *Brouwer* for his part seems to attach great value to the *priority*; he disputes that Koebe was in Karlsruhe in possession of a proof without gaps, while he, Brouwer, at that time had completely proved along his own lines the matter for the case of the boundary circle. [18]

---

[18] About the exchange of letters between Brouwer and Koebe I don't know anything at all.

[153] *Verzerrungssatz.*

That it has come to a conflict is not because of the matter itself, but the cause is rather the contrary characters that have collided here, Koebe's lack of concern about the claims of others, and Brouwer's irritability and passionate vehemence. [19]

Herr Geheimrat *Hilbert*, with whom I spoke today, and who sends you his best greetings, strongly rejected to exert influence whatsoever on either of the combatants; he didn't go into the matter at all and said only 'they are two adult persons, they must know themselves what they do'. It is not clear yet when Brouwer comes to Göttingen, anyway only *after* the Pentecost holidays.—

What I called in our discussions 'completeness of an axiom system' was in mathematized form nothing else but this: of every theorem, in which only are used *such* concepts, as are defined on the basis of those occurring in the axioms, it must be determined *on the basis of the axioms* whether they are true or false. And that is really the ultimate notion of 'completeness' that one can ask for. Each question that is *comprehensible* on the basis of the axioms, must be decidable with their help. If I leave for example the axiom of Eudoxos out of the axiom system for the real numbers, then the question 'Are there infinitely small magnitudes', i.e. is there a number $\varepsilon > 0$ for which every integral multiple $n\varepsilon < 1$, is comprehensible on the basis of these axioms {the concepts: 0, 1, entire number, multiplication, $<$ and $>$ occur in them; entire number $= 1 + 1 + \ldots + 1$}, *but is not decidable.*

About set theory, real variables and differential equations of mathematical physics I will try to collect some material for next time. If in fact, you, dear Herr Geheimrat, will lecture in the next semester about 'The development of mathematics in the 19th century' (and not about projective geometry), then I am of course willing to take part in the corresponding seminar; I will be able to learn there much myself.

Sunday evening I have returned again on foot to Goslar. On the journey home I met Hilb. [(154)]

---

[19]See last page of this letter [ed. - Weyl had added an extensive footnote on a separate sheet:] At the end of last semester I received once from Brouwer a card with the content whether I wouldn't have so much influence on Koebe to help him, Brouwer, 'to get his property back'; Koebe had kept the manuscript that Brouwer wanted to compare with the page proofs for one or two weeks with him, notwithstanding Brouwer's request to send it back immediately. Altogether I have the impression that the present tension between Brouwer and Koebe is caused by such personal frictions, much more than by differences in content.

[(154)]Probably E. Hilb.

I wholeheartedly hope, dear Herr Geheimrat, that your recovery will make further good and quick progress, so that you will not be confined for too long to the solitude in Hahnenklee.

With the most devoted greetings,
most respectfully yours [155]
Hermann Weyl

[Signed autograph – in Klein]

1912-05-22

From L. Bieberbach — 22.V.1912   Köningsberg

Dear Mr. Brouwer [Lieber Herr Brouwer]

Thank you very much for your kind letter. In order to give an as detailed as possible answer, permit me to repeat each time the individual sentences of your letter, and use this as a starting point for my answer.

1. 'To a certain system s of generating substitutions of a group of Schottky type belongs one (as far as the class is concerned) completely determined Riemann surface F equipped with p return cuts. When therefore there belong to the system s several fundamental domains [156] that cannot be transformed into each other by permitted modifications, then this can only be because the p return cuts of the surface in both cases are completed in a different way to a canonical system of p pairs of return cuts.'

The completion to pairs doesn't matter at all. The problem is rather that different fundamental domains belong to the same system of generators. These correspond then on the Riemann surfaces, that are uniquely determined by the group, to different systems of p return cuts, i.e. to two systems, that cannot by mere translation over the surface, be transformed into each other (cf. Dissertation [157] p. 22–23, p. 35–36). One sees this quickest in the case $p = 3$.

----

[155] *Mit den ergebensten Grüssen, Ihr Sie hochverehrender.*   [156] *FB'* in text, replaced throughout by '*Fundamentalbereich*'.   [157] [Bieberbach 1910].

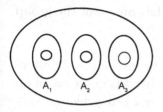

Ring shaped surface a)
Return cuts A_1 A_2 A_3

[..?..] Let the fundamental domain then be something like

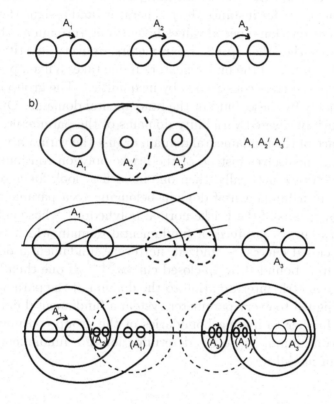

Moreover, in the right punctured hole I imagine pictured the new fundamental domain, which one gets by applying A_2 once, and on the left side analogously. The boundary corresponding to A_2' is indicated. [158]

2. 'But then I don't understand how this non-uniqueness of the fundamental domain in the case of given generators can influence the determina-

---

[158] Brouwer had deciphered the poorly readable above lines, and inderted them at the bottom of the page.

tion of the group by its generators, in other words, the control of the group
continuum by means of the invariants as parameters.

You seem to assume implicitly that by indicating the invariants of finitely
many or also all substitutions of the group, the group itself is determined up
to linear transformations. This theorem seems to me not self evident. Anyway, I can't prove it. In Fricke, the proof in the boundary circle case seems
to rely completely on the fact that only one fundamental domain belongs to
a system of generators. But that is not satisfied. Anyway, here one can make
a change. If one takes for instance the p generators and assigns then three of
the $3p$ coefficient relations special values by norming through a substitution,
then one can take the other $3p-3$ as parameters, or every substitution has 3
[...?...] 2 fixed points and the multiplicator. If one takes 3 fixed points, there
remain $3p-3$ parameters constrained by inequalities. The group is then determined uniquely by these, but not the fundamental domains. One has now
either the multiply covered variability domains of the parameter, hence the
'Riemann' space of the fundamental domain or just the variability domain of
the parameters themselves instead of the Fricke polygon continuum. If one
wants to proceed geometrically, then one must first look for a geometrical
normalization of a fundamental domain belonging to a parameter system,
for example by means of the Fricke normal polyhedron, whose cut from the
sphere *perhaps* always produces a fundamental domain which is bounded
by p pairs of closed curves. — indeed, not every fundamental domain of a
Schottky group is bounded by $2p$ closed curves. [159] If one then has in this
way given a geometric interpretation to the domain of the parameters, that
is, one has assigned to every parameter system a fundamental domain, then
one must go further to finding boundaries of the group continuum within
the polygon continuum, i.e. to the determination of a fundamental domain
of the group of modules.

––––––––

Zusatz [160] Now it remains to show that this variability domain is a continuum; this is handled in Fricke again on the basis of the geometric meaning
of the invariants but in the final analysis on the basis of the unique determination of the fundamental domain by the invariants. To prove the
analogue here seems not to have succeeded until now. Hence the advantages

––––––––––––––––––––

[159] Here Bieberbach refers in a footnote to the '*Zusatz*' on the next sheet – 'cf. see
other side'.   [160] Supplement

of your new continuity proof, which in the present context operates only with neighborhoods, no longer with the full continuum.

———

Thus one cannot, on the basis of the existence of those parameters, conclude the existence of a group *continuum*, and therefore one cannot determine it by these means.

These things, which are certainly a bit vague in their being indefinite and unsettled, were floating in my mind in the case of my note. I did not at all want to write all of that down, just observe that already in the beginning there must be differences from the line of Fricke's proof.

It was not my wish at all to express myself in print about these things, in any case I didn't want to have printed anything about my remarks in Karlsruhe which were by themselves essentially superfluous. It happened only on Klein's explicit 'command'. Also in my dissertation I have restricted myself to what I could determine with certainty. I seems, now that the continuity proof is settled, rather unnecessary to continue on this road, unless it is for quite other purposes — convergence of the Schottky sequences.

At the same time, I send a copy of my dissertation. I believed that I had done so a long time ago. I therefore beg your pardon for this omission.

If you are of the opinion that my note needs an thorough textual change, I ask you for a brief communication.

I will not come to Cambridge. But in any case to Münster. So see you there and cordial greetings.

L. Bieberbach

K.i.Pr. [161] 22.V.1912

NB. On p. 33 of my dissertation I have made a really stupid mistake. The result is correct as can be seen much more easily, namely by showing that the changes of the [?] on p. 32 can always be fulfilled by new circular domains.

[Signed autograph – in Brouwer]

———

[161] Königsberg in Preussen.

1912-05-31

To D. Hilbert — 31.V.1912   Blaricum

Dear Mr. Geheimrat! [Lieber Herr Geheimrat!]

I'll now come next Sunday to Göttingen via Löhne, Hameln, Elze, and I will arrive at 5.38 in the afternoon. My wife has decided to accompany me. We will stay in Hotel Gebhard, and will stay until early Wednesday. We will use the Wednesday and Thursday for a tour of the Brocken, and the night of Thursday to Friday is reserved for the return trip, because on Friday I will be busy in Amsterdam.

For a better preparation of our coming conversation, I enclose two letters which will answer for you the plausible question why I got mixed up at all with Koebe in connection with the publication of my continuity proof. Indeed, the contribution to the continuity proof I presented in Karlsruhe consisted of two parts, of which the first one (the 'invariance of domain') already was submitted in July for publication, whereas with respect to the second (the 'extension of the group set to the set of automorphic functions with m poles'), Koebe claims priority, according to the enclosed letter of Bernstein (cf. the part marked with pencil). Because, moreover, this part didn't seem very deep to me, I hesitated of course to publish it, even though Blumenthal urged me to do so. Finally I sent the manuscript early November to Fricke with the question whether he considered the contents new and worth publishing, and then I received the enclosed answer. The statement about Koebe therein [20] that is marked with pencil complicated the matter so much that I, when a short time later Blumenthal as well as Fricke and also Klein (namely indirectly through Fricke) requested me to publish, I could not possibly do so without, for the sake of more certainty and clarity about Koebe's achievements, getting in touch with Koebe himself, because otherwise there was the danger that Koebe would accuse my publication of being trivial and me of being a plagiarist. In the exchange of letters with Koebe I then received on my very specific questions again and again evasive answers; the only thing I got out of him was the mutual agreement to edit our notes about the continuity proof in mutual understanding. How he then later broke his word and the matter got dragged along, you know.

---

[20]The manuscript was at first incorrectly understood by Fricke, who hadn't been in Karlsruhe, hence the unfounded criticism contained in his letter of December 1.

Well, the rest we'll discuss next week. My wife eagerly looks forward to meeting you again, as I do, and we both greet you cordially.

Your
Egbertus Brouwer.

[Signed autograph – in Hilbert]

12-11-07

From F. Bernstein — 7.XI.1912   **Göttingen**

Dear Friend! [Lieber Freund]

I have suffered the last two months from a severe depression and although I thought all the time of writing you again, I couldn't get myself to make a decision to do so. That has actually been weighing heavily on my mind — it becomes ever more difficult the longer one waits.

Now things are improving in every respect. I feel physically good again and I am happy to be able to do something again. The cure in Wildbad [162] ruined me so badly that I often had to stay in bed for days. Only in Halle I have completely recovered. I got rid of the rheumatism, so much good has at least come out of it.

It was quite a pity that we couldn't meet this vacation. I had so much counted on it.

At our lunch table there have been big changes. We are not in Gebhard anymore. There only a disagreeable physicist has remained, whom we were foolish enough to get stuck with last semester. I eat in the Ratskeller with Försterling, Mrs. Jalli, Paul Hertz and others that you don't known: Defregger, Schwartz, Rusitskya.

Weyl has — incomprehensibly — left us.

You must have received Rosenthal's Habilitation thesis.

Can you tell me perhaps what kind of an impression Borel's rejoinder in the Annalen [163] made on you? Blumenthal has nicely tricked me, because he showed me a totally different manuscript, but not at all the final version, like for example his comparison of our proofs. Now I don't know

---

[162] A *Kurort*, a 'Spa'.   [163] [Borel 1912].

whether in my answer it is clear enough that I think the note of Borel in the Rendiconti too inexact to be counted as a source of a proof. Because I have still some publishable material about the subject, I could once more put his embellishments in the spotlight.

How are you and how is your family? Many greetings to your honored spouse and the young lady your daughter.

With best greetings

Yours truly
Felix Bernstein

[Signed autograph – in Brouwer]

1913-02-06

From L. Bieberbach — 6.II.1913   **Königsberg (Preussen)**

Dear Mr. Brouwer [Lieber Herr Brouwer]

Unfortunately I must trouble you with a probably silly question. Recently I noticed your proof of the Jordan curve theorem in Annalen 69 [164] (If you still have a sufficient number of reprints available, I would be grateful to obtain one, because I have reprints of all your articles except this one.). On page 172 of this you construct a polygon p. Of this you use the property that it has an interior and an exterior. Now it seems to me that to conclude this one must know the correctness of this statement for *every* polygon. For I don't see how for example on page 171 for the polygon π one can obtain evident information of this decomposition property of p by a suitable special choice. With a construction according to the π-recipe one only obtains polygons that necessarily also contain points of N_1.

In this conjecture of mine I was enhanced by the fact that I don't see at which other place the two-sidedness of the Cartesian plane is used, except in this polygon theorem. I miss the proof of the above, which anyway isn't difficult, in your article. I would be very grateful if you would so kind as to put me on the right road.

---

[164] [Brouwer 1910b].

With cordial greetings

Bieberbach

[Signed autograph – in Brouwer]

13-04-16

To D. Hilbert — 16.IV.1913   **Amsterdam**
                                                             Overtoom 565

Dear Mr. Geheimrat! [Lieber Herr Geheimrat]

It is perhaps known to you that I am occupying myself for some time
now with the new edition of Schoenflies' Bericht on set theory. This came
about as follows: For some years now I was repeatedly urged from various
sides to write a book on set theory, because the existing books and encyclopedia articles about this subject are too unreliable and superficial. When
in the summer of 1911 in Göttingen such exhortations were addressed to
me again, and I at the same time learned that Schoenflies was preparing a
new edition of his Bericht, I thought that the desired aim could be reached
with relatively little loss of time on my side, if I was given the opportunity
to check Schoenflies' book during the printing process, and if necessary to
improve and complete it. The difficulty to bring Schoenflies to submit to
my supervision was soon removed, when, Fricke, who knew Schoenflies personally, offered to mediate, on the occasion of visiting him in Harzburg. [165]
Schoenflies then was even most pleased to accept this proposition that was
put to him through Fricke (how Fricke formulated it is however unknown to
me); as a consequence I am involved in a correspondence about the relevant
galley proofs. Meanwhile it turned out that with respect to the method
and intensity of my cooperation Schoenflies and I harbor fundamentally different tendencies: Schoenflies would like to restrict my influence if possible
to improvement of the false theorems and proofs, while I of course aim in
addition at accomplishing completions and more depth. In this struggle I
am the weaker party, because Schoenflies possesses the right of the final
decision, even though he occasionally does make certain concessions out of

---

[165] This refers probably to a visit of Brouwer to Fricke, see the correspondence with
Fricke.

courtesy (or maybe also out of fear that I will desert him because he has seen what a tremendous amount [166] of errors I have picked out for him). The farther the handling of the proofs now proceeds, the more Schoenflies urges me to hurry and the less I can achieve with him, maybe also because he knows he is safe with respect to the cruder errors, so he feels gradually less dependent on me. Nowadays I almost feel the work, once undertaken and therefore to be completed, to be a Sisyphus task. For example, it is a small effort to compose an insertion; but to find then later that Schoenflies, who initially had left the editing work to me, nonetheless wants to 'improve' it later himself, i.e. to insert errors, and also not to be able to put him right, he, who is always in a hurry, and who is, as he admits himself, totally overworked — that is for me an intolerable situation! Also in this way a meager advantage will be reaped which by far doesn't match the efforts I spent, because the general foundation of the book becomes deficient, and difficult questions will remain undiscussed in the book.

Relief would only be possible when from the side of a third party gentle pressure would be exerted on Schoenflies. In this respect I don't want to ask you anything specific, but Schoenflies will next week be, as he writes to me, in Göttingen, and then he will probably spontaneously come to speak with you about his Bericht. Should this happen, then a certain suggestion would come from you to the effect that he should leave me as much freedom as possible, wouldn't it? Given the great respect of Schoenflies for you, such a suggestion would immediately turn out to be very effective, that I am sure of.

Well, I wanted after all just to inform you about the above mentioned state of affairs. Anyway, there is no harm in it that you know about it, and maybe this knowledge enables you to drop a few words in the next weeks, which might be of the greatest help — not to me personally, but to set theory, hence to mathematics, which we both love.

Cordial greetings to you both, also from my wife. When it is somehow possible, I will come myself next week for a few days.

Totally yours [167]
L.E.J. Brouwer.

[Signed autograph – in Hilbert]

---

[166] '*Unmenge*'.   [167] *Ganz Ihr.*

)13-06-16

To D. Hilbert — 16.VI.1913   **The Hague**
                                                                      Haag

Dear Geheimrat [Lieber Herr Geheimrat]

I beg you now for advice. I can become an full professor, and in fact both
in *Groningen*, where on the one hand I will be completely *free* in my professional activities, but where on the other hand I will find a petty provincial
town and probably fewer sympathetic colleagues, and in *Amsterdam*, i.e.
in a lively big city, which always has been intimately connected with my
life, and which is close to my cosy home in Blaricum and to the dunes, and
where I also feel comfortable in the faculty, where however I will be mainly
charged with teaching *Mechanics*. If my official duties were the same in both
universities, then I would of course not hesitate to choose Amsterdam; but
I cannot possibly envisage to what extent an unceasing involvement with
applied mathematics would divert my research from its natural course, and
hence, at the same time, would more or less paralyze it. What is your opinion in this matter? Should I risk to stay where I feel at home, and count
on it that the harmony between my thinking and mechanics will establish
itself completely automatically?

You know me after all, and you have such rich experiences as a researcher.
I wouldn't know anybody whose advice I would seek more than yours, now
that I carry on the most intense struggles of indecision The small-town social
life and the pressure of convention must be terrible in Groningen, and there
is no countryside at all.

Now for something different. Just recently I read that a fourth edition of
your Grundlagen der Geometrie [168] will appear. Have the remarks about
Appendix IV, that I communicated to you in the fall of 1909 (i.e. on the
work from the Annalen 56 [169]) been taken into account? I would anyhow be
happy to help out with the correction of the paragraphs concerned, should
you wish so, and if the authorization for printing [170] hasn't been given
yet.

Unfortunately, the effect of your suggestions on Schoenflies hasn't been
lasting. Just look at the enclosed letter. [171] You will understand how
difficult it is to me to have to read, after all my efforts, words like those

---

[168] Foundations of Geometry.   [169] [Hilbert 1902]. *Brouwer to Hilbert 28.X.1909*, see
CWII p. 102 ff. with Freudenthal's comments.   [170] Imprimatur.   [171] Refers to the revised
Bericht.

marked with pencil in the enclosed letter of May 29. Much indeed, that after an endless exchange of letters, finally found its correct formulation, he now diligently starts to mess up again. He must be *very* overworked, because he makes mistakes any student should be ashamed of.

But I will stay on my post to the end, and patiently continue to teach him and try to save what can be saved at all.

Now many cordial greetings to you and your wife

Your
L.E.J. Brouwer.

Please return the enclosed letters of Schoenflies to me. At the moment I work as member of the examination committee of the Technical University in Delft, and live for the time being in The Hague. At the same time I am sending you a picture postcard that certainly will evoke good memories in you.

[Signed autograph – in Hilbert]

1913-07-04

To D. Hilbert — 4.VII.1913   **Amsterdam**
                                                                        Overtoom 565

Dear Mr. Hilbert [Lieber Herr Hilbert]

I suddenly received an ultimatum from Groningen, [172] and I have decided today to opt for Amsterdam and mechanics. [173] Quod bonum felix faustumque sit! [174]

With Schoenflies things are getting ever worse. When there is not within a few days a complete change in his behavior, I will reach the point that I finally give up the whole enterprise for which I have suspended — in the general interest — all activity of my own for 8 months. I only hesitate

----

[172] Brouwer was offered Schoute's chair.   [173] Brouwer was an extraordinary professor in Amsterdam, so the full professorship in Groningen had its advantages. In July 1913 he was appointed full professor in Amsterdam, after Korteweg had given up his chair, and accepted an extraordinary chair.   [174] May it be good, fortunate and prosperous (Cicero).

because I can't stand the thought that I have wasted my energy for such a long period of time. Couldn't you write him once more a line?

With the best greetings,

Your Brouwer.

[Signed autograph, postcard – in Hilbert]

13-08-16

To A. Schoenflies — 16.VIII.1913   Jungborn (Harz)

Dear Mr. Schoenflies, [Lieber Herr Schoenflies]

Enclosed, I send you your sheets back, and I add my own proof which I have written down in extenso on sheets 1), 2) and 3). It still seems to me the best thing that you just stick to replacing the considerations between the middle of p. 312 (starting from 'Sind also') and the middle of p. 314 ('Mit diesem Resultat'), by my proof. If this is done in the form proposed by me, i.e. introduced by the footnote on sheet 1 above, then we avoid on the one hand the edge with respect to Lebesgue, and on the other hand the reader will *not in any* respect get even a *whiff* of an impression as if at this point your force had failed in some way.

More specifically I have the following objections to your elaborations. In the first place you write in your letter that in the case of two sets one can deduce the theorem about the sum of sets also *directly* from relation (a) on sheet *b*. This would however only be the case when one has already *in advance* the certainty that the sum of two measurable sets is *again measurable*.

Secondly — and this is more important — you assume on sheet *a* that $\{\alpha'\}$ and $\{\beta'\}$ and likewise $\{\alpha\}$ and $\{\beta\}$ and $\{\gamma\}$ are *relations*;[175]. this is not the case; *the difference of two relations need not always to be a relation.*

---

[175] *Beziehungen*

[Ed. supplement]

*[A sheet with some remarks has been preserved with the above draft. The
remarks were apparently meant to be included in the 'Vorwort zur zweiten
Auflage'.* [176]*]*

pointed out[×]

[×] the cooperation of Mr. Brouwer is the more valuable because
Mr. Brouwer's personal views on the foundations of set theory are
in many points in sharp contrast with the guiding principles of this
report.

————*For the preface*————

(before: all in the text) referred, and this though his personal etc.

---

also [177] the word 'auch' [178] must be omitted; for it gives the impression that the 'Besserungen und Richtigstellungen' [179] constitute only
something 'nebensächliches' [180] in my 'Unterstützung' [181]; by leaving out 'auch' they will however appear as the 'wesentliche Inhalt' [182]
of the 'Unterstützing' — and that is also the 'genaue Wahrheit.' [183]

1913-11-08

From É. Borel — 8.XI.1913   **Paris**
Université de Paris,
Ecole Normale Supérieure
45 rue d'Ulm

Dear Sir [Cher Monsieur]

I hope the letter that I have sent you yesterday to the University has
reached you. Reading your first letter, I had interpreted your remarks in the

---

[176] 'Preface to the second edition.' See also Freudenthal's notes in *CWII* p. 367–370, in
particular note 9, p. 369–370.   [177] The following is entirely in Dutch except for the quoted
words.   [178] 'also'.   [179] 'improvements and corrections'.   [180] 'of minor or secondary
importance'.   [181] 'support'.   [182] 'essential contents'.   [183] 'exact truth'.

sense of the explications you give me in the your second one. I hope that
the publication of my course, if it is realized, will satisfy you.

As regards to reprints, I have unfortunately the very bad habit of leaving
the packages mostly unopened for months or even years, because of lack of
time. And when so much time has passed by, most of the time I have
thought about them, and blushed about the corrections or simplifications or
additions for which I plan a new publication, and I don't have the courage
to dispatch the old and obsolete publication. That is why I haven't sent you
that note of the Bulletin de la Societé Mathématique; also I have renounced
completely from having reprints made of some publications, like the Comptes
Rendus, that don't provide them for free.

I send you by the same mail a large package to repair my shortcomings
to you.

Yours very truly [184]
Emile Borel

[Signed autograph – in Brouwer]

14-06-04

From D.J. Korteweg — 4.VI.1914   **Amsterdam**
Vondelstraat 104-F

Amice,

Furthermore, De Vries informed me how much Göttingen takes up your
time, and I understand very well that you don't wish to take talks now upon
you.

My request was in fact solely the consequence of my endeavor to raise
the level of talks as high as possible, and I rather expected that this time
you would ask to be excused.

Your outpouring was less expected by me.

I saddens me much that you like your professorship so little.

However I consider this to be a subjective phenomenon, indeed related
to your great gifts, the way everything is more or less related in a certain
person, but not as inseparable from such gifts.

---

[184] *Votre bien dévoué.*

Methinks this is proved by our physicists who actually are members of foreign Academies, and yet for a long period had not less official worries than you (van der Waals, Lorentz who took over Onnes'[185] lectures for medical students).

Thus it is hard to believe that six lectures a week, partly of an elementary nature, a few examinations a month (and well over four months almost undisturbed vacation) should stop someone from doing scientific work, even of the highest order.

If this indeed is the case with you, then truly there is nothing for it than that you accept as soon as possible a German professorship, and that opportunity will not fail to appear, although I expect that also there inhibiting influences will occur, if you are really *that* sensitive to them.

Another question is whether you, if you can share with me the conviction that the problem must be found *in yourself*, can't do a thing or two to diminish the conflict.

For example, to prepare your lectures in the vacation, so that you are all the time well ahead, and that each time you only need a moment to prepare. That takes away much of the nervous and hurried aspects that are inherent to the teaching new material for the first time.

And then I believe that the more regularly, I almost would say more commonplace, one arranges one's *external* life such as accommodation, working hours etc., the more one's energy increases and the easier and less painful one's *internal* life develops.

I don't know whether you can or want to follow such advice, but you will understand that I feel obliged to give it to you after your poignant outpouring.

I very much hope that you won't blame me for it; if not I would be very sorry, but I felt not free to omit it.

With cordial greetings

Your
D.J. Korteweg

[Signed autograph – in Brouwer]

---

[185] H. Kamerlingh Onnes.

To G. Hamel — 20.VI.1914   Amsterdam

Dear Mr. Hamel, [Sehr geehrter Herr Hamel]

Blumenthal forwarded your letter of June 9 to me; allow me from now
on to write to you directly, to thank you cordially for your interest and kind
help. Your idea to reduce the treatment of practical stability to a 'slow'
withdrawal from the equilibrium position has surprised me very much, but
it seems germane to me, and I share your conviction that moreover the
stability on the smooth rotating saddle must allow for an experimentally
verification.

I think I can shorten the proof even more than in your letter. Let the
general solution of the frictionless equations of motion be:

$$
\begin{aligned}
x &= c_1 e^{\lambda_1 t} + c_2 e^{-\lambda_1 t} + c_3 e^{\lambda_2 t} + c_4 e^{-\lambda_2 t} \\
y &= k_1 c_1 e^{\lambda_1 t} - k_1 c_2 e^{-\lambda_1 t} + k_2 c_3 e^{\lambda_2 t} - k_2 c_4 e^{-\lambda_2 t} \\
\dot{x} &= \lambda_1 c_1 e^{\lambda_1 t} - \lambda_1 c_2 e^{-\lambda_1 t} + \lambda_2 c_3 e^{\lambda_2 t} - \lambda_2 c_4 e^{-\lambda_2 t} \\
\dot{y} &= k_1 \lambda_1 c_1 e^{\lambda_1 t} + k_1 \lambda_1 c_2 e^{-\lambda_1 t} + k_2 \lambda_2 c_3 e^{\lambda_2 t} + k_2 \lambda_2 c_4 e^{-\lambda_2 t}
\end{aligned}
\tag{I}
$$

The state of motion of the material point can be defined on the one hand
by the values of $x, y, \dot{x}, \dot{y}$, on the other hand by the four quantities $\gamma_1, \gamma_2, \gamma_3,$
γ_4, by which we mean the corresponding values of c_1, c_2, c_3, c_4, if we let the
given state of motion correspond to the zero point in time. The system of
values $(x, y, \dot{x}, \dot{y})$ and $(\gamma_1, \gamma_2, \gamma_3, \gamma_4)$ are one to one and homogeneously lin-
early related. To one orbit there belongs the simply infinite manifold of value
systems (by which we in the following mean a real orbit curve) $(\gamma_1, \gamma_2, \gamma_3, \gamma_4)$,
and this has in the stable case the property that every $(\text{mod}.\gamma_\nu)$ is constant,
and hence also $\sqrt{\Sigma(\text{mod}. \gamma_\nu)^2}$. For, one has $\gamma_1(t) = e^{\lambda_1 t}\gamma_1(0)$ and so on.
Likewise $(\text{mod}. d\gamma_\nu)^2$ is constant for two adjacent (also with respect to time)
orbits. For, one has again $d\gamma_1(t) = e^{\lambda_1 t}d\gamma_1(0)$, and so on.

From the equations (I) it furthermore follows that the ratio of

$$
\sqrt{dx^2 + dy^2 + d\dot{x}^2 + d\dot{y}^2}
$$

to the corresponding values of $\sqrt{\Sigma(\text{mod}. d\gamma_\nu)^2}$ varies between two fixed
bounds, which are both different from 0 and ∞. The maximal value of
$\sqrt{dx^2 + dy^2 + d\dot{x}^2 + d\dot{y}^2}$, measured from each point of the one to the 'closest' (i.e. yielding a minimal $dx^2 + dy^2 + d\dot{x}^2 + d\dot{y}^2$) point of the other orbit I

call the path distance of the two neighboring orbits; let h be the maximum value in the whole system of orbits, which maximum certainly exists because of the above, of the ratio between the maximum and minimum value for any two neighboring orbits of the expression $\sqrt{dx^2 + dy^2 + d\dot{x}^2 + d\dot{y}^2}$ thus obtained.

Then, because of the sliding friction, the path distance covered during the time element dt is $< hk\, dt$. The increase of the 'shift' (i.e. the minimum of $\sqrt{dx^2 + dy^2 + d\dot{x}^2 + d\dot{y}^2}$ for the orbit hence likewise is $< hk\, dt$ and the increase of the shift between the times t_0 and t is $< hk(t - t_0)$, which proves your theorem.

Please allow me one more remark. You write in your letter the equations of motion in the following form:

$$\ddot{x} - 2\omega\dot{y} + ax = 0$$
$$\ddot{y} + 2\omega\dot{x} + by = 0$$

and you state the *conjecture*, that in case $a, b > 0$ the friction causes convergence to a position of rest. The fact that standstill is reached already after a finite time follows from the existence of the energy integral of the frictionless motion:

$$H \equiv \frac{1}{2}(\dot{x}^2 + \dot{y}^2 + ax^2 + by^2) = c$$

In fact, for the changes of H caused by the friction the following holds:

$$\frac{dH}{dt} = -k\sqrt{\dot{x}^2 + \dot{y}^2}$$

Hence the positive definite form H can only decrease, which proves the stability of the motion with friction. Furthermore the total orbit length must be finite, because $dH/dt = -k$. Now if the end point of the orbit lies at a finite distance from the equilibrium point, then close to this end point only unboundedly decreasing velocities occur, the direction of which must approach the direction of the attracting force, because of the attractive force ($X = -ax, Y = -by$) and the friction force, and it cannot cross that direction, but on the other hand it must reach it, because otherwise the acceleration would converge to a finite limit, whose direction would be different from the limit of the velocity, so that the velocity could not be exhausted by the integral of the acceleration. Hence the velocity has in the end point the direction of the vector $-(a\bar{x}_x + b\bar{y}_y)$, so the point *approaches* during the end of the orbit the equilibrium position ($x = 0, y = 0$). Hence the limiting value of the attractive force can't be equal to the friction force, because otherwise the components of velocity and acceleration in the direction of the limiting tangent would have the same sign in the neighborhood of the end of the orbit!

Hence in the neighborhood of the end point the friction force dominates the attraction force with a finite surplus, so after a distance ε from the end point is reached, the end point itself will be reached within a time element of order ε.

But if the end point of the orbit lies in the equilibrium position itself, then $x, y, \dot{x}, \dot{y}$ will finally all become vanishingly small. At that moment however the friction force dominates, and the resting position is reached in a vanishingly small time.

Again thanking you, yours sincerely, [186]

Your
L.E.J. Brouwer

[Signed autograph, draft – in Brouwer; also in *CW II* p. 684–686, with Freudenthal's comments]

14-07-13

From D.J. Korteweg — 13.VII.1914   **Amsterdam**

Vondelstraat 104-F

Amice,

With reference to your letter one pragmatic remark.

Would you please postpone your official discharge as member of the prize contest committee until both problems that have been entered by you or with your cooperation, have been dealt with? [187]

I will then delete them from the program of 1915; but the possibility exists that answers are submitted, and methinks you will see the reasonableness of my request, which amounts to you being then able to function as first reporter.

---

[186] *Mit nochmaligen Dank, in grosser Hochachtung.*   [187] The Dutch mathematical society offered regularly prize problems, that were judged by the above mentioned committee. These prize problems have generated some outstanding research.

Another proposal is that you *stay on as* a member and that I take on me the commitment to have you report only on the problems posed by yourself (and not as second or third for the others).

I think I can assume responsibility for this, taking into account the heavy duties that await you in your function as editor of the Annalen, [188] because I feel convinced that also the other committee members will prefer this over your full discharge. With respect to a possible successor as chairman, this letter can serve as a guarantee; that's the reason why I will write down the P.S. that I am adding on a separate sheet.

Please, a yes or a no on both these matters.

Greeting

Your

D.J. Korteweg

---

13 July '14

P.S. Whereas in different circumstances I would be very pleased with your prestigious appointment as editor of the world's foremost mathematical journal, my heart isn't in it for more than one reason.

First, I view the work, with which you are being flooded from Göttingen, as a very serious and *enduring* obstruction against continuing your own independent work, and yet *that* is what you will be judged by in the long run, also in Germany.

Second, I foresee that consequentially you will more and more withdraw from the life of the Dutch mathematical community, even though just the opposite attitude is expected from a Dutch professor, and rightly so in my opinion. That this is a great disappointment for me, is less relevant; however the fact itself would be very regrettable for the further development of this life.

Third, I fear that you will search for the cause of diminished fertility in a place where it is not, or only for a small part: in your *professorship,* and that you will consider this more and more to be a real nuisance.

At first I had planned to discuss this point in more detail, and raise, among other things, the obvious objections to your proportion. [189] but your comparison of a Dutch professorship with six hours of lectures, and

---

[188] Brouwer was made associate editor (*Mitarbeiter*) of the *Mathematische Annalen* in the summer of 1914.   [189] This is rather vague, it probably refers to the comparison below of the duties of the Dutch professor and of the family doctor in the country.

some of them of a very elementary nature (while the others leave you great freedom in the choice of subjects, plus four months of vacation) [190] to the position of a country physician with a busy practice, makes me, on further reflection, feel that this would be not be successful because in my view you have lost here all sense of proportion.

One thing I have to admit in order not to be unfair to you. Namely this, that your elementary lectures seem to be for *you* a great problem, because they make you impatient and seem to make you temporarily unfit for other work.

I wish I were younger and had more time ahead, to do in addition to the obligations I haven taken on, to which I have recently adapted my position, [191] something for you in this matter; but that is not possible and one cannot very well demand that from De Vries.

Moreover, I can't really understand that this problem should be in the long run insurmountable for you.

––––––––––

In my opinion you are just as grossly exaggerating, in calling the situation of mathematics in the universities deplorable, and in seeing in the Dutch environment an obstacle for the development of a gifted young mathematician.

I totally disagree with that. It seems to me that this Dutch environment, consisting of the Academy, [192] universities, and the Wiskundige Genootschap [193] (lectures; prize contests; the Revue [194] which, as it were, presents each beginning mathematician with the worldwide constantly developing mathematics, in which he will have to take part if he wants to accomplish anything; an almost complete journal collection) is by no means the deplorably insufficient environment you claim it is.

Methinks even that if you consult your own experience, you will have to recognize that you found in it many beneficial stimuli, a great freedom in the choice of your field of study, and for the rest nothing but recognition and encouragement.

In my opinion one should not overestimate the influence of the environment, neither in the positive, nor in the negative sense, because after all every mathematician of any importance has to take his *own* education

––

[190] In the original text there is punctuation instead of brackets.   [191] Korteweg, who was only four years away from his retirement, had exchanged his chair for an extraordinary professorship, so that Brouwer could become a full professor.   [192] KNAW.   [193] Dutch Mathematical Society.   [194] Revue semestrielle

in hand and find his *own* way; but in the emergence of you, and recently
Schouten, I see the proof that this environment is at any rate not unsuitable
at all.

I really believe that every gifted mathematician has the opportunity to
fulfill his capabilities, which naturally are rarely *very* extraordinary.

People of this kind are rare and therefore because of the laws of probability small in number, hence distributed irregularly over the different breeding
grounds. In the Netherlands one may not expect to find always representatives of them in each field. I was overjoyed when it appeared that now
mathematics got its turn, and I would be very sorry if in your case this
would become a disappointment because of you being absorbed in editorial
work, even though this is of the highest level.

Meanwhile, let us hope for the best now the matter is the way it is.

[Signed autograph – in Brouwer]

_____

1915-06-11

From H.A. Lorentz — 11.VI.1915   Haarlem
                                                             Zijlweg 76

Amice,

When we strolled through Amsterdam after the last meeting of the
Academy [195] and we came to speak about the mathematics vacancy in
Leyden, it was on the tip of my tongue, that I would like nothing better
than that you yourself could come to the decision to exchange Amsterdam
for Leyden. I didn't mention it because the faculty hadn't met yet.

Now we had a meeting and it turned out that it was the unanimous
wish that you would, if possible, occupy the vacant post; we all consider this
of the greatest importance for the flourishing of the faculty.

More in particular Kluyver, De Sitter, Ehrenfest and I would very much
appreciate to be able to collaborate with you. You could be certain, that
you would be welcomed warmly and with open arms.

Kluyver and I would like to come over sometime to discuss the matter, in
order to explain the intentions of the faculty and answer questions from you
to our ability. *Preferably* tomorrow, Saturday evening; we can be at your

_____
[195] KNAW.

place at about eight o'clock. We could also come, if that suits you better, on Sunday afternoon or Monday evening. Let me know, if you please, when you can receive us.

With friendly greeting

t.t.

H.A. Lorentz

[Signed autograph – in Brouwer]

Editorial supplement

[*How serious the Leyden offer was, is illustrated by the following letter. Apparently Ehrenfest was willing to make considerable sacrifices to attach Brouwer to Leiden.*]

H.A. Lorentz to P. Ehrenfest — 9.VI.1915   *Haarlem*
                                                            Zijlweg 76

Amice,

You make us a very generous offer, and I believe that we must accept it if there is no other means to relieve Brouwer from mechanics; that is, in the case that the Keesom plan would still meet too much resistance in the faculty, or the minister would after all not be inclined to satisfy our wishes. But I would regret it very much if this should have to be the solution. You have devoted yourself now with heart and soul for almost three years to theoretical physics and you had as a professor very good results; I would regret it if this fortuitous activity would suffer from an larger number of lectures.

On a few other points in your letter I must answer as follows.

a. Your plan to give Keesom only crystallography and in this manner humor M., I would think is very good, but I consider it unrealizable now.

Indeed, in present circumstances one can only obtain some money when it's a matter of great and urgent importance, like in the case of

Brouwer joining us. The problem of crystallography has no relation
at all with that.

[. . . .]

Now it seems to me that the best thing is to mention tomorrow your
declaration of willingness, because it shows that in any case the mechanics course is provided for. We don't have to discuss it right now in
more detail, but we can negotiate first with Brouwer. If he is prepared
to do so, then we can resume the discussions in the faculty, and then
I would not yet want to give up right away on the Keesom plan (i.e.
the first plan).

With cordial greeting, t.t.

H.A. Lorentz.

[. . . .]

[Signed autograph – in Ehrenfest]

1915-06-19

To P. Zeeman — 19.VI.1915   **Blaricum** [196]

Dear Colleague, [Hooggeachte Collega]

On behalf of the Faculty of Mathematics and Physics of the State University of Leyden I am invited to accept a chair in geometry there, with the
promise that the Faculty will completely consent in my putting into practice
the view that discharging the task of a professor consists more of dedication
to one's own scientific work and being accessible for independently working
students that look for guidance and information, than of regular lectures on
routine theories that since long have been expounded clearly in books. The
oral explication of the invitation was summarized by colleague Ehrenfest

---

[196] 'To Prof.Dr. P. Zeeman, Chairman of the Faculty of Mathematics and Physics of
the University of Amsterdam.'

with the words: 'So materially nothing more is asked from you than being there.' [197]

Now there are three circumstances that make the Leyden chair offered to me preferable over my present Amsterdam working environment, unless it would be possible to obtain certain encouraging assurances from the Curators.

First, I have since long experienced that the 'Leyden' interpretation, of the task of a professor, alluded to above, is not shared generally in the Amsterdam Faculty of Mathematics and Physics, which has become for me, who as youngest member of the Faculty doesn't feel strong enough to go my own way against other currents, the reason that I am, since my accession to office, handicapped in my scientific work in a very discouraging manner. Only from an encouragement from the Curators, I could derive the strength to carry out my task in full accordance with my own insights and conviction, which, incidentally, would be beneficial not only to my own scientific work but also to the interests of the students, and where there should be no fear at all that I would take the above quoted words of colleague Ehrenfest too literally as a guideline.

Second, the three mathematical sciences (mathematics, mathematical astronomy and theoretical mechanics) are represented in Amsterdam by a weaker teaching staff than in the State Universities, because we have for those subjects only two full and one extraordinary professor available, and in Groningen en Leyden one has three full professors for them. In Utrecht mathematical astronomy is joined with practical astronomy, and theoretical mechanics with theoretical physics, so that a precise comparison isn't possible; but because there one has two full professors for pure mathematics (without astronomy and mechanics), one must consider the strength there roughly equivalent to that of each of the two state universities.

Third, I, who cannot feel in good health for more than a few days, in the low-lying Dutch towns, and who has never been capable of intense mental work, while residing in such towns, would as a Leyden professor be able to live outside of the municipality without further formalities, for example in Noordwijk or Wassenaar, whereas as an Amsterdam professor I have to use two houses, one in Amsterdam because of the municipal ordinances, the other in Blaricum, where I am obliged, in the interest of my work and my health, to seek refuge for several days per week.

This disadvantage which is for me connected with Amsterdam could also be remedied, if I would be permitted to live outside the municipality. Disad-

---

[197] Ehrenfest was quoted in German.

vantages for the regular course of teaching or my contact with the students should not be feared, because as a rule I am accessible at the university for students during the four days per week that I spend in Amsterdam for office activities, and for more extensive discussions people already now know very well how to find me in Blaricum, where my home has a telephone connection and is easily reached within an hour from Amsterdam. [198]

I already had a conversation on the above matter with our President Curator. He indicated to me that he personally was not unsympathetic to my viewpoints and wishes, and he said that he planned to bring up this matter next Friday, June 25, in the meeting of the Curators. It doesn't seem unlikely to me, that our President Curator would appreciate an explanation from your side before that time, and it is this consideration that has led to this letter to you.

Sincerely yours,

Your servant
L.E.J. Brouwer

[handwritten note on top of first page:] 'This letter back to P. Zeeman, please.'

[handwritten note (Zeeman) at the bottom of last page:] 'mathematical reading room!! + entry for books (f 500 per year).'

[Signed autograph – in Zeeman]

1915-09-18

To C.J. Snijders [199] **— 18.IX.1915**   **Blaricum**
                                                        Loevesteyn [200]

Copy

Excellency [Excellentie]

In the conviction of acting in the national interest, I take the liberty to call Your Excellency's attention to a branch of practical mathematics,

---

[198] The distance is about 25 km.   [199] To 'His Excellency the Supreme Commander of Land and Sea Forces at 's Gravenhage', i.e. to general C.J. Snijders   [200] One of Brouwer's houses in Laren-Blaricum.

which recently found application in the armed forces of several countries, but which is, as far as I know, not applied in the Netherlands army; I mean *photogrammetry*. Especially, it is the flying service, the usefulness of which is considerably raised by connecting itself to a photogrammetric service. Indeed, only by means of photogrammetry is it possible to obtain exact maps and profiles of the recorded terrain from aerial photographs (which are strongly deformed, and about which moreover generally neither the correct location, nor the correct orientation of the aircraft during the shooting of the photo is known). And also, only by means of photogrammetry is it possible, if one possesses a map of the terrain in peacetime, to indicate on such a map the location of the means of war (such as batteries or trenches), observed and photographically recorded by the airmen, in order to fire at them with a chance of success, also if there is no opportunity to perform range shots. I am sending Your Excellency a brochure as an enclosure, in which are expounded the basics of photogrammetry and its methods in the developmental stage of 16 years ago. I will be glad to provide extensive and more recent literature, and I am also prepared to give oral comments on the above.

I have the more readily proceeded to writing this, because it seems to me that an efficient photogrammetric service can be established at fairly small costs and in a rather short time.

Hoping that Your Excellency will excuse my frankness as motivated by the national interest, I sign with due reverence

(sgd.) L.E.J. Brouwer,
Member of the Royal Academy of Sciences,
Professor of Mathematics in Amsterdam

Enclosed: Jahresber. der M.V. VI,
Heft 2, containing a photogrammetric report of Finsterwaldes [201].

[Autograph, copy – in Zeeman]

---

[201] [Finsterwalder 1899].

1915-10-12

From C.J. Snijders [202] **— 12.X.1915**   **'s Gravenhage**
                                                          General Headquarter

Copy

To Professor L.E.J. Brouwer [Aan den Hooggeleerden Heer L.E.J.
Brouwer]

Member of the Royal Academy of Sciences
Blaricum, house 'Loevesteyn'.

Returning the *Jahresbericht, etc.* enclosed in your missive of last September 18, I have the honor to communicate to you the following.

In the department of aviation the need has until now not been felt to reduce images obtained of terrain to particular images, that are suitable for performing measurements upon.

By the use of our excellent topographic maps on a scale of 1:25,000 on which the minutest details are indicated, it is possible to use a photograph taken from an airplane for marking precisely each added fortification, trench, etc. on the map.

Nonetheless the commander of the aviation department has turned his attention to the study meant by you; the results obtained in this matter in Austria by Schimpfling are very encouraging indeed, so there will be no hesitation to proceed to the establishing of a photogrammetric service in the aviation department, if the need will be felt.

Meanwhile I thank you for the pains taken by you to draw attention to this matter.

The General,
(sgd.) Snijders

O.V.I. No. 3363
(Div. G.S. No. 9565
Attachments: a booklet,
Subject: Photogrammetry

[Autograph (Brouwer), copy – in Zeeman]

---

[202] General C.J. Snijders, commander in chief of the Dutch army.

915-11-04

From W. Blaschke — 4.XI.1915   Leipzig
                                                                    Fockestrasse 51

Dear colleague, [Sehr verehrter Herr Kollege]

I have recently spoken with Dr. Ackermann, an owner of the publishing
house B.G. Teubner, and I approach you at his request. It concerns the
following. Your fundamental geometric articles are not easily accessible. On
the one hand they have appeared, scattered here and there, furthermore the
Dutch language presents difficulties to some, and finally your articles are
very succinct.

It would therefore be much in the interest of science, if you could decide
to present your researches together in one book. Mr. Ackermann would be
very glad if you would give preference to his publishing company, which
can be considered to be the strongest mathematical publishing house next
to Gauthiers-Villars, and with which I myself always have had the best
experiences.

As far as I can recall some of your remarks, you are not really opposed
to the idea of writing a book. Perhaps you could some time write to me or
directly to Teubner, what you think about it. Let us hope, that the time
will not be too far off, that there will also again be opportunity for peaceful
meetings of mathematicians.

With best greetings

Your
W. Blaschke

[Signed autograph – in Brouwer]

15-11-19

To W. Blaschke — 19.XI.1915   Leipzig
                                                                    Fockestrasse 51

Dear Mr. Blaschke [Lieber Herr Blaschke]

A few years ago Blumenthal had already asked me to edit a book for
his series appearing with Teubner: 'Fortschritte der mathematischen Wis-

senschaften in Monografieen'. At that time I thought I could not yet promise it; now $^{\langle 203 \rangle}$ my circumstances have changed somewhat, and I believe I can already make a promise. You can inform Mr. Ackermann on my behalf, but please point out to him that Blumenthal has prior rights to get the book for his series. The title would be something like: 'New investigations in topology'. $^{\langle 204 \rangle}$ I intend to also incorporate the work of others (Tietze, Carathéodory, Lebesgue, Sierpiński).

Your letter suddenly reminded me that you have already asked me before Christmas of last year for a report for the Fortschritte $^{\langle 205 \rangle}$ on my paper: *Eenige opmerkingen over het samenhangstype η.* $^{\langle 206 \rangle}$ I am now really very sorry that I was so much occupied by exams, that I simply have forgotten to answer you. I wholeheartedly beg you to accept my apologies for that. It would have been, by the way, in my interest to answer you immediately essentially as follows:

The content of the article is given in the Revue Semestrielle des Publications Mathématiques $^{\langle 207 \rangle}$ XXI 2, p. 99. The $^{\langle 208 \rangle}$ essential point is contained in the theorem formulated there in the last three or four lines. This theorem was mentioned in a conversation in Cambridge $^{\langle 209 \rangle}$ between Borel and me about sets of measure zero; neither of us had a proof at the time. Following that we each published independently and simultaneously a proof; I [did] in the article: 'Eenige opmerkingen enz.', $^{\langle 210 \rangle}$ Borel in the paper: 'Les ensembles de mesure nulle', $^{\langle 211 \rangle}$ which appeared in the Bulletin de la Societé Mathématique de France, 41 (1913), p. 6–14. But the Borel proof is not correct because on p. 9 he only ascertains that the ratio *of the dimensions of the successively constructed domains* lie between $(1 - \varepsilon_1) \ldots (1 - \varepsilon_n)$ and $(1 + \varepsilon_1) \ldots (1 + \varepsilon_n)$, but that the ratio *of the* $^{\langle 212 \rangle}$ *corresponding ordinate differences of the corners of the domains* fail to do so; this entails that it can very well happen for a certain n, that the Borel conditions cannot be fulfilled, so that the construction fails. This is not just a gap in the proof that can be filled, but a real error which makes the whole proof collapse. I would really appreciate very much if in a report about both articles for the Fortschritte the above matter could be elucidated. Is that still possible? I

$^{\langle 203 \rangle}$ partially crossed out part: 'things have changed a little, and I would gladly follow Blumenthal's invitation'. $^{\langle 204 \rangle}$ *Neue Untersuchungen über Analysis Situs.* $^{\langle 205 \rangle}$ *Jahrbuch über die Fortschritte der Mathematik*; Author's review of [Brouwer 1913a] in JFd.M 44, p. 556 (also in *CW II* p. 405). $^{\langle 206 \rangle}$ Some remarks about connectivity type η. [Brouwer 1913a, Brouwer 1913c]. $^{\langle 207 \rangle}$ The Dutch reviewing periodical. $^{\langle 208 \rangle}$ The draft is in telegram style – the sentence started with 'aber' ('but'). $^{\langle 209 \rangle}$ International Congress of Mathematicians, August 1912 $^{\langle 210 \rangle}$ [Brouwer 1913a]. $^{\langle 211 \rangle}$ The sets of measure zero. $^{\langle 212 \rangle}$ Words have been crossed out, so that the sentence is not quite clear.

have already explained the error to Borel himself, shortly after his article appeared. [213]

Did you receive any news from Weitzenböck? In the first year of the war I received a few cards from him, now, however, nothing more for several months.

At Pentecost I have visited Study.

Cordial greetings

from your
Brouwer

[Signed autograph draft/copy – in Brouwer; also in *CW II*, pp. 410–411, with Freudenthal's comments]

916-02-07

To P. Ehrenfest — 7.II.1916   Blaricum

Dear Ehrenfest [Waarde Ehrenfest]

Your last letter still is on my desk, and only now it occurs to me that you wanted to ask me *a few more questions*, but that you wanted to await permission. Of course you can rest assured that you can put those questions to me, even though I can't say in advance whether I have time to study them.

I have searched the literature for the second question in your last letter, but I have not succeeded in finding the answer, which by now also for me is also *of the highest interest*, so if you find the solution somewhere, you would do me a great pleasure by telling it to me. I only can write you this about it:

Let $ds^2 = \sum_{h=1}^{n} \alpha_{hh} dx_h^2 + \sum_{h,k=1}^{n} 2\alpha_{hk} dx_h dx_k = \sum_{h=1}^{n} \beta_{hh} dy_h^2 + \sum_{h,k=1}^{n} 2\beta_{hk} dy_h dy_k.$

Substitute $dx_h = \sum_{k=1}^{n} \frac{\partial x_h}{\partial y_k} dy_k$, then it follows:

$$\beta_{hk} = \sum_{\mu=1}^{n} 2\alpha_{\mu\mu} \frac{\partial x_\mu}{\partial y_h} \cdot \frac{\partial x_\mu}{\partial y_k} + \sum_{\mu,\nu=1}^{n} 2\alpha_{\mu\nu} \left(\frac{\partial x_\mu}{\partial y_h} \cdot \frac{\partial x_\nu}{\partial y_k} + \frac{\partial x_\mu}{\partial y_k} \cdot \frac{\partial x_\nu}{\partial y_h} \right)$$

---

[213] *Brouwer to Borel 7.XI.1913*; see also Freudenthal's comments in *CW II* p. 407, 409.

i.e. for all values of $x_1 \ldots x_n$ the points

$$\left(\frac{\partial x_1}{\partial y_1}, \frac{\partial x_2}{\partial y_1}, \ldots, \frac{\partial x_n}{\partial y_1}\right), \left(\frac{\partial x_1}{\partial y_2}, \frac{\partial x_2}{\partial y_2}, \ldots, \frac{\partial x_n}{\partial y_2}\right), \ldots, \left(\frac{\partial x_1}{\partial y_n}, \frac{\partial x_2}{\partial y_n}, \ldots, \frac{\partial x_n}{\partial y_n}\right)$$

must form a polar simplex with respect to the manifold

$$\sum_{h=1}^{n} \alpha_{hh}\xi_h^2 + 2\sum_{h,k=1}^{n} \alpha_{h,k}\xi_h\xi_k = 0,$$

if all coefficients $\beta_{hk}(h \gtrless k)$ will cancel.

Which condition is equivalent to this, that the n points

$$\left(\frac{\partial y_1}{\partial x_1}, \frac{\partial y_1}{\partial x_2}, \ldots, \frac{\partial y_1}{\partial x_n}\right), \left(\frac{\partial y_2}{\partial x_1}, \frac{\partial y_2}{\partial x_2}, \ldots, \frac{\partial y_2}{\partial x_n}\right), \ldots, \left(\frac{\partial y_n}{\partial x_1}, \frac{\partial y_n}{\partial x_2}, \ldots, \frac{\partial y_n}{\partial x_n}\right)$$

must form a polar simplex with respect to the manifold

$$\sum_{h=1}^{n} A_{hh}\xi_h^2 + 2\sum_{h,k=1}^{n} A_{hk}\xi_h\xi_k = 0$$

For $n > 3$ the n functions y have to satisfy $\frac{n(n-1)}{2}$ hence *more than n* partial differential equations, which is generally not possible. However, for $n = 3$ the 3 functions y must satisfy 3 partial differential equations and *the existence proof of Cauchy works for these partial differential equations.*

Yet if we choose *for $x_3 = 0$* $y_1, \frac{\partial y_1}{\partial x_1}, \frac{\partial y_1}{\partial x_2}, y_2, \frac{\partial y_2}{\partial x_1}, \frac{\partial y_2}{\partial x_2}, y_3, \frac{\partial y_3}{\partial x_1}$ en $\frac{\partial y_3}{\partial x_2}$ as arbitrary functions of x_1 en x_2, then for $\frac{\partial y_1}{\partial x_3}, \frac{\partial y_2}{\partial x_3}$ en $\frac{\partial y_3}{\partial x_3}$ can be found such functions for x_1 en x_2, that for $x_3 = 0$ the partial differential equation is satisfied (namely, it amounts to the determination of a polar triangle of a conic section, every vertex of which must lie on an arbitrary given straight line, and such a polar triangle can always be found).

So the problem is indeed possible for $n = 3$; If I have written on my last post card, written in haste, trusting my memory, that the problem is generally impossible, *also for $n = 3$,* be so kind as to return that card to me as it is discrediting for me.

Many greetings from home to home. How about the appointment of Van der Woude?

Brouwer

[Signed autograph – in Ehrenfest]

1916-05-06

To P. Ehrenfest — 6.V.1916   Blaricum

Dear Ehrenfest [Waarde Ehrenfest]

Hereby I return to you my letter of February 7. The post card with the incorrect information that preceded it I have destroyed, I am sorry if you think this narrow-minded and humorless, but after a certain experience with the German mathematician Koebe I have made it a firm *principle* for myself, firstly to be extremely careful with scientific correspondence and secondly to always try to get back into my possession any letters written by me from which scientific discredit might be extracted. This is a cool intellectual habit, which everyone who had an experience like mine would have adopted, and which is not accompanied by any mental affect of fear or remorse or such. I hope you will recognize the justification for such a habit in some cases, and that you will not have to withhold your respect for me on this account.

Many thanks for the bibliographic references concerning Einstein; since then I heard a talk in the Academy [214] by Lorentz on the subject, which deeply impressed me.

Furthermore I hope that you will accept my apologies for the delay in writing this letter; it was a consequence of being overloaded with correspondence. I had considered your last letter as 'not urgent' and consequently had put it on a pile where it had to wait its turn.

Cordial greetings also to your wife and from mine

t.t.
Brouwer

[Signed autograph – in Ehrenfest]

16-09-16

To the Belgian Government — 16.IX.1916 [215]

A QUESTION FOR THE BELGIAN GOVERNMENT

From conversations with Flemings residing hereabouts and belonging to various directions of domestic and foreign politics, it has become apparent

---

[214] KNAW.   [215] The present letter was published as an open letter in the Dutch weekly '*De Nieuwe Amsterdammer*'.

to me that among them there is well-nigh unanimity regarding the following facts that are in my view not at all generally known:

> *In Belgium there is no law regulating the official language used in university education.*

Hence a Belgian government that does not make Dutch the language for teaching in the University of Flanders and French in the Walloon university, cannot appeal to any law as an excuse for this violation of the natural rights of half the Belgian population, and bears personally full responsibility for this injustice. Hence the Flemings, who have been watching the German government violating Belgian justice for two years, cannot in the least be required to turn a blind eye to the eighty years of violation of Flemish rights by the Belgian government.

Furthermore *international law requires the German occupying force to maintain public life in Belgium to the best of its ability while respecting the national laws.* Because the national laws remain silent about the official language of higher education, the German authorities are not at all obliged to imitate the violation of justice committed by the successive Belgian governments, and according to international law it would even be *obligatory* to make the Ghent university Flemish, were it not that . . . because of a decision of the Belgian government at the beginning of the war, higher education in Belgium has been *suspended*, and its reinstatement by the German occupying force would only be legitimate if the interest of public order made this mandatory, which can be doubted with good reason.

Also the Flemings whose political attitudes are not foremost dominated by indignation about the German invasion, seem to have to refrain from any support of the German authorities in their efforts to make the Ghent university Flemish. Why do nonetheless many feel strongly inclined to give such support? *Because they distrust the Belgian government*, and they fear that after the war it will swiftly forget that the army defending Belgium was four fifths Flemish, and that it will violate Flemish rights as before.

Such a distrust may be insulting for the Belgian government; but it cannot deny that its past gives some cause for it, because when a bill was proposed (not by the government itself, which would have been proper, but by the members of Parliament Franck, Huysmans and Van Cauwelaert) to regulate the official language of higher education (a bill which in fact only had nothing but a negative tenor, namely to forbid *by law* further violations of Flemish rights in the future), it found before the war unconditional support with – if I'm right – only two of the ten members of the government.

And therefore the question: *Why doesn't the Belgian government make it easy for the Flemish to determine their position with respect to the German authorities in the matter of making the Ghent university Flemish, which touches so intimately upon their existence as a people, by openly declaring that the Flemish rights will not be violated again after the war if they can help it; namely that they have unanimously decided to take up the Franck-Huysmans-Van Cauwelaert bill after the reinstatement of Belgium?*

Thus it would provide the proof, to the satisfaction of all the Flemings, that it has the moral courage to refuse unconditionally not only to deliver the whole of the Belgian people to the German urge for expansion, but also to refuse to deliver half the Belgian people to the French urge for expansion.

L.E.J. Brouwer

[Printed – in 'De Nieuwe Amsterdammer'.]

917-04-16

From A. Schoenflies — 16.IV.1917   **Mönichkirchen**
                                                      Hotel Windbichler

Dear Mr. Brouwer [Lieber Herr Brouwer]

On your recommendation we are reading here (my wife and I) the new novel by Meyrink, The green face! [216] After the first chapters I wanted to ask you whether you recommend me to train myself in time for the profession of magician and swindler, so I can find a reliable livelihood in The Hague in the new time after the war — now I know that I would be broken, atomized and blown away horizontally in the great cyclone. Unless you might in the meantime have learned vigilance, and I would have enjoyed the same undeserved fate as your friend, like Pfeill as friend of Hauberrisser. [217]

By the way, did you read Gerhart Hauptmann's *Emanuel Quint, der Narr in Christo*? [218] It touches in part on Meyrink's novel, but is much more of value and to be taken seriously, compared to Meyrink's mixture of

---

[216] Gustav Meyrink, *Das grüne Gesicht* (1916). This novel plays in post-war deca-dent Amsterdam.   [217] Pfeill and Hauberrisser are characters in *Das grüne Gesicht*.
[218] Gerhart Hauptmann was awarded the 1912 Nobel Prize for all of his work, but mostly for his novel *Der Narr in Christo Emanuel Quint* (1910) (The Fool in Christ: Emanuel Quint).

talmudic and Hegelian wisdom, of besotted dialectics, and a little understanding of his own!

   Cordial greetings

   your
   A. Schoenflies.

[Upside down at top of the letter:]   Cordial greetings
                                        your Emma Schoenflies

[Signed autograph, postcard – in Brouwer]

1917-06-09

To G. Mannoury — 9.VI.1917   **Laren**

Dear chap [Beste Kerel]

   Thank you for your history of mathematics and for the specification of
your hours. To exclude misunderstandings (for it has become clear that
Korteweg obtained a wrong idea from your letter about this) I repeat once
more that if a mechanics course by you is realized next semester, you will
only get the third year as listeners; maybe you can take in the second year
too, when it has advanced far enough in analysis in the course of De Vries,
which was also the case with my mechanics course that I started in 1915.
   Enclosed I send you 8 times 12 copies in four languages of the manifesto plus the statutes of the International Academy of Practical Philosophy
and Sociology. [219] Since the time of drawing up the manifesto the Board
of Directors for which at the time only four members were designated, is
extended with L.S. Ornstein (professor of physics in Utrecht) and G. Mannoury. You declared yourself willing to do so, didn't you? That we have
never invited you for another meeting, was only because we were certain
that you were incapable of attending. We hope that this will change after
the summer, and that you may perhaps now already find time to send some
copies of the circular to Dutch or foreign acquaintances, so as to get letters
of approval, and as a preparation for the appointment of representatives in

----

[219] Note the expanded name, cf. [Schmitz 1990] p. 223.

other countries. In that case add the names of Ornstein and Mannoury in ink, both in the statutes and to the signatures of the manifesto, and correct the sole remaining printing error in the German manifesto, where an umlaut was left in 'beinflusst'.

Your
Bertus Brouwer.

[Signed autograph – in Mannoury]

)17-10-01

From F.M. Jaeger — 1.X.1917   **Groningen**

Amice,

Many thanks for your letter and the effort spent. Today I have informed Schoute [220] about the matter, and I now let you know in the name of both of us that we would very much appreciate it if you would go ahead and talk to Zeeman about the matter, and if you would for instance tell him that we (and you as well) would be pleased to confer with him and Lorentz sometime at the end of October about that matter, if convenient before the general session [221] in a separate meeting. Personally I think that Mr. Korteweg's objections will rather be purely theoretical. Probably he doesn't know the military environment and has a much too exalted opinion of the amount of initiative among military authorities. For three years now the directorate of the army has had the chance to improve the army by means of the adjoined intellect. The result has been nil, simply because of the total lack of initiative. About the boundless bigotry in those circles I could tell you far worse stories.

Hence: *nothing* can be expected from common sense or initiative of the army administration, a fortiori not in a time of panic. So it *has* to come from our side. Wouldn't it perhaps be good to call Lorentz and Zeeman and also Mr. Lely [222] in conclave? He very much detests the military muddle, and he knows its spirit, or rather the total lack of any spirit, and maybe

---

[220] the Groningen meteorologist.   [221] KNAW.   [222] Minister of public works (1913–1918), see *Brouwer to Lorentz, 16.II.1918.*

he could achieve something for us with Cort van der Linden, [223] or in the Cabinet.

In any case, to me too, it seems that right now the Academy really can achieve something *good* in this matter, and that the best thing is to take action as soon as possible. —

In November my Lectures on the Principle of Symmetry will appear. I shall honor you with a copy. For, although it is not in the first place a mathematical work, it will probably, —if its mathematical shortcomings are kindly passed over,— give you some pleasure in a related field, and it would give me pleasure, if you would think well enough of it to introduce it to your younger pupils as something that might be if some use for their general education. In pure mathematics I am just a plebeian; possibly there is something informative for them in the Applications of the theory. If the gentlemen want a meeting still before the end of October, perhaps it is best to have that on Saturday morning.

> With friendly greetings
> tt
> Jaeger

[Signed autograph – in Brouwer]

1918-01-09

To J.A. Schouten — 9.I.1918   Laren

Copy

Dear Sir [Weledele Heer]

I have informed you at the time about my view that the mental attitudes of the two of us are not suited for mutual understanding. At the same time I asked you only for a message, whether you wanted your duplicate manuscript [224] back from my archive on legal grounds.

The letter that was subsequently received would have been opened in the Christmas vacation, were it not that I heard from my friend Ornstein

---

[223] Liberal prime minister (1913–1918).   [224] See *Brouwer to Klein 19.IX.1919*.

(from whom you have earlier tried to find out my more intimate feelings in a manner that, as I assume, is permitted according to your morals, but that I find highly improper) that you have again taken a step with him in this matter 'to avoid squabbling' (!) Consequently the opening of your letter has been left undone, and neither will the letter received today from you, be read by me.

As I have meanwhile ascertained not to have a strong legal position in the matter of retaining your duplicate manuscript, I will now have a copy made at my own expense, and then I will return to you the copy belonging to you.

And now I urgently request you to leave me alone in the future. In my capacity of member of the Academy, Member of the board of the Dutch Mathematical Society [225] and Editor of the Annalen I always have felt obliged to reserve a large part of my time in the interest of young mathematicians at the beginning of their career, and you have profited amply from this. In return I demand no gratitude or apologies for the efforts made(even though words of to this effect from others never were entirely absent, when they took up my time in the same manner as done by you), but I do demand the strictest possible respect for the method that I consider correct in discharging this demanding task. And by your failing in this respect — also after the hint given to you — you have automatically put an end to any availability of my time for you (even for reading your, to me incomprehensible, letters).

Sincerely yours [226]
(w.g.) L.E.J. Brouwer

[Signed autograph, copy – in Brouwer]

•18-02-04a

To M. Buber 4.II.1918ᵃ   Laren

Dear Sir, [Hochgeehrter Herr]

The executive committee of the *Internationales Institut für Philosophie* has instructed me to answer your letter to our member Mr. Borel [227] of

---

[225] *Wiskundig Genootschap.*   [226] *Met verschuldigde gevoelens.*   [227] Henri Borel, the sinologist.

March 17, 1917, in which you raise a fundamental objection to our manifesto. Hence I beg you to take into account the following:

In several cases the occidental word has indeed in addition to its material value also a spiritual value, but the latter is always subordinate to the former, and while the first has attained a certain and lasting orienting effect on the activity of the community in the sense that it stimulates the separate individuals to hinder each other as little as possible in their pursuit of physical certainty and material comfort, and possibly also even to support each other, the latter lacks any influence on the legal relationships (except insofar it is abused for deviously committing injustices); consequently its effects are weak, temporary and localized.

Words that have an exclusively spiritual value and that are suitable for orienting the community towards inhaling and exhaling the world spirit [228] and towards observing the Tao, don't exist in the occidental languages; should these exist, their effect would be paralyzed by the mutual physical hatred of people that live too close to each other, which has roots in the mutual distrust of the purity of their birth, and which obstructs the pursuit of material comfort of the separate individuals only to a small degree, but to a high degree obstruct the inhaling and exhaling the world spirit. The introduction of the first word with exclusively spiritual value into the general human understanding will as phenomenon be inseparably connected with the insight that this physical hatred is intolerable, and will immediately give rise to legal rules about human procreation.

But a possibility for this introduction will only be created, when the 'mystery of the emergence' of this word has taken place not in the isolated individual, but in the mutual understanding of a *community* of clear feeling and acutely thinking people that furthermore are materially not too close to each other.

Yours truly, [229]

Prof. Dr. L.E.J. Brouwer

[Printed text – in *Comm. of the Intern. Inst. for Phil.* **1**, 1918; cf. [Brouwer 1918c]]

------

[228] *Weltgeist.*   [229] *Mit vorzüglicher Hochachtung.*

Editorial supplement

[*The following argument is at the heart of Buber's objection to the signific enterprise* [230] *, see* [Brouwer 1918c]:]

> Word creation, the making of a word, is for me one of the most mysterious events of spiritual life, indeed I admit that in my view there exists no *essential* difference between what I here call word creation and that which has been called the appearance of the Logos. The emergence of a word is a mystery, which takes place in the inflamed and receptive soul of man who is poetically creating, discovering the world. Only such a word that has been begotten in the spirit, can originate in man. Therefore, in my view, it cannot be the task of a community to make it. It rather seems to me that a society, such as the one planned by you and your friends, may only aim at *purifying* the word. The abuse of the great old words can be fought, the use of new ones can not be taught.

)18-02-15

From H.A. Lorentz — 15.II.1918   Haarlem

Amice,

After our last conversation we have considered in the board-to-be of the 'Scientific Committee' [231] in more detail how it can operate, and more in particular which subcommittees will have to be formed from ordinary and extraordinary members. You will recall that the committee will have the right to co-opt extraordinary members, a form that has been chosen because it seemed undesirable that one should have to ask for a decision of the Minister, each time when the need was felt for cooperation of experts in some field. I hardly need to add that the activity of the extraordinary members will be appreciated as much as those of the ordinary members. For the question which persons were to be proposed as ordinary members and which not, the crucial factor was mainly the size of the task that we had in mind for them; also most of the proposed ordinary members are

---

[230] *Buber to Borel 17.III.1917.*   [231] Wetenschappelijke Commissie van Advies en Onderzoek; Scientific Committee of Advice and Research

experimentalists or technologists, i.e. people that have at their disposal the resources of a laboratory, or of factory, that they control.

It is not necessary that I give you the full list of subcommittees (for nutrition, clothing, fuels and minerals, agriculture, animal food, etc.), but I think that I can inform you already now that it seems to us that there should be one for survey-photographs obtained by airplanes; naturally we put our hopes on you in this respect. Moreover, that we thought of this point, we owe to what you have done already in this matter, the importance of which I have emphasized right away in my conversation with the Minister.

So we would appreciate very much if we could include you as extraordinary member in the committee and also if you would take a seat in the mentioned subcommittee. To this subcommittee would furthermore belong Dr. Schoute of the Meteorological Institute, who has already often been in the air, and myself. I am not informing Schoute yet, but with you it is of course a different case.

I myself would be very pleased if I could be of use to you in your work in any way and if I could contribute, so as to help that justice is done to it. Reading what you have written already about the subject gave me the impression that I would be in the right place in this subcommittee, whereas in the main committee with its experimentalists and technologists, I probably will have the feeling not quite to fit in. I would have stayed entirely out of the matter if I were not chairman of the Academy [232] and if the Minister hadn't explicitly insisted that the function of Executive Committee would be in the hands of some members of the Academy.

In view of this I could not shirk my responsibilities and naturally I have had a great part in the preparation, so there is much for which I am responsible. From the outset I had in mind an arrangement such as we now are going to get and of course I immediately have been considering who might be the ordinary and extraordinary members. The task that we would like to see you take on, is the one I had intended for you from the beginning. I imagined that you could accept that and so fulfill your duties vis-à-vis the nation, without too much disruption in your scientific work.

With cordial greetings from house to house

t.t.
H.A. Lorentz

[Signed autograph – in Brouwer]

---

[232] KNAW.

To H.A. Lorentz — 16.II.1918   Laren

Dear Mr. Lorentz, [Hooggeachte Heer Lorentz]

In the past weeks I have sincerely tried to acquiesce in my non-apppoint-ment in the Scientific Advisory Committee, [233] but I can't succeed. And on the contrary, this incident brings me more and more out of balance. Therefore I cannot act otherwise than asking you kindly to look at the affair from my point of view, based on the following exposition.

Since more than two years I have the ambition to establish a photogram-metric service in the army, and the fact that my original motives were to be found mostly in the danger, considered rather great by me, that I would yet be called up (and I know from experience what this means to me); and that this danger now has been reduced to minute proportions, in no way dimin-ishes my wish to continue the work, once I had initiated it, in this direction, until my goal has been reached. In view of this I have started on a series of articles about photogrammetry in the 'Aeronautical journal', [234] and in the meantime it becomes ever more urgent to carry out of experiments; so for some months now I have been looking for an opportunity to have these carried out under my direction, where in the first place I recalled to my mind that in foreign countries members of the Academy regularly receive commissions, also from the Ministry of War, and in the second place I have kept in mind Article 2a of the rules of our own Academy. My wishes and aims in this matter I have mentioned for the first time in a conversation with Jaeger at the Academy meeting of September 1917. For me, and as I think, also for Jaeger, the main issue was that because of Article 2a of the Rules members of the Academy who wished so, should be given the opportunity to give directions to the Government in the interest of national defense, and have experiments performed in the interest of the fruitfulness of these direc-tions. Moreover we spoke as a side issue about the desirability of exempting the members of the Academy from ordinary home guard duties.

As the Board had already indicated at an earlier occasion that I shouldn't bring up important matters directly in the plenary Academy meeting, but that I should do so first with the Board, I thought I should act thus in this matter as well, and I turned to colleague Zeeman in the first days of October

---

[233] Wetenschappelijke Commissie van Advies en Onderzoek, Scientific Committee for Advice and Research.   [234] Luchtvaartorgaan.

1917; he expressed his full agreement with my plans. In the interest of the Academy's prestige he found it, in view of earlier experiences, necessary indeed that the occasion were used to point out to the Government the rights and the place of the Academy, and he was quite prepared to pass my expositions on to the other Board members, and to ask their cooperation in the fulfillment of my wishes; I did not hide from him my fear of obstruction by the Vice-Chairman, [235] where it concerned a proposal originating from a proposal of mine.

A few days later already, I heard from Zeeman that you completely agreed with us (with the only exception, that in the side issue you rather wished the exemption from home guard duties to be extended to all professors), because no opposition from the side of the Vice-Chairman had been noted.

Subsequently the meeting of the Academy Board with Minister Lely, [236] Jaeger, Schoute and myself took place, and neither there nor in the Extraordinary Meeting of the Academy in November 1917 anything whatsoever happened, nor was any word spoken that could give me reason to suspect that you or Zeeman had changed your opinion in any respect, and hence, in accordance with all unwritten laws of human relations, would not do so; further that the initiator, whose ambition it was to be adopted in the meanwhile conceived Committee of Advice, would indeed be included.

And that, if any obstruction was met, you would warn him for the purpose of designing a joint plan of resistance, the more so where you had in the memoir that you submitted to the minister (as appears from the reading of it at the meeting) specifically mentioned, as an example, the field where I could in particular be active in the committee.

Instead of this, and without any prior warning, I am told two months later by Zeeman, casually, and without any accompanying clarification, that the Committee is all set and that I am not included. And when I protested to you after receiving this staggering message, I got from you no other consolation than the suggestion of the possibility that for the purpose of proposing the mentioned activities, I could be placed in a subcommittee.

Apart from the order of the probability that this possibility becomes reality, and apart from the question whether I could do any productive work in this position (what certainly would *not* be the case if Korteweg is the only mathematical member of the Committee itself, and if therefore my work would more or less fall under his responsibility). Finally, apart from all personal paternity rights, the dignity of the Academy doesn't tolerate in my

---

[235] Korteweg.   [236] Minister of public works (1913–1918), see also Jaeger to Brouwer, 1.X.1917.

opinion that in a Committee established on the initiative of the Academy itself, the mission of which is part of the regular task of the Academy, a member of the Academy should have to forego a place to which he aspired, and to withdraw to the second rank for the benefit of an outsider.

I have elaborated extensively, but I wanted to be clear and complete. I hope that I have succeeded, that you won't hold my frankness against me, and that I can look forward to an answer from you. If you want to allow me in this matter to have a conversation with you, then I would gladly use the opportunity.

Sincerely yours

Your
L.E.J. Brouwer.

[Signed autograph, draft – in Brouwer]

───────────

Editorial supplement

[sheet with Brouwer's handwritten remarks] [237]

> If the *cause* as mentioned by Lorentz (in letter of 15.2.18) for the 'form' of the organization (consisting of ordinary and extraordinary members), and the *criterion* stated by him for ordinary and extraordinary members is correct, the extraordinary members should have at least an advisory vote in the main committee.

───────────

●18-05-23

From C. Carathéodory — 23.V.1918   **Göttingen**
                                                    Friedländerweg 31

Dear Mr. Brouwer, [Lieber Herr Brouwer]

Many thanks for your letter, as well as for sending me your article [238] and also the article of Van der Corput. Concerning the latter, there is a

---

[237] Both sheets in Brouwer Archive.   [238] In view of the topic (Lebesgue measure), probably [Brouwer 1918b].

series of reasons why we won't print it in the present form. Part of these reasons you will find in the enclosed letter of Landau. If even Landau, who has spent the last five years almost exclusively on these problems, can't understand the article in spite of his great diligence, then something must be wrong. The second reason is purely formal: already for several decennia the Annalen have the fundamental rule not to print dissertations (I believe that the only dissertation that has appeared in the Annalen was the one of Hurwitz). Whereas parts of dissertations have very often been printed (e.g. Erhard Schmidt's investigations on integral equations). The third reason, which has to do with paper shortage, is purely personal. About three months ago Noether sent a long article by R. König which he has accepted, and of which you probably will have seen the galley proofs. Then six weeks ago a second article by the same author, which he also accepted and which was even longer. I protested against that on purely formal grounds, namely that we now have per year only 26 sheets available and that it is impossible that we spend almost one third of that for one author without harming the other authors. Hilbert and Klein supported me and I am now expecting any day that Köning will withdraw his article. It would now be an insult to Noether, when we immediately would accept such a long article as the one of Van der Corput. The solution that Landau proposes, that part of the article appears in the Liechtenstein journal [239] and the rest in the Annalen, seems to me one that should satisfy all parties concerned, and I hope that you also agree with it, or that you make another suggestion. We can reserve up to 40 pages for Van der Corput, I think. However, in the present size I estimate it to be over 100 pages — that is more than the number of pages that all your own discoveries have demanded.—Three weeks ago a small article of four pages arrived that I found *very amusing*, entitled 'on Brouwer's fix point theorems' by an unknown Hungarian. [240] I wrote to him that he might add the proofs of a few theorems that he only stated and that we would probably accept the article. Now it turns out that it is a fourth semester student. Isn't that amusing?

With many greetings

Yours truly [241]
C. Carathéodory

---

[239] *Mathematische Zeitschrift*; Lichtenstein was the editor in chief of the journal.
[240] *Über die Brouwerschen Fixpunktsätze*; [Kerékjártó 1919].   [241] *Mit vielen Grüssen – Ihr sehr ergebener.*

P.S. Klein would like to have articles about the theory of gravitation in the Annalen also. Maybe you can stimulate some Dutchman who does this kind of thing (for example De Sitter) to produce something. Of course it should not be too long.

[Signed autograph – in Brouwer]

918-11-25a

To D. Hilbert — 25.XI.1918[a] [242]

Dear Mr. Hilbert, [Lieber Herr Hilbert]

May the hale heart of your fatherland overcome the present crisis; and may the German lands soon blossom in exceptional ways in a world of justice. [243]

That wishes you

Your Brouwer.

[Signed autograph – in Hilbert]

918-11-28

From A. Denjoy — 28.XI.1918   **Utrecht**
Stationsstraat 12[bis]

Dear Mr. Brouwer, [Cher Monsieur Brouwer]

Infinite thanks for your kind idea to congratulate me with the great events the history of my country is going through now. Our joy is made by

---

[242] Identical message to Klein (in Klein Archive).   [243] November was a fateful month for Germany; after the armistice (11.XI.1918) the emperor abdicated and fled to Holland. Revolution was in the air, etc. The future was bleak and uncertain.

all we have suffered, by all we have feared and by realizing now that those sad days and the threat of shameful slavery seem to be over.

Yours cordially, A. Denjoy

A. Denjoy.

[Signed autograph – in Brouwer]

1919-02-16

From J. Noordhoff — 16.II.1919   **Groningen**
                                                                      N.V. Erven P. Noordhoff's,
                                                                      Boekhandel en Uitgeverszaak
                                                                      Oude Boteringestraat 12

Dear Professor, [Hooggeleerde Heer]

In reply to your letter about taking over your work '*Grondslagen der Wiskunde*', [244] I am pleased to confirm the preliminary promise that Mr. Wijdenes has made to you in Amsterdam, that I appreciate it very much to undertake the marketing of the available copies of the work and that I would even more be pleased if I could succeed to sell a great part of the stock of this work by sufficient advertising, so that you can be found willing to work on the manuscript of a second printing of the '*Grondslagen*' in order to have it published in the series of Mathematical books which is published by me, after consulting Mr. Wijdenes. At first it was Wijdenes' idea that if the stocks of your '*Grondslagen*' were small, the copies could be put aside and a new printing could be undertaken right away. But now that it turns out that the now available copies are 240 in number, it is in this expensive time of paper and printing, a pity to make the available copies worthless by printing a new edition right away.

I would like to suggest that you henceforth commission me with the selling of the available copies of your '*Grondslagen*'. I will try to see to it that by good advertising the sales of your work increase and I propose that you let me henceforth do the accounting on the following conditions:

You receive each year in the month of January a statement with the available number of copies. The sold copies will be credited to you for half

---

[244] *Foundations of Mathematics*, Brouwer's 1907 dissertation.

the price. As the price is f 2,90, as I believe, you will receive f 1,45 per copy sold. As soon as it appears that by good advertising, the stock has greatly diminished, we can confer further about the manner and time of publishing a second printing. If you think it is desirable to have a second edition printed sooner, I would be glad to talk things over. You will receive a fee for the reprint of f 40,- per sheet of 16 pages, in the format and type of the works of my series, known to you. This fee of f 40,- per sheet is paid when a part or the entire work is published. Of course I leave the possible publication of the second printing entirely up to you, but I do inform you that it is my opinion that the sale of a new book always is better than that of a book that is already a few years old.

Sincerely your obedient servant [245]
Noordhoff

[Signed typescript – in Brouwer]

_____

)19-02-26

From F.M. Jaeger — 26.II.1919   **Groningen** [246]

Amice,

If the Board *explicitly* wants to stipulate in its proposal that we *remain member of the existing Association internationale*; that we will *not* become a member of the interallied firm, and keep our complete freedom to act; and if furthermore the notorious 'justified feelings' would disappear from the document,— then I could agree with the proposal, at least in the essentials, even though I think that in that case, that League of Nations in the background is rather superfluous. The statutes of the interallied confederation are a faithful *copy* of the now published project of the so-called League of Nations; the leitmotif of both is how the victors play the boss. It seems to me that on such a monstrosity we cannot base a missive, as required in this

_____

[245] *Hoogachtend Uw dienstwillige dienaar.*   [246] In this letter Jaeger discusses the issue of joining the Conseil Internationale de Recherche; its secretary, A. Schuster had invited the KNAW in a letter of 19.IX.1919. The sentiments in the Academy were mixed. Brouwer and the Groningen group led the opposition. For more information, see [Van Dalen 1999, Van Dalen 2005] section 9.1, 13.4; [Otterspeer and Schuller tot Peursum-Meijer 1997].

case — exactly because that foundation has been, as you quite correctly have remarked, *condemned already beforehand*! The gentlemen in Paris and London have disgraced themselves severely vis-à-vis Science, and if we co-operate with their plans, we disgrace ourselves *with* them, and even more so, because for us there isn't even the 'excuse' that we are in a state of war psychosis...

Now I am quite certain that the Board of the Academy will *not* approve of the conditions mentioned in the beginning of this letter. Indeed the odds are 99:1 that the interallied will emphatically reject a proposal in which we state that we remain member of the old Association. I believe that the chances for such a proposal will be even less than for the Groningen project,— although I don't swear obstinately by the latter, and would be pleased to give it up for something better. But the proposal of the Board *is* not something better, but in *my* view it is something much worse, namely a petition based on the veneration of success, to those that have committed an injustice, and a document that will make us lose face *and* in the eyes of the Allied *and* of the Central Powers.

This matter,— as so many,— concerns for a large part issues of instinctive feelings about morality; and precisely for *that* reason we said last Saturday that in *our* opinion a 'compromise' between the two viewpoints wasn't possible. I still believe so,— unless Lorentz or one of the other gentlemen can convince me that his or their point of view is morally superior to ours. As far as mixing politics with pure scientific matters is concerned, I still cannot see that.

Meanwhile,— we still have time to let our thoughts mature in this difficult business; *before* anything else, one should try to make op one's own mind about the value of the moral motives that will have to determine our position in this.

With friendly greetings [247]

tt
Jaeger

[Signed autograph – in Brouwer]

---

[247] *Met vriendschappelijke groet.*

To A. Hurwitz — 10.VI.1919   Laren

Dear colleague [Hochgeehrter Herr Kollege]

May I ask you a favor? I would like next month to make a trip to Switzerland. Now I hear in the Swiss General Consulate in Amsterdam that I would have the best prospect to get permission for this journey, if I would indicate as the purpose for the trip 'discussion of scientific interest' and if at the same time my entrance request would be supported by some Swiss colleague in the same field by a letter directly to the Federal Center of the Aliens Police [248] in Bern. Would you be willing to lend me your support and, perhaps get also another colleague from Zürich to sign? If Weyl would still be in Zürich, he would be the most suitable person to cosign, because I actually do have scientific topics I want to discuss with him. [249]

I thank you cordially in advance for any help. Also, I thank you once more for your message a couple of months ago concerning regular Riemann surfaces, which led me quickly to the result that the enumeration in question hasn't been explicitly carried out for any genus except zero. Subsequently I have devoted two Comptes Rendus notes to the question for genus one (Sessions of March 31 and April 28), [250] but unfortunately I have not yet received any reprints.

If my trip comes about, it will be a great pleasure for me to meet you. With warm greeting

Yours truly [251]
L.E.J. Brouwer

[Signed autograph – in Hurwitz]

---

[248] Eidgenössische Zentralstelle für Fremdenpolizei   [249] The foundational discussion of 1919 in the Engadin resulted in Weyl's conversion to Brouwer's view point, see [Van Dalen 1999], section 8.6.   [250] [Brouwer 1919c, Brouwer 1919b].   [251] *Ihr ganz ergebener.*

1919-06-28

To D. Hilbert — 28.VI.1919

Dear Mr. Hilbert! [Lieber Herr Hilbert]

I don't know whether these lines can bring you any consolation, but I set great store by declaring to you on the day of the signing of the Peace Treaty [252] that, seen from Holland, the Allied Powers have, through the peace extorted today, taken upon themselves a guilt, that is certainly not less than the combined guilt of those (whoever they actually were!), that started this war.— My sincere thanks for your letter from Switzerland. How glad would I be to meet you again soon, if it were somehow possible.— At the end of the day, we scholars are after all in a fortunate situation, because such a large part of our realm of thoughts is completely independent of political nonsense.

Cordial greetings from house to house

Faithfully yours, your [253]
Egbertus Brouwer.

[Signed autograph, picture postcard – in Hilbert]

1919-09-08

From F. Klein — 8.IX.1919   Göttingen

Dear colleague, [Sehr geehrter Herr Kollege]

Your letter of September 5 arrived just now. We have not missed anything yet, and everything can be arranged somehow.

There was an inadvertence in the procedure of Teubner – who had the proofs of Kerékjártó and the following ones typeset till the end of the issue – in which Cararathéodory possibly had been involved. Probably the opinion was that from your side only a few words would have to be corrected. Now there are larger changes and these, of course, cause under the present circumstances out of proportion large costs. To what extent, I cannot guess

---

[252] Peace treaty of Versailles (28.VI.1919).   [253] *In Treue Ihr.*

for the time being. Anyway, I want to ask you and the other editors that in the future you send only manuscripts to Teubner that are completely ready to print, so that more extensive corrections will be avoided.

Carathéodory wrote to me from The Hague that you have doubts as to the acceptance of Schouten's paper. Fortunately I haven't yet undertaken anything definite in this respect. I have only generally voiced the opinion that it would be more practical if articles that relate to Lorentz and Einstein would not be given such a prominence in the Annalen as was done so far. Indeed the constraints in the printer's shop have become less, even though they have not been overcome. When you could do something in this direction, I would be grateful.

For the rest Cara [254] will have also spoken to you about the vague plans regarding the long term future of the Annalen, that float around. Scientific publishing in Germany now has to deal with quite different conditions than before (where the only thing that is so worrying is that nobody can say whether the reshaping in the external circumstances already have come to an end). In about a month an extensive conference with Teubner (more specifically also because of the Encyclopedia) will take place here in Göttingen. Let's hope we really can find a solid foundation!

Yours truly [255]
Klein

[Signed autograph – in Brouwer]

19-09-19

To F. Klein — 19.IX.1919   **Laren (near Amsterdam)**

Dear Geheimrat [Hochgeehrter Herr Geheimrat]

In reply to your letter and your card, I first of all inform you that I recently have rejected a large article by Schouten about the application of his 'direct analysis' to the theory of relativity for the Annalen, in the first place because the author doesn't understand the art of presentation and in the second place (which is more important) because, in short, his achievements

---

[254] The generally adopted abbreviation of 'Carathéodory'.   [255] *Ihr ganz ergebener.*

consist of wrapping up results already found by inventive authors into a new (but thick and opaque) attire. In addition, the quotations are very complete in inessential points, but very incomplete in the essential points, so that the superficial reader of these articles gets a wholly false impression of their value. What is lacking in Mr. Schouten is, by the way, not talent but erudition and moderation, so I don't exclude at all the possibility that in the future he will turn into a good mathematical author.

Because I don't consider myself a prominent expert in this field, the rejection of Schouten's article (which certainly has also been recorded in the editorial archive of the Annalen) has only taken place after I had sent the manuscript to Study and obtained his advice. To his negative assessment for my information, Study has added, among other things, the following words with respect to the author: 'I don't expect that a factual discussion with such a muddled head would be of any advantage to him.' Weitzenböck too, whom I see as the second representative authority in these matters, completely shares the unfavorable judgment about the publications of Schouten, and refers to the latter's '*Grundlagen des Vektor- und Affinoranalysis*' [256] as 'that horrible book that he has committed'. [21]

I myself am, by the way, to be blamed to a certain degree, that at the time I have prematurely called the attention of the Annalen editors to Schouten, because in the summer of 1913 I sent the article 'On the classification of the associative number systems' [257] (since then published in vol. 76) to Blumenthal, with the recommendation to publish in the Annalen after checking the salient point of the contents, namely the 'principle of continuation of self-isomorphism' for novelty, because this novelty (which I could not judge myself) was crucial for the value of the article. I believe that Blumenthal then sent the manuscript to Hölder, who gave the definitive approval, and only later it turned out that the mentioned principle had been much earlier explained by Cartan, and in much more transparent form.

As far as the misunderstanding (fortunately with no serious consequences) with respect to the printing of the article of Kerékjártó is concerned, I had pointed out when I sent this article to Carathéodory that, apart from the final corrections by me concerning the content, it needed a drastic reworking of the language, and that I was willing to do this myself, if necessary, but that I rather left this to a German to get a perfect result. Only from the proof sheets that I received in Switzerland, I learned that until now no such

---

[21]Please consider the information about these words of Study and Weitzenböck as confidential.

[256]Foundations of vector and affinor analysis.   [257]*Zu Klassifizierung der assoziativen Zahlensysteme.*

rewriting had been done, so I have taken it on myself. However, in this matter Carathéodory (with whom I have been spending several very nice and cozy days in my house) is probably not to be blamed either, because from certain signs of incoherence in the correspondence of the two of us, it appears that some letters or cards must have been lost.

As far as the dating of my article that appears in vol. 80 issue 1 is concerned, as a matter of principle I never date my publications (apart from very special exceptional cases), and in this case I even have (for reasons that are by no means secret but somewhat laborious to describe) a special objection to it; hence I would like to ask you to agree that at the end of the article I leave out place and date.

The Bernstein quoted in the introduction is indeed Felix Bernstein of Göttingen; I have nothing against inserting an F or the entire first name at the relevant place.

With cordial greetings, also from Carathéodory

your revering [258]
L.E.J. Brouwer

[Signed autograph – in Klein; part of draft in Brouwer]

019-10-18

From J. Nielsen — 18.X.1919   **Hamburg**
                                                      Abendrothsweg 50 II

Dear professor, [Sehr geehrter Herr Professor]

It has pleased me very much, that you approve of the contents of my article. With respect to the required changes I await your communications.

The justification of the theorem in question in my dissertation — § 4 can be read without connection with the preceding — is of course not sufficient. I realized that at the time (1912), but had to finish my dissertation quickly and after that I was so absorbed by other work that I didn't come back to it. Therefore, when this summer in Göttingen the problem surfaced again at the occasion of a discussion about the paper by Mr. von Kerékjártó, I was all the more eager to use the opportunity to put the proof in order.

---
[258] *Ihr verehrender.*

The presentation in the dissertation is now perhaps useful as a convenient illustration for the topological core of the idea of the present proof.

Allow me to enclose on this occasion two reprints from the Mathematische Annalen of last year that deal with infinite groups. Now I am most of all involved in making group theoretical principles useful for topology. More specifically I have been trying already for a long time to find the group of mapping types for a surface of genus $p > 1$. The solution of this problem will also give a necessary condition for the solution of the fixed point problem in the most general case. At the moment I am making some progress. When I would be allowed to submit to you at some later moment a communication, I would owe you my warmest thanks

Sincerely yours [259]

J. Nielsen

[Signed autograph – in Brouwer]

1919-10-21b

To F. Klein — 21.X.1919[b]   **Laren**

Dear Geheimrat, [Hochgeehrter Herr Geheimrat]

Carathéodory has not returned here yet. He should be back on the 15th, but he informed me at the last moment that because of a sudden trip of Venizelos [260] to London, he is forced to stay in Paris until Venizelos will have returned.

Enclosed I send you the note by Mr. Wolff, for the Annalen, about which I wrote to you recently.

I have started my discussion with Nielsen; it may possibly have to take quite some time. If this should realize its aim (that is the establishing of a flawless and a best possible direct proof) in a satisfactory manner, then I must be able to count with certainty on it that the author is not at the same time going to negotiate with the managing editors about the publication of his article; only because this principle of Carathéodory was strictly maintained with respect to Kerékjártó, I was at last able to get

---

[259] *Ihr sehr ergebener.*   [260] Eleuthérios Venizélos, Greek statesman and diplomat, at the time Greek representative at the Paris Peace Conference.

something good out of that young Hungarian; and only because Blumenthal, when dealing with Juel, was in this respect too tolerant, a load of confused nonsense by the latter author has been published in the Annalen.

It is only because in the present case the author has, as I believe, close personal relations with Göttingen, that it would be for me, for certainty's sake, most welcome to have your guarantee that the article, that was sent to me for refereeing will anyway be accepted only by me, in order to preclude in advance any possibility of vain effort.

Please accept my apology for my frankness and thank you very much in advance for your possible assurance.

As always, your devoted [261]

L.E.J. Brouwer.

[Signed autograph – in Klein]

19-11-09

From F. Klein — 9.XI.1919   Göttingen [262]

Dear professor! [Hochgeehrter Herr]

The difficulties that publication of the Mathematische Annalen with Teubner recently have met, and that became so clearly visible to all, because of the competition of the journal of Springer, now have culminated in a crisis.

On September 30 an elaborate discussion took place here in Göttingen, in the presence of Mr. Ackermann, [263] between Giesecke representing Teubner, and von Dyck, Hilbert and me. In particular Hilbert has emphasized that we must insist on publishing one volume per year in peacetime strength; through attracting more mathematical physics the business can very well hold its own next to the mathematische Zeitschrift. The next day von Dyck made me the proposal, that he would withdraw from the board of chief editors in favor of a representative from mathematical physics,— I have answered him that we only can accept this offer if he would join the ranks

---

[261] *stets Ihr verehrender.*   [262] Letter to the editors.   [263] of Teubner.

of associate editors, [264], this already for the reason, that no semblance of dissension within the editorial board could come up in the eyes of the public.

Meanwhile volume 80, number 1, which Carathéodory concluded before his travel abroad, has been finished under my supervision by Teubner. Over and above this, we still have a few short manuscripts:

1.) Sternberg, Asymptotische Integration gewöhnlicher Differentialgleichungen — objected to.
2.) Bögel, Stetigkeit von Funktionen mehrerer Veränderlicher — under revision.
3.) Rademacher, Ueber partielle und totale Differenzierbarkeit.
4.) Ostrowski, Existenz einer endlich Basis bei Systemen.
5.) Nielsen, Fixpunkte bei topologischen Abbildungen. For refereeing with Brouwer.
6.) Frl. Noether, Zur Reihenentwicklung in der Formentheorie.
Typesetting of this manuscript has not yet commenced.

So far the matter seemed to proceed in a normal fashion, until I received a letter, dated October 27, from the publishing company, saying that Teubner couldn't bear the exceptional expenses which were demanded by publishing the Annalen in the yearly extent requested by us, and that he left it to us to look for a new publisher. Indeed, at the meeting of September 30 it had been mentioned that Springer, as stated orally at one time or another, might be willing to take over publication of the Annalen from Teubner.

In essential agreement with von Dyck, and, as soon turned out, Blumenthal, Hilbert and I have subsequently written to Springer in this sense and from him we immediately received a telegraphic and a written commitment that left nothing to be wished.

'Thanking you for your letter, I wish to express my special joy about the trust that can be read from your proposal. I am very happy to take over the publishing of the Mathematische Annalen. In this willingness of course is included the wish to do everything to make it possible that this journal, famous of old, can be successfully continued. I commit myself explicitly to agree with a size of the journal that allows the publication of all eligible articles ...'

---

[264] The Mathematische Annalen had a board of chief editors, called *Herausgeber* and a board of associate editors, called *Mitwirkenden*.

Thus the transfer of the Annalen to Springer's publishing company may be considered concluded, and I only have to ask the gentlemen of the editorial board that weren't part of the two negotiations to remain faithful to the Annalen; when no cancellation is received by Hilbert or me within fourteen days, we will assume the agreement of each of the gentlemen.

Details can only be negotiated orally with Springer. For this we have in mind November 26, because then Blumenthal will be here while passing on his way through. Our plan is that Blumenthal gets from now on a position such as Liechtenstein has at the Mathematische Zeitschrift, where we assume that in the long run the occupation of Aachen [265] will no longer hinder the necessary business traffic with the rest of Germany.

Finally it must be noted that Mr. Ackermann writes to me in an long letter that he had heard only afterwards about the letter of October 27 of the Teubner firm to me, and that he regrets very much the course of the events.— Further also, that the lines above have been written in full agreement with Hilbert, whom I have furthermore asked to initiate all steps that are conducive to the further organization of the Annalen.

Your truly [266]
(signed) Klein.

[Typescript, (copy) – in Brouwer]

19-11-10

From A. Schoenflies — 10.XI.1919   Frankfurt a. Main

Dear Mr. Brouwer [Lieber Herr Brouwer]

Because this year your trip through Frankfurt a/M could not materialize, we hope all the more for the next year. Then you can certainly activate your affection again for our Harz forests!— Though I don't know whether the Engadin [267] will attract you even more. Anyway I also hope to see you one of these days.— Today one more thing. I still have an elaboration of the note in the Göttinger Nachrichten (1912, p. 605), which you have entrusted to me. Should I perhaps send it back to you, or do you wish me still to keep

---

[265] by the French.   [266] *Ganz ergebenst.*   [267] Switzerland.

it? The reason that I write is partly that you may perhaps not think of it.—
After having survived the anniversary of the revolution (November 9, 1918),
I hope for a steady improvement; if only the Entente, in fact La France,
doesn't aggravate it too much. But it is indeed one of my axioms that I
believe in the victory of common sense, which is inherent *in the things*; this
alleviates the difficult time for me.

With cordial greetings from house to house

Your
A. Schoenflies.

[Signed autograph, postcard – in Brouwer]

1919-12-04a

From B.G. Teubner — 4.XII.1919[a]   **Leipzig**
                                                              Poststrasse 3

Dear sir! [Sehr geehrter Herr]

From your kind reaction in writing of November 22, I see that Herr
Geheimrat Klein has already informed you about the matter of the Annalen,
which is for myself exceedingly unpleasant. The gentlemen of the editorial
board have demanded from me that I increase the extent even over what
it was before the war; and this would go with an annual subsidy from the
publisher of 15-20,000 Mark. He is not able to accept this for a Journal;
taking into account the economic situation in which he has found himself,
because of the circumstances caused by the war and the revolution, and
especially in consideration of the fact that I, not only for the Annalen, but
also for other mathematical enterprises, have in the course of many decennia,
made sacrifices, running into hundreds of thousands, it is most regrettable
that the mentioned gentlemen of the editorial board, did not take this into
consideration, and that no venue was sought to enable the continuation
of the Annalen in my publishing house where it now has appeared for 50
years. Rather, after my statement that I could not be required to increase
the subsidy substantially in the present circumstances, which also wasn't
done for the editors of other scientific journals, they have seen fit to contact
without further ado the Springer firm, which of course sees in the takeover of

the Journal a special advertising object for the expansion of its mathematical publishing. Your request provided me with an occasion to reveal to you the reasons that lie at the basis of the discontinuation of the publishing of the Annalen by my publishing house, because it is of course of importance to me that outsiders don't get the wrong ideas about this. In any case I will also find an occasion to make these reasons public.

Issues 1-6 of the 28th volume of the Jahresbericht der Deutschen Mathematiker-Vereinigung has already appeared and they were dispatched to you on the 13th of last month. I hope that meanwhile they have come into your possession.

The first correction of your article [268] was sent to you on the first of the month, because it wasn't for the first issue but for the next one.

Sincerely yours [269]

Ackermann

[Signed typescript – in Brouwer]

)19-12-29

To A. Schoenflies — 29.XII.1919   **Berlin** [270]

Christliches Hospiz St. Michael

Wilhelmstrasse 34

Dear Mr. Schoenflies [Lieber Herr Schoenflies]

You will be surprised to get from here a letter from me. It so happens that I simultaneously received an offer from Göttingen and from Berlin (or more precisely both faculties have put me first on their list of possible candidates), and for that reason I am now here to confer with the ministry. The decision to come to Germany or stay at home will be a very difficult one; in itself I would be most happy to come, indeed, the university facilities in Berlin and Göttingen are tremendously better than in Amsterdam, and here I could expect a much wider scope and also much more frequent stimulation, on the other hand I am afraid that under present conditions I would have to go back considerably in pecuniary respect, because it seems to me that

---

[268] [Brouwer 1919d]   [269] *Ganz ergebenst.*   [270] In the margin of the letter Brouwer had made in pencil a list of expenses; hence the document may well be a draft.

a university professor without financial means can hardly subsist here with a family. Therefore I should in any case for the time being have to give up the idea of investing my small capital of eight to ten thousand guilders in a villa in the German countryside; because I will definitely need the interest of that for my cost of living, as this would produce almost 10,000 marks at the present low exchange rate for the German currency. In addition I hope to be able to negotiate a fixed salary of 25,000 mark, so together with the tuition fee I can get up to a total annual income of almost 40,000 mark, from which, by the way, about 5000 mark taxes and 2500 mark rent must furthermore be subtracted. Do you think that a family of four persons (five years ago we have in fact adopted a friend of my daughter as foster daughter [271]) on this basis can exist in Berlin without having to worry about food, and so that I can still also buy the necessary books? Your advice would be valuable for me, because the colleagues here will understandably depict the circumstances rather too favorable, than too unfavorable, just because they like to get me immediately. I will certainly stay here for another eight to ten days; and I would appreciate very much to get an answer from you to Berlin; for the rest it appeals strongly to me, if the railway situation makes it possible without too much discomfort, to make the return trip via Frankfurt. I didn't hear from you about your Swiss debt after my letter to that effect; I am curious whether the method I proposed to you suited you and whether you have used it.

Meanwhile wishes you and your family a happy New Year

Your
L.E.J. Brouwer.

My address is as is printed at the head of this letter.

[Signed autograph (draft?) – in Brouwer]

––––––––––––

––––––––––––

[271] Cor Jongejan.

Chapter 4

1920 – 1929

From H.A. Lorentz — 25.I.1920 **Haarlem**
Scientific Committee
for Advice and Research

Amice,

It seems to me too that, replying to the enclosed communication, you could draw attention to our Subcommittee for Photogrammetry. On my part I would be happy, also in view of the plan to establish this new committee, to have a talk in the next two weeks with the Minister of Education and maybe others about the Scientific Committee for Advice and Research. But can I say then, that I have learned about the plan from the Minister of War? The letter that you sent me was marked 'secret'. [1]

Perhaps you can ask the Commander of the Aviation Department whether you are allowed to inform me, as chairman of the Scientific Committee for Advice and Research, about the plan without mentioning the names of the persons considered for the Committee. The 'secret' probably will refer to those names and I don't have to know these in order to bring up the matter.

You will understand that I have heard with great interest about your nominations [2] in Berlin and Göttingen. I am delighted about the great ap-

---

[1] The letter is not extant. It is not unreasonable to assume that it is another copy of the letter of 19.I.1920.   [2] Lorentz erroneously writes 'appointments' (*benoemingen*).

preciation of your merits thus shown from the side of you German colleagues. I understand very well that especially the Berlin proposal has, apart from its many drawbacks, its attraction, and that you must seriously consider it. But I very much hope that you will come to the decision to stay in the Netherlands.

With amicable greetings

t.t.
H.A. Lorentz

[Signed autograph – in Brouwer]

1920-02-04

To Mayor Amsterdam — 4.II.1920   **Berlin S.W.**
                                                                  Hospiz St. Michael
                                                                  Wilhelmstrasse 24

Dear Mayor, [Hooggeachte Burgemeester]

Should my wishes find a favorably reception with the curators, [3] would you then perhaps have the kindness inform me about it by a few words to the above address? In view of the way I have been received here in Berlin, and the courtesy shown to me, I would appreciate very much to convey my decision orally to my Berlin colleagues (especially if it is unfavorable for them).

Assuming that I can stay in Amsterdam, I would like to make yet another proposal to you, namely that Mayor and Aldermen try to find a way to authorize me already now to put the credit of f. 10,000 for the reference library [4] at my disposal. For I believe that I can *now* and *personally* make purchases in Leipzig that will be two to three times cheaper, than they would have to be *later* and *from Amsterdam*. It would be simplest if the city of Amsterdam or one of its institutions had an account with a German bank, and that it would be prepared to transfer money in German currency to the account of German booksellers, following my instructions. In that case I

---

[3] The mayor was, ex officio, president of the board of curators of the University of Amsterdam.   [4] *handbibliotheek.*

could conclude the transactions irrevocably by payment in cash, and check the shipping to Holland in person.

With my apologies for the trouble I cause,

Sincerely yours [5]
L.E.J. Brouwer

[Signed autograph – in GAA]

)20-02-12

From Mayor of Amsterdam — 12.II.1920   Amsterdam

Dear Professor, [Hooggeleerde Professor]

I received your letter of February 4 last only yesterday evening. I am very much pleased that I can inform you that the Curators have declared to be prepared to transmit your wishes to the City Council, and as the City Council has yesterday said that it does not object to consent to your wishes, trusting that you will be retained for Amsterdam and the fatherland. To avoid misunderstandings and to confirm our conversation on January 26 last I mention your three wishes below. [6]
I. Your annual salary will be raised to the maximum of f. 10,000, effective January 1, 1920.
II. An amount of f. 10,000.- is made available for buying back volumes of mathematics journals,
III. The number of teaching staff is increased by two lecturers, for teaching the undergraduate students. [7]

I discussed with Prof. Hendrik de Vries the possibility that, to save expenses, the lecturer's positions could be combined with teaching a not too large number of hours at the Gymnasium [8] or one of the high schools [9] I informed the Council that a solution in this direction would be looked for, without committing myself.

---

[5] *gaarne Uw dienstwillige.* [6] These three desiderata are the basis of the promise of a 'Göttingen in Amsterdam'. [7] In Dutch *'candidaten'* i.e., students who have passed their first university examination after about two years of study. [8] A secondary school with Latin and Greek. [9] *Hoogere Burgerscholen*, secondary schools without Latin or Greek, but with a strong science program.

The council members have been bound to secrecy in this matter under reference to article 43 of the Municipal Law.

Concerning your proposal to open a credit for the acquisition of journals – I cannot possibly consider this at this moment. Before taking the required steps in this matter, I would in the first place need a statement from you that you will remain at the University of Amsterdam. Moreover, purchases in Germany require extreme caution, *especially with respect to the required export permit*. The biology department of your faculty has experienced a few months ago a great disappointment in this domain. So if it is necessary to go to Leipzig, then this always can be done later, after the necessary arrangements have been made with the financial experts of the city.

Sincerely yours [10]

Your
T. [11]

[Initialled autograph draft – in GAA]

1920-02-21

To Mayor of Amsterdam — 21.II.1920   Laren

Dear Mr. Mayor, [Hooggeachte Burgemeester]

Having received your letter of the 12th of this month at this address, I have the pleasure to confirm once more in writing that I fully agree with the contents of your letter, and once more to thank you for having made it possible by your efforts to make me remain in the fatherland. Also with respect to the board of curators, I beg you as Chairman to accept my gratitude for their cooperation.

Sincerely yours [12]

L.E.J. Brouwer

[Signed autograph – in GAA]

---

[10] *Met de meeste hoogachting.*   [11] J.W.C. Tellegen.   [12] *Gaarne Uw dienstwillige.*

920-03-00

From Brouwer et al. to KNAW [13] — III.1920   Amsterdam

The undersigned propose for the Foreign Membership of the Royal Academy of Sciences, Mr. Jacques Hadamard in Paris, without any doubt the most versatile, astute and fertile of the living French mathematicians. Among the very diverse domains of research in which Hadamard had a key role in the last 30 years, the undersigned mention the theories of analytic continuation, entire functions, orbits in mechanics, wave propagation, vibration modes of plates, distribution of prime numbers, functional calculus, integral and integrodifferential equations, and calculus of variations.

The undersigned are of the opinion that the place left vacant by Poincaré among the Foreign Members of our Academy cannot be filled better than by the man who also was his successor in the Section de Géométrie de l'Académie des Sciences in Paris. [14]

D. Korteweg
H.A. Lorentz
W. Kapteyn
J.C. Kluyver
Jan de Vries
J. Cardinaal
Hk. de Vries
L.E.J. Brouwer

[Signed autograph – in KNAW]

––––––––––

20-03-25b

From J.A. Schouten — 25.III.1920[b] [15]   Delft
Rotterdamsche Weg 2⁵

Dear Sir, [Weledele Heer]

In polite reply to your letter of the 20th, I inform you that the promise contained in my letter of November 20, 1919, copy enclosed here, does admit

––––––––––

[13] In Brouwer's handwriting.   [14] Hadamard was appointed in Paris in the year of Poincaré's death, in 1912. He was duly appointed in Amsterdam.   [15] Erroneously dated 1919.

no other interpretation than the relinquishing of one of the *manuscripts* in the state it has been submitted during the summer of 1917 to you. Concerning the state of the *binding* and the *manner of binding* no promise whatsoever has been given by me. For your further information, the manuscript intended for you was divided for my own convenience into two parts of a more convenient thickness. My plan was, as already announced to you, to send you consecutively both parts, each of course neatly bound. In my humble opinion I thus would have completely fulfilled the promise I made, because I promised the *manuscript*, not the *binding*, and a manuscript doesn't change of course by an artifice as mentioned.

Meanwhile I have taken proper notice of your statement that there is no possibility of restitution of the manuscript in parts. So you refuse acceptance of the manuscript now offered to you in a completely respectable form and completely as agreed upon, just because of the fact that this was bound in two volumes instead of in one. As you have no grounds at all for demanding that I reunite the manuscript in a single binding, I consider myself relieved of the obligation to satisfy the promise made by me at the time.

I record that even after a not particularly polite request from you in November 1919, I have immediately *kindly promised* the manuscript to which you didn't have any *legal claim*. Furthermore, I have out of *kindness* informed you telegraphically about the contents of a letter which you hadn't read yet, in order to save you the costs of having a manuscript retyped, of which the possession was already promised to you weeks ago. So on my side there was no lack of consideration and patience. Where you have reacted since the second half of 1917 to this, for reasons as yet unknown to me, with unkindness and with misrepresentations, it cannot be expected from me that I am forthcoming with respect to a legally unsupported *demand* now formulated by you, where a *request* would have been more appropriate.

Sincerely yours [16]

J.A. Schouten

[Signed typescript – in Brouwer]

---

[16] *Hoogachtend.*

From O. Blumenthal — 1.IV.1920   Kreis Heinsberg
Waldhotel Wasserberg

Dear Brouwer! [Lieber Brouwer]

First of all I want to tell you, while awaiting your promised letter, that
I am away from Aachen for 14 days. I am staying quite close to the Dutch
border, namely close to Dalheim, which is the German border station on
the line between Roermond and Mönchen Gladbach. [17]   *It would mean a
great deal to me to meet you soon.* The forest here is very beautiful, maybe
that attracts you.

I come back to the dating problem. You misunderstood the agreement
of the editorial board. It has after all been laid down that the date of
acceptance will be shown. It is the intention of the proposer, that the date
should not be a ground for priority claims of the author, but it would give the
public a possibility to check how much time elapses between acceptance and
publication. The date of acceptance, not of submission, was chosen because
we are afraid of disputes with the authors in case of returning manuscripts
for revision, about what the date of submission is: the author thinks the
date of reception of the manuscript that was returned later, and the editor
thinks the date of reception of the manuscript that is ready for printing.
When 'date of reception' is chosen, the author has more rights, and in case
of 'date of acceptance' the editor. I admit that one can disagree about the
efficiency, and I am quite ready to enter an argument with you.

Your misunderstanding originates from the following: as implementation for the agreement that the date of acceptance should be shown, I have
proposed that *in general the acceptance day should be the day that the print-
ready manuscript is in the incoming mail.* In that way it should, in the interest of the author, be prevented that an editor will have a paper unnoticed
with him for months, which could also happen.

You admit that apart from the date of acceptance there is justification
for an author-date. Nonetheless, I would, also now, give the editor the right
to reject an author-date that seems unjustified. This, as opposed to my
earlier view.

On the other hand there are also cases where there is justification for an
author-date next to a date of *reception*. I have just now seen such a case: an

---

[17] nowadays 'Mönchengladbach'.

article was for three months at the Mathematische Zeitschrift and then was sent to me by Lichtenstein with the request that we take it: not because the article was bad, but because the journal already had an article by the same author. In this case the author-date undoubtedly is justified.

Maybe we will find in face to face discussion a solution, which is correct for all cases. In writing one gets involved in complications. As Clemenceau once said: Je suis dans l'incohérence, j'y suis, j'y reste. [18]

Best greetings and 'auf Wiedersehen'!

Your
O. Blumenthal

[Signed autograph – in Brouwer]

1920-05-06a

From H. Weyl — 6.V.1920[a]   **Zürich**

Dear Brouwer! [Lieber Brouwer]

Finally I have sent the long promised [object] off to you. [19] It should not be viewed as a scientific publication, but as a propaganda pamphlet, thence the size. I hope that you will find it suitable to rouse the sleepers; that is why I want to publish it. I would be grateful for your opinion and comments. Did I enclose everything that you let me have only as a loan? If not, please reclaim it; the lecture on Formalism and Intuitionism [20] was already in my possession in the old days; at that time I did not pay attention to it or understand it ...

At the moment the matter of the appointment is finally approaching a decision. The reason for the delay was Berlin; and after Herglotz apparently turned it down, I have been offered Berlin in addition to Göttingen. The day after tomorrow I depart. I feel rather loosely tied to Zürich. Neither for mathematics, nor for myself I can realize here something. I'll write to you

---

[18] I am in [in a state of] incoherence, here I am, here I stay. The second part is in fact a famous quotation by itself, namely of general MacMahon in the Crimean War (1855).   [19] Manuscript of the 'New Crisis' paper, [Weyl 1921].   [20] [Brouwer 1912a], [Brouwer 1914]

about the result. Today a couple of cordial greetings from your

Hermann Weyl

[Signed autograph, postcard – in Brouwer]

Editorial supplement

D. Hilbert to H. Weyl — 16.V.1920 *Göttingen* [21]

Sontag

Lieber Weyl,

[..]

I heartily wish that you can improve your financial situation in Zurich as far as your wishes go. Should you, however, decide for Germany, then it is not clear to me why you should prefer Berlin. What I can quite understand with Brouwer and Landau—Brouwer wanted just temporarily to stay in Berlin, and to get familiar with Berlin and the nimbus to be appointed in the capital, were his motives, and Landau has his roots in Berlin and also the financial basis, which cannot be replaced by any salary, necessary for Berlin— does not apply to you: moreover, you can in a few years time obtain a transfer to Berlin, when later the extremely unpleasant and not to be envied circumstances in Berlin have been improved.

With best greetings to also to your wife,
your
Hilbert

[Signed autograph – in Weyl]

20-05-06b

To H. Weyl — after 6.V.1920[a] [22]

Your unreserved scientific assistance has given me an infinite pleasure. The reading of your manuscript was a continual delight and your exposition,

---

[21] Only the for Brouwer relevant part of the letter is reproduced.   [22] This draft is poorly readable.   Some sentences have been left unpolished or unfinished.   C.f. [Van Dalen 1995].

it seems to me, will also be clear and convincing for the public ... That the two of us have different opinions some side issues, will only be will only stimulate the reader. However, you are completely right in your formulation of these differences of opinion; in the restriction of the objects of mathematics you are in fact more radical than I am; however, one cannot argue about this, these matters can only be decided by individual concentration.

Referring to your expositions on the concept of a continuous function I would like to draw your attention to my concept of a completely defined function of the continuum. I mean by that a law which assigns to each point of a point species that locally coincides [23] with the continuum a further point of the continuum. Such a function can very well be discontinuous without being in any manner generated by putting together continuous functions on separated continua; one can, by the way, operate with them in many ways (one can, for example, integrate them in certain cases without having information about their continuity or discontinuity).

Apart from our points of difference, I have the following remarks:

To the non-existence proofs (to which belong for example the cardinality theorems on p. 13 and 43 of my first treatise [24] and also the Hilbert finiteness theorem for complete systems of invariants in his first proof) you don't devote any space in your enumeration of mathematical judgments. On p. 3, l. 8 (and likewise on the analogous place on page 13 of 'The Continuum' the meaning of the word 'Sachkenntnissen' [25] is obscure to me.

It seems to me that the whole point of your paper is endangered by the end of the second paragraph of page 34 [26]. After you have roused the sleeper, he will say here to himself: "So the author admits that the real mathematical theorems are not affected by his expositions? Then he should no longer disturb me!" and turns away and sleeps on. Thereby you do our cause an injustice, for together with the existence theorem of the accumulation point of an infinite point set, many a classical existence theorem of a minimal function, and also the existential theorem of the geodetic line without the second differentiability condition, loses its justification!

The statement you formulate on p. 37, l. 3–6, which by the way, as you know, contradicts my opinion, should be explained a bit more in detail. It seems to me that also the reader who has followed you closely so far, will have problems with this passage. Your discrete function and mixed function to me seem, just as well as the continuous function, to be contained in my

---

[23] A notion from [Brouwer 1919a]: A locally coincides with B if $\neg \exists a \in A \forall b \in B(a\#b) \wedge \neg \exists b \in B \forall a \in A(a\#b)$. [24] [Brouwer 1918a]. [25] Factual knowledge. [26] [Weyl 1921] p. 66

spread concept. My spread law can very well give in advance for every choice sequence the certainty that after it once has generated a sign, it will henceforth generate again and again nothing. [27]

I am tremendously curious about your decision between Göttingen, Berlin and Zürich. May you see clearly and make the right choice. That won't be easy for you!

I can keep the copy of your manuscript you sent me, right? You don't have to send me back anything. Because some of my reprints have been printed anew, I want to ask you to inform me which of the following publications of mine you have at present:

1. Intuïtionisme en formalisme (Dutch)
2. Intuitionism and formalism (English)
3. De onbetrouwbaarheid der logische principes (Dutch)
4. Het wezen der meetkunde (Dutch)

I can now supplement the possibly missing items. Once more, many sincere thanks for the joy and satisfaction that your text has given me, cordial greetings also to your wife and 'auf Wiedersehen'!

Your Egbertus Brouwer.

[Draft handwritten – in Brouwer]

20-07-26

From H. Dingler — 26.VII.1920   **Munich**
                                                      Clemensstr. 47-III, München

Dear Professor! [Sehr geehrter Herr Professor]

Please accept my warm thanks for your kind, rich package; I have immediately occupied myself with the extremely interesting reading of the material. I was very much interested to find in you a strong inclination also towards epistemological problems, that have been occupying me already for many years (until now I was only familiar with your more mathematical articles). I have yet to get acquainted with your set theory. For the time

---

[27] This is part of Brouwer's definition of 'spread', see e.g. [Brouwer 1981] p. 14, [Van Dalen 1999] p. 314.

being I don't quite understand how you want to get around the fundamental theorem of the excluded middle. However, something like that is after all certainly possible, just as it is possible to construct non-Euclidean geometries. I enjoyed very much your demand for a constructive (I would rather say 'synthetic') set theory. [1] Fortunately the holidays will start in a few days, and then I'll find more time to go into your valuable writings.

Thanking you again,

Sincerely yours [28]

H. Dingler.

[Signed autograph – in Brouwer]

1920-08-07

To F. Klein — 7.VIII.1920   **Bad Harzburg**
                                                                   Krodothal 4

Dear Mr. Geheimrat, [Hochgeehrter Herr Geheimrat]

Refereeing an article by Schouten amounts in my opinion to first translating the cumbersome and worthless symbolism into common language, [29] then sifting from among the great mass of trivialities thus obtained, the few theorems that matter, and finally figuring out on which places, unquoted by the author, these theorems, insofar they are correct, have appeared earlier in the literature. Then the result, certain from the beginning, is the rejection.

---

[1] Also the definition of a set by a law, I find *highly* sympathetic in the case of the higher sets, as well as your set theoretical theorems (until now only in the formulation; I have yet to learn to understand more closely the meaning and proof).

[28] *Mit verbindlichen Empfehlungen und nochmaligem besten Dank, Ihr ergebenster.*
[29] Although Brouwer was no admirer of formalisms — see e.g. Brouwer, Intuïtionism and Formalisme (inaugural address, University of Amsterdam, 1912), [Brouwer 1913b] p. 84 — he would not object to efficient notations. Schouten's formalism, however, was more than he could take. Brouwer was not alone in this view, cf. *Brouwer to Klein 19.IX.1919* and [Van Dalen 1999] section 8.3. Klein did not share Brouwer's negative opinion (see below); he asked Weyl for a second opinion, (15.I.1920), but Weyl was not forthcoming, cf. [Van Dalen 1999] p. 298.

But to carry out the justification of the rejection in a logical and matter of fact manner, demands not only a large and unrewarding investment of time, but also a library that completely contains the newest literature, hence for this reason already I am unable to undertake this assignment here in Harzburg. Also, on the other hand I by now feel justified, after having protected the Annalen already a few times from the embarrassment of accepting a Schouten article (indeed, the article about the classification of associative number systems [30] has been accepted by Hölder), to waste no more time and effort on this author, and to restrict myself to declining any responsibility for the publication of his productions.

I apologize for expressing myself somewhat bluntly, but I see no other possibility to express my point of view clearly in any other way.

As regards Haalmeyer, in the past months he has submitted his article two more times to me; both times it seemed to me capable of improvement and I have handed it back to him.

In my further publications about topological groups I will probably have more often the opportunity to refer to the 'Theorie der automorphen Functionen' by Fricke and you, especially where I prove the topological equivalence of the topological and linear infinite discontinuous groups.

With many greetings

Yours truly and cordially [31]
L.E.J. Brouwer

[Signed autograph – in Klein]

20-08-20

From G. Mittag-Leffler — 20.VIII.1920   **Tällberg** [32]

Dear Colleague, [Tres honoré Collègue]

Are you still in Amsterdam? I have been told that you have accepted to become the successor of Carathéodory in Berlin, but I don't know any details. If that is the case, could you not think of Frédéric Riesz as your

---

[30] [Schouten 1918].   [31] *Ihr wie immer hochachtungsvoll und herzlich ergebener.*
[32] Letter forwarded to Villa Friedwalt, Krodotal 4, Bad Harzburg (envelope).

successor? It is improbable that you could find a worthier one. He is now in a very unhappy position, being fired from Koloszwar (Klausenburg), because he couldn't give courses in Romanian.

I allow myself to send you three brochures of mine, and I would be happy if you would always send me reprints of what you publish yourself.

Please accept the expression of my great respect and my admiration for your beautiful works,

Yours truly, $^{\langle 33\rangle}$

Mittag-Leffler

[Signed typescript – in Brouwer]

1920-08-28

From J. Wolff — 28.VIII.1920   **Groningen**

Amice,

Enclosed I submit to you an article, $^{\langle 34\rangle}$ for which I give you full authority: if you think it good enough for the Academy, $^{\langle 35\rangle}$ then you would do me a very great pleasure to present it. In case of the least doubt I ask you urgently not to present it, and then I'll hear from you about it some time.

As an extension of the notion of limit, I assign to each set V_δ depending on $\delta > 0$, with $V_{\delta'} < V_\delta$ if $\delta' < \delta$, a limit set (L), which is the intersection of the closed hulls $^{\langle 36\rangle}$ of V_δ. In that way one can for example speak about the limit set $^{\langle 37\rangle}$ of a function 'in a point'. Usually one only considers the extremal elements, those are the two limit functions. Wouldn't it be nice to classify functions according to the nature of their limit sets? As appears from my article, functions for which the limit sets are all points or continua must form an important class, see for example p. 4 § 10.

I have meticulously checked in the Revue $^{\langle 38\rangle}$ whether functions have been studied at all according to this program and I come to the conclusion that this is not the case. I think it is interesting to examine the kinds of

---

$^{\langle 33\rangle}$ *Agréez, je vous en prie, tres honoré Collègue, l'expression de ma haute considération et de mon admiration de vos beaux traveaux.* $^{\langle 34\rangle}$ Possibly [Wolff 1920]. $^{\langle 35\rangle}$ KNAW $^{\langle 36\rangle}$ The author uses here a German term. $^{\langle 37\rangle}$ in Dutch: *limesverzameling.* $^{\langle 38\rangle}$ *Revue semestrielle.*

continua that occur in the complex plane near the differential quotients of a function: if they are all points, then the function is holomorphic.

With friendly greetings

t.t.

Wolff.

[signed autograph – in Brouwer]

)20-09-07

To H. Weyl — 7.IX.1920   **Bad Harzburg** [39]

                                                                 Krodotal 4

Dear Weyl [Lieber Weyl]

As a supplement to my postcard from Switzerland, first the following: in building up mathematics in Amsterdam I don't pursue at all the plan of establishing there an intensive lecture and seminar business, but only to bring together a circle of people whose mathematical work is mainly a stimulating and controlling side phenomenon of their general spiritual development, in other words people who feel themselves to be more or less the thinking organ of the community and who unabashedly relegate the directly tangible academic teaching activities to the second place, after this calling. (Indeed, I see the drive for mathematical knowledge — which is fundamentally different from the joy of solving mathematical problems — as a characteristic of a mental attitude that safeguards a free and wide view on the most diverse moral and practical domains, which is considerably superior to the prevailing view.) To this I add that we mathematicians in Amsterdam have secured in the last years a very large degree of academic freedom and that we use this in the above sense. Moreover, we are respected in our faculty (of natural sciences) and our subject is held there in an esteem that is free of skepticism. However, in the other faculties (maybe with exception of the medical faculty) we have more or less the reputation of Bolshevists.

---

[39] The unmentioned topic of this letter is Brouwer's attempt to get Weyl to accept a chair in Amsterdam.

Concerning assistants: I have one who manages the reading room and who has worked out a few of my lectures. My colleagues don't have and don't want one. You certainly can get one, as soon as you want. Besides the salary (which by the way is expected to be increased again in the near future) there are *no* tuition fees. [40]

When you prefer, you can live of course in a suburb — like I do, for example at the North Sea in Zandvoort, about 30 minutes by train from Amsterdam. Rent and taxes together will amount to between 1000 and 2000 guilders; both are substantially lower in suburbs than in town.

My colleagues De Vries (Vossiusstraat 39) or better even Mannoury (Koninginneweg 192), who has four children of school going age, will be able to inform you precisely about schools; just ask them specific questions. The schools in town are excellent, of those in the suburbs I have heard less praise, but also there they are certainly bearable. In Amsterdam there is even a German school, but I don't know anything about its quality.

As far as your official language is concerned, you have automatically permission to teach for two years in a language other than Dutch; this permission will then be extended when needed, I believe one year at a time. Ehrenfest lectured already in Dutch the second year he was in Holland. For Denjoy it will be already the fourth year that he lectures in French in Utrecht.

At the end of the week I'll be in Holland again, and from there I'll come to Nauheim. [41] Please write the rest to me in Laren. Our lectures start again on the first of October.

Please recommend me to your wife and accept with my wife's greetings a warm handshake from

Your
Egbertus Brouwer

[Signed autograph – in Weyl]

---

[40] I.e. money paid directly by the students to teachers.   [41] Where the 1920 Naturforscherversammlung was to be held.

A. Denjoy to O. Blumenthal — 4.X.1920   Utrecht [42]

Stationsstraat 12bis

Sir, [Monsieur]

Back from a holiday and from the Congrès International des Mathémati-
ciens held in Strasbourg from 22 to 30 September, I find at my return in
Utrecht a postcard written in pencil, which you were so kind to address to
me from Mr. Brouwer's house.

Despite the warm memories I have kept ever since our meeting in Rome,
I do not believe this is the time to renew our personal relations.

As long as the governments that belong to the League of Nations have not
arrived at an unanimous decision about the admission of Germany, mathematicians from my country will keep, I believe, their reservations with
respect to colleagues from yours.

The visible reasons for a renewal of the conflict between our two countries
are from gone. One must have seen the devastation of certain regions in
the North and North-East of France and measured the amount of work
and expenses necessary for rebuilding to realize that the people of France,
heavily reduced as they are in their means of production, will not consent
in assuming that task alone, and in exonerating yesterday's enemies, less
tested than they are.

In case Germany would rise to escape her obligations and France would
have to resort to force in order to submit her, the initiative, taken already
by a French scholar, to ignore prematurely all reservations with regard to
a German colleague, that position, taken in an offhand manner, would be
regarded as thoughtless and irresponsible.

I am in no doubt that such is the opinion of all of us French mathematicians. If it is seen with disapproval across the border, that is of no concern
to us, if only because the war has strengthened us in our firm belief in our
better judgment, despite the low esteem in which it used to be held in the
old days.

More often than not the past six years have proved us right. It is not the
French way of thinking that did not stand the test of the facts. Accordingly,
we will continue to give it credit, even if it results in contempt again: costly
as the effects of it are for us, they cost others even more.

---

[42] This letter has been reproduced in *Brouwer's to the Minister of Education*
27.IX.1922.

I conclude. The day the French government will believe to have received
proofs of good will on behalf of your government and will judge them to
be sufficient as a justification for the restituting to Germany the rank of an
ordinary nation; that day, too, I will no longer see any fundamental obstacles
to engage in, or renew, relations with those of my German colleagues at any
rate who will not have provoked, by their resounding manifestations, the
personal feelings of resentment of scholars belonging to the Entente.

Meanwhile, I obey the orders dictated by the attitude of the government
of my country.

Yours sincerely, [43]

A. Denjoy

[Typewritten signed original – in Brouwer]

───────────

1920-10-17

To A. Denjoy — 17.X.1920 [44]   Laren

Sir, dear colleague, [Monsieur et cher collègue]

Our colleague Mr Blumenthal has shared with me your letter of 4 October, by which you thought fit to answer a postcard he had sent you whilst
staying with me.

I have no doubt that you realize the consequences of that incident for
our personal relations: the laws of hospitality oblige me to see the attitudes
taken towards one of my guests as engaging me personally. Allow me to tell
you that my opinions on the political responsibility of scholars (especially of
us, members of the academy of a neutral country) are diametrically opposed
to yours.

Rest assured, Sir, dear colleague, of my due respect. [45]

(signed) L.E.J. Brouwer.

[typescript copy – in Brouwer]

───────────

[43] *Je vous prie d'agréer, Monsieur, l'assurance de ma considération distinguée.*
[44] Reproduced in the letter from *Brouwer to the minister of education 27.IX.1922.*
[45] *Soyez assuré, Monsieur et cher collègue, que je vous portie la grande estime qui vous est due.*

From A. Denjoy — 20.X.1920   Utrecht [46]
Stationsstraat 12bis

Sir and dear colleague, [Monsieur et cher collègue]

I take pleasure in acknowledging the receipt of your letter of 17 October. I deeply deplore its conclusions, if not, the very natural course of action as regards Mr Blumenthal, you claim to be the reason.

First of all I have my reservations on whether it is right for a host to take remarks addressed to someone staying under his roof, as being directed against himself.

Undoubtedly you would not allow me to take offence if it is your pleasure to welcome at your home people I cannot possibly meet. So why would you be offended if I decline the offer of a conversation on behalf of one of your guests?

Recently you went to Nauheim to attend the Conference organized by the Germans, at the same time as the Strasburg Conference, the scientific interest of which presumably was not less great than that of the first. If you had written me from Nauheim I would have replied with my customary cordiality but the Germans would have been gravely mistaken if they had believed that my sympathies towards their guest was also extended towards his hosts.

If, in due time, when the suspicions towards Germany will be lifted officially, you wish to do your best to bring together scholars that were former enemies, your actions would be seen with great interest. But take my cordial advice and believe me when I predict that it will be more effectual if you take more and better notice of what is going on in the scientific circles of the Entente and if you also heed more carefully the indispensable precept not to impose to friends of one camp your sympathies for those of the other.

The French do not like orders — neither to give nor to receive them. For four years we have received orders of a different force but we preferred not to listen. Eventually, on the contrary, when our orders were finally heard, nobody failed to obey.

Your letter contains a phrase which I cannot help being somewhat disturbed by, namely, on the 'political duty of us, members of academies of the neutral countries.'

---

[46] Reproduced in Brouwer's letter to the minister of education 27.IX.1922

If last April there had been among the mathematicians staying in Holland, anyone deserving better than I do your votes and those of your colleagues, nothing should have prevented you from electing him. If on the other hand I seemed to be the least unworthy candidate for being one of yours, I do not believe your appeal asked anything from me beyond accepting it.

Accordingly, I do not believe to be ungrateful for the honor that has been done to me, if I acknowledge no other duty towards the Academy than to contribute to her works with all my efforts.

I am convinced that if the Paris Académie des sciences had you as a correspondent or foreign associate, she would not ask more than that from you.

If the post of member of a Dutch Academy would involve obligations of a different kind, especially of a political nature, I would be weak enough to have no hesitation between a foreign title, however honorific it is, and my being a simple French citizen, which does not lend itself to being compromised.

In the same domain there is one rule, though, to which I would believe to be bound. If a section of the Academy would address a letter of rebuke to a German or Austrian learned society, elementary tact would prevent me from lending my French name to a manifestation that could damage the reputation of Dutch science beyond your borders.

You have my approval if you believe that the *neutral* members of the neutral academies can be useful in bringing together scholars from different European countries, provided that for the time being you limit yourself to establish and maintain contact between their *works*. Some weeks ago you asked me, on behalf of some Germans whose name I do not know and about whom I do not care, some of my latest articles. I have never ceased to be willing to send you copies of those as long as I have enough reprints. I have no objection whatsoever to contributing in that way to lessen the lack of publications the Germans complain about.

But it would be counterproductive to strive, prematurely, for the reunion of the authors themselves.

Public opinion attributes to scholars, more than to the majority of other individuals, like businessmen for example, a kind of national character, which must make scholars very cautious when it comes to lending their personality for informal contacts, which could be criticized by sensible patriots.

Towards Germany, grooved by ambivalent tendencies, France feels, among other things, like a self-conscious neutral party confronted with a conflict in which others are engaged but in which she could be dragged along.

For more than four years France has resisted under difficult conditions. For a few months more we can have the patience

to tolerate a government that I wish will make the transition from a state of war to a state of peace.

There should be no doubt that the feelings expressed in this letter do not diminish the great esteem in which other mathematicians hold you and in which I join without any reservation. Believe me to be, Sir, dear colleague, your devoted,

Sincerely yours, [47]

A. Denjoy

[Signed typescript – in Brouwer]

20-10-27

To A. Denjoy — 27.X.1920   Amsterdam [48]

Sir, dear colleague [Monsieur et cher collègue]

Thank you for your letter of 20 October. You will no doubt agree with me, if I don't see the usefulness to continue a discussion on hospitality (neither on the consequences of my own extended to Blumenthal, nor on the acceptability of the hospitality extended to me by Strasbourg for reasons of the accident of the place of my birth), if I of course do not dream of interfering with your political views as French citizen; and finally if, that in case you are interested in the way I think of the tribute we scholars ought to pay to opinion (whether it be in the country of our birth or in that in which we are active or indeed in the world at large), I limit myself to sending you the official report of the session of our Academy of 31 October 1914 (p. 828).

As regards the way you see the role of an ordinary member (the question is not about correspondents or foreign members) of the Amsterdam Royal Academy (especially with respect to art. 2, sub c.,[2] of the Rules of the

---

[2]It is because of this article, that M. Blumenthal, citizen of a country that has friendly relations with the government of The Netherlands, quite naturally turned to you, an ordinary member of the Academy of Sciences of The Netherlands, when there was reason to talk to you about certain scientific matters.
(signed) L.E.J. Brouwer

[47] *Croyez-moi, Monsieur et cher collègue, votre tout dévoué.*   [48] Reproduced in the letter from *Brouwer to the minister of education 27.IX.1922.*

Academy), I must admit my surprise, but, all things being considered, in the present situation this is a matter of concern only to you and the Dutch government.

Needless to say, my dear sir and colleague, on one hand that I infinitely regret the circumstances that remove me from a man of your worth; on the other hand that those circumstances do not diminish the feelings of respect I have for you.

(signed) L.E.J. Brouwer

[Typescript, copy – in Brouwer]

1920-10-29

From A. Denjoy — 29.X.1920   Utrecht [49]
                                                              Stationsstraat 12bis

Mr. A. Denjoy [50]
Utrecht
Dear Sir [Den Heer A. Denjoy]
I do not wish to keep the letter below, which I am really not able to take cognizance of.
I urgently request you to direct no further letters to me.
In the mean time, please be assured of my sincere respect.
L.E.J. Brouwer.

Sir, dear Colleague, [Monsieur et cher Collègue]

I am not going consult article 2 sub c of the rules of the Academy. [51] I would be surprised if the statutes of this society allowed her ordinary members to break off even their epistolary relations and obliged them to open the doors of their apartments to any visitor who has nothing to do with that Academy.

'In the present situation', you say, 'my way of seeing the role of an ordinary member (I had only hypothetically assimilated you to a correspondent or foreign member of the Paris Académie des Sciences, given the fact that

---

[49] The original letter was returned to Denjoy; it is likely that the notes on this typed duplicate were made known to Denjoy.   [50] Note on top of page in Brouwer's handwriting.
[51] KNAW

only French citizens can be ordinary members), 'is a matter of concern only to the Dutch government' and myself. [3]

I will not scrutinize the mysterious meaning of those sybilline words. I am perfectly tranquil. The Dutch government will not expel from her academy a Frenchman to punish him for having declined for the time being the invitation of a German to meet him.

Your government will have no wish to please, by an incident of this nature, the enemies of the good relations between our two countries.

As far as I am concerned, I will scrupulously avoid widening our differences as long as no qualified person interferes.

French public opinion — which is all that matters to me, my letter to Mr Blumenthal has clarified that point, French public opinion is already not too well disposed towards Holland. It is felt too clearly that certain people here would have seen the disappearance of France and her civilisation as a minor accident. They would not deplore it if the world had become German. The aggression of 1914, four years of German crimes, on land as well as on sea — all that would be no more than peccadilloes. It is in nobody's interest to confirm the belief of my compatriots — which for that matter is not exact — that all Dutch people think this way.

Your letter shows me that you acknowledge only a vague attachment to Holland, created by the accident of place of birth. Are you not exaggerating your indifference? If the Belgians or the English just had been invading your country, and had pillaged and destroyed the wealthiest region, from Rotterdam to Amsterdam, killed 300,000 young men, maybe you would have felt enough aversion towards the aggressor to make you feel Dutch.

Your obligations towards what happens to be the place of your birth do not allow you to visit a conference at Strasburg but they do allow you to visit one at Nauheim. I would have understood if you declined both the German and the French invitation. Your duties towards Holland entail rigors and accommodations that strike me as strange.

Except for the kinds of countries you mentioned, one recognizes also the country of affinity, a category the existence of which you will find difficult to challenge.

---

[3][Brouwer's note in margin; in the pamphlet it is a footnote:] 'This quotation is incorrect and must *perhaps* explain the unreadable sequel. If you really think that the Dutch Government has nothing to do with the manner in which Rules established by the government are interpreted by an official involved, then I don't want to quarrel with you about that either. What I meant to write was not more than that *except* the official involved and the Government, *certainly no third party* needs to bother with it, and apparently you agree with that.
LEJB.'

But above all you forget the country of nationality. To belong to a nation implies charges but also advantages. Any man should see it as an honor — and for any man it is also wise — to be attached to a people under all circumstances.

I dare to congratulate myself for being able to reunite in one country those of nationality, of affinity and of birth.

It is no coincidence that my origins are in Gascogne. Given the fact that for many generations all my ancestors have been living in that corner of France, it would have been against all the odds had I been born elsewhere.

I have no hesitations in feeling myself a member, and a very humble member, of one family, together with all those who have made my language, incomparably superior to any other because of its rigor, its precision, its immaterial and energetic vigor. That language is perfectly apt to give expression to certain spiritual meanings I see in myself [het Frans begrijp ik strikt genomen niet — er moet een transcriptiefout gemaakt zijn]. And it does not easily lend itself to translate confused mental dispositions that my nature dislikes but in which many a foreign soul finds pleasure.

I can recognize myself in the aversion of French intellect from vainglory from charlatanry and from appeals to superficial curiosity.

Among the dominating traits that are most characteristic for the French people is that I quite enjoy to rebel with all that is in me against characters opposed to mine.

I know of no people with a greater inclination to criticize themselves and greater aversion from admiring themselves.

There are no others on whom arguments of noblesse have more effect and contemptible reasons less impact. They are not like those to go to war hoping to come back rich.

All these affinities determine my impression that I am not a Frenchman by accident.

Your respect touches me, but I have never asked for it. Less respect for me, and less antipathy for my country would be more to my satisfaction
Sincerely yours [52]

(signed) A. Denjoy

[Typescript copy with notes signed by Brouwer – in Brouwer]

---

[52] *Veuillez agréer, Monsieur et cher collègue, l'assurance de mes sentiment dévoués.*

921-01-01

To H. Weyl — 1.I.1921   Laren [53]

Dear Weyl, [Lieber Weyl]

Many thanks for your letter, which just arrived.

When a while ago your telegram came, my disappointment was, to my own surprise, very great; and from that it became clear to me how much I would have liked to have you here. I was already prepared for this negative result, and when no message was forthcoming for such a long time, I had, as I thought, completely reconciled myself with it; the effect of the definitive message shows me that I succeeded only quite imperfectly in this reconciliation. Well, let us hope that you have made the right choice for you and your family, and that the matter will bring all that is beneficial and desirable to your work environment in Zürich.

Perhaps we will succeed in accomplishing that the position intended for you will be offered to another young person with a very outstanding reputation, and as such Bieberbach (whom I would have liked the best) cannot very well be considered, because he is too little known beyond the narrow circle of mathematical professionals, and he also doesn't have a completely unchallenged name as initiator. His prospects for Berlin seem to be fairly good, if however he cannot hold his own here too, he should at least get Leipzig or Hamburg (assuming that Blaschke goes to Berlin; I already wrote to you about that earlier).

For your position, in case the vacancy will be maintained, Birkhoff, who is also known to people in the fields of astronomy and mechanics and who is especially considered a star of the first magnitude, is now the first who comes to mind, and he moreover belongs to a Dutch family (both his parents were born here) and who can perhaps be won over, because in America he isn't yet a full professor.

If it doesn't come to that, and if one returns to the original arrangement of two extraordinary professors or lecturers then I will ask you maybe a few specific questions about Polya. Anyway, I will keep you informed.

What Klein means by a 'reconciliation' between Schouten and me, is not clear to me. For Klein there can be only one relation between Schouten and me, namely that I have rejected papers by Schouten, that were given to me to referee; but Klein knows that this was because of the plagiaristic

---

[53] 'Laren, New Year's Day 1921'.

character of these articles, and not at all on personal grounds. Anyway, recently I haven't seen any publications by Schouten. As nothing is more pleasing than to revise an unfavorable judgment about an author as soon as there is an occasion, I would like to ask you to indicate to me the place in the journal of the 'positive achievement' of Schouten that you mentioned (or rather to let me borrow the journal issue or a reprint for some time); if I then find about the relevant mapping problem a new theorem that hasn't been copied from somewhere (e.g. from Cartan Bull. Soc. Math. 45, p. 57–81), then I would be glad to recognize and appreciate it.

Now dear chap, here's to your health and that of your family for 1921. May the mountains bring you health and vigor. I long to stay there: I don't feel well at all the last few weeks and every day that I don't lie down for a few hours I have a considerable temperature every evening. By itself that is no reason to worry, because I have more often such periods, but if I stay like that for a longer time, then I will request a vacation and then I must go Switzerland for a few weeks. Then we will be able to meet very soon again. Otherwise hopefully in next summer or at the next congress.

Cordial greetings to you and your wife, also from mine, and believe in my faithful friendship

Your
Egbertus Brouwer

[Signed autograph – in Weyl]

1921-01-17b

To A. Schoenflies — 17.I.1921[b]   **Laren**

Confidential

Dear Mr. Schoenflies [Lieber Herr Schoenflies]

No doubt, you will have received these days several recommendations regarding the Bieberbach-vacancy that has arisen with you: may I also for my part direct your attention to a colleague, with whom I am certain you would make an excellent choice? I am thinking of Blumenthal, with whom I have been in contact for roughly a decennium, and whom I have learnt to appreciate more and whom I have learned to appreciate more and more during this period, and also in more and more aspects. In particular I am convinced that he hardly has an equal among our confreres in all-round mathematical knowledge, in energy for work, in helpfulness and moreover in honesty and decency. The activity in which I have been able to observe him first hand (apart from personal contacts), so that the above mentioned opinion has become firmly rooted in me, is the publication of the Annalen (in which I was involved as a silent assistant of the editorial board in 1911 – 1914, and from 1914 on officially as an editor), which before the war was for three quarters in his hands, and similarly from the beginning of 1919 on. By far the larger part of the refereeing was done by him, either alone or together with a specialist engaged by him for this purpose; and if the Annalen of Klein and Hilbert have stood their ground in the first ranks of the mathematical journals, then it owes it in first place to the untiring, unselfish and expert work of Blumenthal, and this work must be valued all the more because it requires on the one hand considerable talents, and on the other hand it brings no honor at all, because for the wider public it takes place completely in the shadows. That nonetheless Klein and Hilbert never got Blumenthal a university chair, [54] I can only explain by the Machiavellian principle 'le premier devoir des rois, c'est l'ingratitude', [55] and in addition Blumenthal's excessive modesty (he never tried to get a professorship himself) has played a role. How much Blumenthal formed the core of the editorial board of the Annalen is shown by the years 1914 – 1918, during which the journal was most dangerously ailing because of Blumenthal's military service, and it would have certainly succumbed, if, immediately after his return, Blumenthal would not have given all his energy to it, so that the old 'Standing' was recovered in a few months.— Furthermore I have been able to observe in the last months that Blumenthal is realizing the injustice done to him, and is starting to become embittered; I believe that his wife feels the injustice even more keenly than he does and she longs to get away from Aachen.

Now, my conclusion is as follows: if you should get Blumenthal to Frankfurt, then an old debt of the mathematical community to Blumenthal will be settled, and on the other hand you would get the headquarters of the Mathematische Annalen in Frankfurt and moreover an enthusiastic and very energetic colleague, and whose modesty will moreover preserve in undamaged form the leading position taken by yourself.

---

[54] Blumenthal was a professor at the Technische Hochschule (Institute for technology) at Aachen, which did not count as a university.   [55] The first duty of kings is ingratitude.

In any case, I hope you don't blame me for writing the above to you: I saw it as my duty.

With most cordial greetings from house to house

Your
L.E.J. Brouwer.

[Signed autograph – in Brouwer]

1921-01-31

From R. Weitzenböck — 31.I.1921   **Graz**
                                                            Glacis-Strasse 59

Dear Brouwer, [Lieber Brouwer]

Thank you very much for your letter of the 24th of this month! First of all I want to thank you most cordially for having thought of me and for doing so much work in advance.

In this matter I might then first say the following. In the seventh year of my marital state, I have, in November 1920, become settled halfway, meaning that as full professor [56] I have a corresponding position, and I can in the local circumstances live my life with my family and with my work. I don't want to give up this situation without sound reasons, and more in particular I don't want to start more or less at the bottom.

What is a lector at your university? Does it correspond to our Extraordinarius (extraordinary professor)? Are they civil servants with normal retirement rights? For example, when I would become lector and then die after half a year, would my wife then get anything like a pension? The legal position with respect to the board of the university will not cause me any special discomfort. I believe that once I am over there, it will be straightened out in due time.

The teaching duties you indicated (number-forms and theory of invariants) would *suit me very well*, and would delight me. How many hours per week would be considered there? Of course I would commit myself to lecture in Dutch after at most two years. Also I would, in case the matter is

---

[56] *Ordinarius.*

arranged, apply for the Dutch nationality. My brother wants, as far as I know, now to do likewise.

Now about the question of money. Can one live in your country with 6000 fl [57] in roughly the following manner: (with family) three or four rooms and the rest – in particular a bathroom, in a modest neighborhood or outside of town, a housemaid, every day enough to eat, but otherwise no special demands for clothes and amusement. Maybe once a year a trip to here? For me two things matter most: in the first place to be with my family free of all the petty material worries that one has here every day and every hour – roughly to the degree of the way we lived before the war. And in the second place to be able to dedicate myself in peace and quiet to science and finally get around to intensive work.

If you believe that the sofar mentioned considerations in their two main points can be arranged, then I am happy to agree with your propositions, and will come to you. Let me repeat the two main points: the first one is the status of a lecturer (I am yielding secure ground here!) and the second concerns the material aspects.

In this letter I include a curriculum vitae and a publication list for the case that you think you can pursue the matter further for me, and that you need data about me.

Of course there are many other questions to be settled, as you mention with justification. More specifically I raise two matters, namely the question whether housing situation is also so bad in your country and the matter of moving. The latter must of course be payed for me, I could not afford this myself.

Please inform me if you know more about these things. Especially when you are in the position to tell me something about the point of view of the faculty and the board of curators. Could it not be to the advantage of your university to point out that Delft has created so many new positions for mathematicians, and thus make a fourth professor possible? I want on no account to view your country as a so-called 'milk cow'. But with us it is, as in Germany, now a custom that the rich foreign countries buy us up. And then I am thinking in the first place of that what is most precious, namely human material. I believe I have told you already in Nauheim [58] that my brother has during his vacation in Europe recruited engineers and building technicians for your colonies, or that he has recommended them to The Hague. As I have heard, some 30 of these have been hired until now.

---

[57] *florijnen* – i.e. Dutch guilders.   [58] The Nauheim conference of 1920, cf, [Van Dalen 1999].

I have witnessed the thing here, at least in part, and I can assure you that your government really got good people.

My wife and I would be *very glad*, if you would pay a visit over Easter to our beautiful Styria and we could have you with us for some time.

Best greetings and recommendations from house to house

Your [59]

R. Weitzenböck

[Signed autograph – in GAA]

1921-02-01a

From H. Kneser — 1.II.1921[a]   **Göttingen**
                                                          Annastr. 2 II

Dear Professor [Hochgeehrter Herr Professor]

Departing from a problem of differential geometry 'in the large', I had conceived some time ago the idea of examining the topologically different types of families of curves [60] on closed surfaces. In this context I consider the family of curves given by an everywhere regular differential equation satisfying the Lipschitz condition (but other assumptions that are invariant under topological mappings would also suffice). The results are the following:

1) The surface is a one or two-sided annular surface.

2) If the family of curves doesn't contain a closed curve, the annular surface is two-sided and it can be mapped one to one and continuously onto the square $0 \leq x \leq 1, 0 \leq y \leq 1$ with identification of opposite points $(0, y$ and $1, y; x, 0$ and $x, 1)$, in such a way that each line $y = \gamma x + c$ with fixed irrational γ and arbitrary c corresponds to a curve from the family, up to at most countably many lines, that each correspond to a complete band of curves from the family.

3) One obtains all other types by joining bands that are bounded by closed curves of the family, whose types I might describe by figures:

---

[59] *Ihr ergebenster.*   [60] *Kurvenscharen,* see [Kneser 1921].

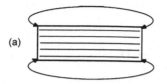

(a)

(only closed curves);

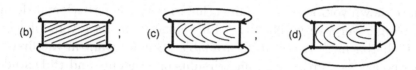

(b)   ;   (c)   ;   (d)

If one joins finitely or countably many bands (a) and (b) together, finitely many bands (c) and one or two bands (d) to form a closed surface, then one obtains all possible types.

Theorem 1 is hardly new; for the others I have found in the literature only Bohl, Acta Mathematica 40. [61] In particular your articles about vector fields seemed to pursue different aims.

However, now I would like to investigate whether an extension to more dimensions is feasible. Here the question about the behavior of a topological mapping of a surface onto itself in case of unbounded iteration, becomes important, especially the question about the properties of such maps that don't have a fixed point for any iteration. Hence I must in any case study your articles about mappings of surfaces in the Amsterdam Verslagen [62] and I would be very grateful to you if you would send them to me, and actually I would like to use the Dutch edition rather than the English one.

Sincerely yours [63]

Hellmuth Kneser

From 1.III to 1.V my address is: Breslau 16, Hohenlostrasse 11.

[Signed autograph – in Brouwer]

---

[61] [Bohl 1916].   [62] Proceedings KNAW.   [63] *Mit den besten Empfehlungen, Ihr ganz ergebener*

1921-02-14

From A. Schoenflies — 14.II.1921   **Frankfurt am Main**
                                                          Grillparzerstr. 59

Confidential

Dear Mr. Brouwer [Lieber Herr Brouwer]

I have shown your letter on Blumenthal to both Bieberbach and Hellinger. Our staffing is very, very difficult. A full replacement for a versatile character as Bieberbach [64] is not at all available. But we want in the first place a man who exerts a great scientific stimulus on students and PhD students, which was the main strength of Bieberbach, and which hardly anybody can emulate. This does not merely include command, even comprehensive command of the subjects, but also an agility to take up and formulate problems, which is a characteristic of Bieberbach.

In the first place we think of Lichtenstein and Polya. I am afraid that the government won't even approach Lichtenstein, who is just now going to Münster. In the case of Polya there is perhaps a personal obstacle. If neither Lichtenstein nor Polya are nominated, then we are so to speak desperate. We have thought of Radon and Rosenthal, and also we seriously consider the name Blumenthal, however as a last resort. But I myself will also leave soon: according to the law I will go into retirement on October 1. In my opinion Blumenthal would then be a very good replacement. But we cannot completely ignore Hellinger and Szàsz either— you see, the situation is complicated in every respect.

On this occasion I would like to allow myself a wish for Amsterdam. Hanna, who was here last Sunday — I gave a Rector's ball on Saturday and I had invited her —, told me that after Weyl declined your offer, you no longer have a position for a full professor, but that you think of establishing two extraordinary professorships. If that is so, and if you haven't made your choice yet, then I would like to recommend Szàsz most warmly. For, he is in the first place an arithmetician and a number theorist, and that's what is still lacking you in Amsterdam. He would also constitute an excellent and at the same time necessary completion of your mathematical circle. He is capable, has many interests and as far as I can judge, he is also a good teacher. We find him here very pleasant, and he is also someone whom we would miss if he weren't there. So you may wonder why I recommend him so warmly.

---

[64] Bieberbach was appointed in Berlin.

It is only because of the uncertainty of the situation here. I don't know if
it would be possible to give him a promotion here. That would in any case
only be possible by calling attention to him for the arithmetical completion
of the completeness program. [65] And exactly for that reason I mention him
out to you too. He is a completely honorable decent personality. As you
know, he has published much, though his papers don't always go so deep.

That is what I would like to give to you in consideration.—
With cordial greetings from house to house

Your
A. Schoenflies

[In the margin:]— Hanna has told us the most wonderful stories about the
stay at your home; please accept our thanks for all the lovely hospitality!

[Signed autograph – in Brouwer]

21-04-10

From A. Fraenkel — 10.IV.1921   **Amsterdam**

Dear Professor, [Hochgeehrter Herr Professor]

Returning from a trip to London to visit the siblings of my wife, I found
here, just now, your kind lines from Italy. I hope that the article from
Crelle [66] that I sent to you in the beginning of March has got or will,
inspite of your absence, get into your hands. Because towards April 20 we
want to return to Marburg, we will postpone our visit to you to another time.

Yet, I should like to use the opportunity to make a few remarks about
the treatise by Schoenflies '*Zur Axiomatik der Mengenlehre*' [67] that was
communicated by you to the Amsterdam Academy. I have addressed these
remarks already about a quarter of a year ago to Schoenflies himself. The
heart of Schoenflies 's article is his treatment of the problems of comparability (p. 794 and p. 808). In more detail: first of all it should be remarked
that the proof for the possibility $(dd) = (a)$ on p. 808 has gaps, because the

---

[65] Probably connected with Sasz' research in the area 'Completeness and Closure'; he
published on completeness of function systems   [66] [Fraenkel 1921].   [67] On the axiomatics
of set theory. [Schoenflies 1920].

system treated there is not only in contradiction with axiom I on p. 803, but also with the very fundamental axiom II on p. 787. Incidentally, this second contradiction can be easily remedied by changing the system treated by Schoenflies into another one. But apart from these single gaps: the alternative 'either $(dd) = d$ or $(dd) = a'''$', from which the comparability of sets follows, i.e. according to Hartogs, eventually the axiom of choice, seems to me a totally arbitrary requirement that goes beyond the principle of choice. If one wants to go beyond Zermelo, then one should allow both possibilities without excluding one axiomatically, and then one will doubtlessly discover that one cannot do without the axiom of choice.

Allow me to venture my opinion beyond this main point of the Schoenflies article, to the effect that an axiomatic foundation with such an extensive axiom system (and moreover with a relatively large number of undefined fundamental concepts) has only little value, when the independence proof is not provided. With Zermelo this is completely different, because with his very few axioms it is *evident* that they are *more or less* independent. But the axiom system of Schoenflies will show itself without any doubt to a large extent reducible when the independence is tested, much to his advantage; the comparison with the axiom systems of geometry fails, because set theory is much less complicated than geometry (and must therefore restrict itself also to very few fundamental notions).

My interest in this matter was stimulated by an investigation completed in the beginning of this year, [68] in which I axiomatically developed the theory of cardinal numbers and cardinalities on the basis of ten axioms, whose independence I completely proved. Meanwhile I have also looked more closely into the independence of Zermelo's axiom system. Unfortunately the work on Gauss' algebra [69] doesn't leave me as much time for my own research as I would like.

In case this letter is forwarded notwithstanding my counter-indication, then I wish you and your honored spouse a really pleasant further journey. With the best greetings of my wife and me I sign

yours truly [70]
A. Fraenkel — Marburg

[Signed autograph – in Brouwer]

---

[68] [Fraenkel 1922].   [69] [Fraenkel 1920].   [70] *Ihr ganz ergebener.*

To R.C. Mauve — 11.IV.1921   Firenze

Greetings from he Piazza Michelangiolo, where twenty years ago I have so often eaten with you both. Since that time nothing has changed here; maybe not even a single new house has been built; there are only new nameplates under many of the most important paintings; there must be a new Director of Fine Arts; for Da Vinci and Michelangelo there is almost nothing but 'già attribuito'; [71] in your house lives a Dottore Medico Chirurgo; on the little omnibus of the Porta Romana you still pay the old Lumps[?]; that meanwhile the prudent Italians have won a war is a scream; it's swarming here with Dutchmen; besides many colleagues, I met here in the streets within one week Eisenloeffel, De Winter and Spigt. [72]

Bye!

Brouwer and Lize Brouwer

[Signed autograph, postcard (picture) – in Collection v.d. Noort]]

To Algemeen Handelsblad [73] — 27.XI.1921   Laren [74]

As a supplement and for clarification of the fragmentary report in your Evening Edition of 26 this month [75] of an incident occurring during a session in the meeting of the Academy, the undersigned would like to remark the following:

The minutes [76] of the Ordinary Meeting of October 29, 1921 will inform the international readership of the Works of the Academy, that on that date the Ordinary Members of the Division of Mathematics and Physics, convening publicly, on the instruction of the Dutch Government and at the expense of the state, have deviated from using the official language of the

---

[71] Formerly attributed.   [72] J. Eisenloeffel (1876 – 1957) silversmith; A.J.J. (Janus) de Winter (1882 – 1951), painter.   [73] A newspaper based in Amsterdam.   [74] Published as a 'letter to the editor' in Algemeen Handelsblad 27.XI.1921   [75] See below.   [76] in Dutch 'zittingsverslag'.

Kingdom of the Netherlands, [77] The undersigned, wishing to be relieved, before the mentioned readership, of being partly responsible for this act, sent the Meeting the motivated notice of absence, which was printed in your Evening Edition of October 29, and mentioned again in the Evening Edition of the 26th of this month; and pointed out to the Secretary in an enclosed letter, that this notice was meant as a public protest, and intended for the Minutes of the Session.

When it appeared yesterday at the reading of the minutes that the notice of absence had not been presented by the Board to the Meeting, and that the inclusion in the Session Proceedings of October 29 as intended by the undersigned would not be realized, the undersigned requested to be allowed to read the motives of his absence during the last meeting to the meeting, in order to effectuate the release of the co-responsibility, as requested by him at least by means minutes of the Session of November 26.

The refusal of this request, which implies a violation of elementary minority rights of the undersigned, will have as a consequence that the readers of the Works of the Academy will get the incorrect impression that the national character of the Academy can be disregarded without serious objection from its midst.

L.E.J. Brouwer.

[Typescript, copy – in Brouwer; handwritten draft also in Brouwer]

Editorial supplement 1

[Handwritten private note of Brouwer, 26.XI.1921, written on the backside of the envelope of a letter to Hk. de Vries]

Jaeger, Winkler, Eykman, Magnus, v.d. Hok, v. Everdingen, Lorentz, Julius, Jaeger, Haga

voted against incorporating my protest in the session proceedings [78]

In favor only H. de Vries and myself

---

[77] Denjoy lectured in French. The lecture was announced in the convocation of the KNAW, *Bolk to Members KNAW, 24.X.1921* as 'De Heer Denjoy zal een mededeeling doen, getiteld *'Recherches récentes sur les séries trigonométriques'*.' [Mr. Denjoy will present a communication, entitled: Recent investigations on trigonometric series]   [78] of a KNAW meeting of 24.IX.1921.

Board Members Went and Bolk did not express themselves in voting
(however, the former had declared earlier to be in favor, the latter
against)

Editorial supplement 2

[*Report in the Algemeen Handelsblad of 26.XI.1921 in the section Science
(Wetenschap)*]

Royal Academy of Sciences

In the meeting today of the section of the Academy of mathematical
and physical sciences, prof. Brouwer asked, after the minutes were
read, that a letter he sent to the Section would be read.

The chairman prof. Went said the this letter was dealt with in the
extraordinary meeting.

Prof. Brouwer remarked that the letter was directed to the ordinary
meeting, and he maintained his request to read, appealing to the meeting.

Only one member supported prof. Brouwer, and after this he asked to
be allowed to make a statement.

Secretary prof. Bolk pointed out that prof. Brouwer's desire already
had been satisfied, because he had published the letter in the 'Handelsblad'.

Prof. Brouwer then asked that in the minutes and in the proceedings
of the ordinary meeting it should be noted that he was refused to read
a note to the meeting. ......

22-00-00

L.E.J. Brouwer, Note on Weitzenböck   1922 [79]

R. Weitzenböck

Weitzenböck wrote in 1908 at the age of 23 years as an officer the book
*Complex symbolism, an introduction to the analytical geometry of multidi-
mensional spaces,* [80] which appeared in the Schubert Collection; a book

---

[79] 1922, or later. Probably part of the appointment procedure.   [80] *Komplexsymbolik,
eine Einführung in die analytische Geometrie mehrdimensionaler Räume.*

which is partly a compilation, but which also contains many new things and which has remained both in set-up and subject unique in its kind, even though unfortunately some minor errors occur in it. With this book Weitzenböck's transition from a military career to a mathematical career became visible, even though he remained active as an officer in the Austrian army for several years afterwards. Between 1908 and 1912 about ten articles follow in the tracks of this book.

Since 1912 he occupied himself mainly with the problems originating from Klein and Study, of finding complete systems of invariants for figures of several classical transformation groups (more in particular the projective and affine groups and the group of motions). All these problems, which had withstood the efforts of Klein and Study (except for the simplest cases) were completely conquered by Weitzenböck in a series of about 25 articles (which mostly appeared in the *Wiener Berichte*), and not only for the projective and affine groups and the group of motions, but also for the Galilei-Newton-group (the group of classical mechanics). The basis of this series of investigations is formed by the so-called fundamental theorems of the symbolic method, which have been given their definitive form by Weitzenböck, and which reduce the invariants of an arbitrary system of algebraic figures to the ones of a certain *linear* system of figures, and which moreover enumerate in the first place all possible types and in the second place all possible rational relations between these types of the above mentioned invariants. Weitzenböck has done this work between 1913 and 1919, where one must keep in mind that he was in active service, [81] almost until the end of the war.

In recent years Weitzenböck has extended his invariant theoretic methods to differential invariants, and he has been able to give among other things quite a number of applications to the theory of general relativity. More in particular he has enumerated all possible mutually independent simultaneous second order differential invariants of a tensor of the first and of a tensor of the second rank (in four-dimensional space); there are only six of these; thereby he has made an important contribution to the questions about the Hamiltonian in general relativity.

At the moment Weitzenböck is indeed the foremost authority on invariants in the world. His article in the Enzyklopädie *Neuere Arbeiten der algebraischen Invariantentheorie. Differentialinvarianten* [82] and his book *Invariantentheorie* (Groningen, Noordhoff, 1922) must also be mentioned.

---

[81] in German '*im Felde*'.   [82] New publications on the Algebraic Theory of Invariants. Differential Invariants.

Furthermore Weitzenböck has shown to be an excellent academic teacher, and in daily contact he is a man of rare simplicity, sociable, honest and co-operative.

LEJB

[Initialled autograph – in Brouwer]

22-04-21

From T. Ehrenfest-Afanassjewa — 21.IV.1922   Leiden

Dear Mr. Brouwer! [Sehr geehrter Herr Brouwer]

Many months ago I have written you a begging-letter, and I have not received any answer at all. It concerned your articles for professor *B. Kagan in Odessa. However difficult it is for me*, I must repeat my request, because Kagan is a person who really deserves that one does something for him. Now he is in the greatest misery — hunger and lack of even the most primitive things in clothing — and yet the first things he begs for are — books, necessary for the continuation of his scientific work. He manages not only to work very hard himself, but also to interest people around him in scientific work.

In the latest letters he explicitly asks for your articles about the foundations of mathematics.

Would you please for once be so kind as to either send me the things, so that I can pass them on (nowadays it is very easy by mail), or tell me that you cannot or will not do it. Then I will try to buy them for Kagan. Until now I didn't do that because I must save as much as possible for shipments of food for our Russian friends and relatives.

With best greetings, also for your wife

T. Ehrenfest-Afanassjewa

[Signed autograph – in Brouwer]

1922-04-26

To T. Ehrenfest-Afanassjewa — 26.IV.1922   Bad Harzburg
                                                              Krodotal 4

Dear Mrs. Ehrenfest [Verehrte Frau Ehrenfest]

Please excuse me for not having answered your previous letter; since the
Dutch Academy of Sciences is acting as n-th pair of oxes to draw the loot
wagons of the Parisian Shylock gang (and has its members that don't agree
with this humiliation without any protest, scolded by the Shylock lackeys)
I have lapsed into such a state of disillusion and apathy that most of the incoming letters remain unanswered. This may explain and excuse that I have
let it come to a reminder from you. But that you imagine the possibility
that I possibly don't *want* to concede to your plea, and that you might request a confirmation of this, which would compel you to *buy* my articles (are
reprints then for sale at all?), adds to the numerous incomprehensibilities
that nowadays pour down on me.

Perhaps, however, you didn't mean it literally this way, and did you
expect just my wholehearted promise, which I now make, that I will deal
with the subject of your letter immediately after my return home (in so far
as no overly paralyzing events or situations await me at home).

With best greetings from house to house

Your L.E.J. Brouwer

[Signed autograph, draft – in Brouwer]

1922-10-10a

From A. Dresden — 10.X.1922^a Madison (Wisconsin)
                                           University of Wisconsin, Madison
                                                Department of Mathematics
                                                      2114 Vilas Street

Dear Professor Brouwer, [Waarde Professor Brouwer]

I am planning to write an article for the *Bulletin of the American Mathematical Society*[83] in connection with Weyl, Mathematische Zeitschrift;
Hilbert, Hamburger Abhandlungen etc. [84]

---

[83] [Dresden 1924].   [84] [Weyl 1921], [Hilbert 1922].

The main point will be to explain in more detail your criticism of the logical foundations. Among my colleagues there is quite some interest in this, but it isn't easy for them to get a clear idea about it. If I want to clarify it, I have in the first place to be certain that I understand it. So I take the liberty to ask you for clarification concerning a few points in your 'Foundation of set theory', [85] part 1.

On p. 3 you define the concept 'element of a spread' [86] as a 'sequence of signs'. Why is then in the case of finite groups of signs or of sequences of type ζ, [87] only the single sign an element?

Why do you speak about 'digit complexes' [88]? Does that mean only 'group of digits, among which the 0 may occur'?

On p. 4 it says that the 'spreads' are a special kind of 'species of the first order' [89] — but then they are properties, not laws. Is the idea that the 'spreads' are a kind of 'elements of species of the first order' [90]?

Could you give a few examples of the 'species'.

What is your view on the Kronecker program of arithmetization; and do I understand your view correctly, when I say that an indirect proof is only permitted when one first has *proved* that of the two cases between which the proof has to decide, at least one occurs?

And what is the meaning of the n on line 7, p. 80 of the JDMV, V. 23? [91]

What would you think of a translation into English of 'The unreliability of the logical principles' in the Tijdschrift voor Wijsbegeerte, 1908?

I enjoyed meeting your brother this summer. Won't you come to America for a trip? For the time being I don't see a possibility to come to Holland.

Sincerely yours [92]
Arnold Dresden.

[Signed autograph – in Brouwer]

---

[85] *Begründung der Mengenlehre.*   [86] *Element der Menge.*   [87] ζ is the sequence of natural numbers.   [88] *Ziffernkomplexe.*   [89] *Species erster Ordnung.*   [90] *Elementen der Species erster Ordnung.*   [91] [Brouwer 1914].   [92] *Steeds gaarne de Uwe.*

1922-11-24

To G. Mannoury — 24.XI.1922   **Laren**

Dear Gerrit [Beste Gerrit]

In the document of Van Ginneken, [93] I read for the first time a formally pronounced ruthless negation of the only thing that attracts me to significs: the hope of the creation of linguistic social means of reform, independent of all existing (in my view mostly obsolete) formation of groups, and by people that in a neutral and humanitarian community would rise above their respective groups. Indeed, this view has in our circle been relegated more and more to the background, but I have patiently allowed that to happen, firstly by acknowledging my learning capacities in this matter, and secondly in the expectation that the community I hoped for would finally be established and would function, notwithstanding all difficulties.

I must now definitively give up this expectation after the experience that one of my fellow members now derives inspiration even from the rejection of my (unaltered) principle, and the consequence of this can be none other than my resignation from our circle. I am even of the opinion that it would tend to unfairness and lack of character if I would under these circumstances keep publishing our joint manifest, knowing it is followed by Van Ginneken's postscript.

Notwithstanding the above, I have the feeling that there is something that ties us four more to each other than to others, but it seems that this je ne sais quoi cannot be admitted into the realm of conscious reality.

With a handshake [94]
your

[Carbon copy – in Brouwer]

-----

[93] Van Ginneken's statement as a member of the Signific Circle contained a plea for understanding and communication within small coherent groups, in his case the Roman Catholic Community. See [Brouwer 1937] and [Schmitz 1990] p. 425.   [94] *Met handdruk.*

To F.A.F.C. Went [95] — 25.XI.1922   Laren

Dear Colleague, [Hooggeachte Collega]

After ample considerations I have come to the opinion that it is rather difficult for me to ask *on my initiative* a fellow member of the Academy to put himself at disposal for nomination into a Committee of advice concerning my address to the minister. [96] This fellow member would involuntarily start to feel himself to be my advocate and this would be improper, because I can, nor may be a party in the treatment of this matter in the Academy. So I withdraw my remark about this point, made in the October meeting.

However, I can say to you that most particulars that could shed light on my conflict with Denjoy are known to our colleague Hendrik de Vries; also, that if our colleague Winkler would like information from my side, I would be completely at his service.

With friendly greetings

Your
(w.g.) L.E.J. Brouwer

[Signed autograph, draft – in Brouwer]

22-12-16

To G. Mannoury — 16.XII.1922   Laren

Dear Gerrit [Beste Gerrit]

I suddenly notice that you have convened me tomorrow morning (in my mind it was the 24th), and because tomorrow Corrie is going abroad with her sister, for which still a lot has to be organized, there is a big chance that

---

[95] Chairman section Physics KNAW.   [96] See *Brouwer to Minister of Education 27.IX.1922* and [Van Dalen 1999] section 9.3.

I can't combine that with my presence at our meeting. So in case I have to be absent again, I want to answer *now* your letter of a few weeks ago, also for the others.

What attracts me to significs has always been and remained: the hope for the creation of linguistic social means of reform, independent of all existing group-forming, and by people that in a neutral and humanitarian community would rise *above their respective groups*. Indeed, this view has been relegated more and more to the background in our Circle, but I have allowed that to happen, firstly allowing for the (until now not realized) possibility that I see the light in this matter, and secondly hoping that the community that was before my mind's eye would still in the end be established and would function, notwithstanding all difficulties.

In your encyclopedia program [97] I hardly find anything at all of my ideal expressed, and an even stronger indication of the solitude of my path is given by the document of Van Ginneken, which not only rejects my principle, but that even derives inspiration from this rejection. [98]

Much more important than the professional and recreational activity of a philological or psychological character, which makes up your encyclopedia, is for me the fulfillment of primary humanitarian duties with signific basis, such as the struggle for the morality of international science, into which I have been driven [99] in the last few years, and which I have seen absorb a great deal of my mental powers. The world needs the spiritual struggle of practical significs more than the accompanying linguistic and psychological theories.

Greetings to Van Eeden and van Ginneken, and a handshake from

your
Bertus.

[Signed autograph – in Mannoury]

---

[97] See [Schmitz 1990] p. 430. This program was proposed by Mannoury when the original plans failed.   [98] See [Schmitz 1990] p. 425.   [99] This is an oblique reference to Brouwer's efforts for the re-establishing of international scientific cooperation and organization, cf. [Van Dalen 1999] Ch. 9.

)23-04-00

To the Dutch Mathematical Society [100] — IV.1923 Amsterdam

Bijlage F. [101]

M.M.H.H.,

Hereby we propose to you to notify the 'Union, etc.' [102] in Paris:

That the Wiskundig Genootschap has joined the Union at the
time in the expectation that this association would develop into
an international association,
that however until now no events have occurred that justify the
hope that this will be the case in a foreseeable future,
and that the Wiskundig Genootschap therefore sees itself forced,
in view of its very limited financial capacity, to resign its membership.

G. Mannoury
Hk. de Vries
L.E.J. Brouwer

[Typescript draft/copy – in Brouwer]

23-04-18

From A. Fraenkel — 18.IV.1923 Amsterdam

Dear Professor, [Hochverehrter Herr Professor]

as I didn't find you today in the University — I only there heard that
the lectures hadn't started yet — I allow myself to say goodbye in this way.
I have received your very friendly card with the proofs that were recently
taken to Laren; [103] I believe that I may infer, if I don't hear anything to the
contrary, that you will essentially agree with the proof sheets sent through
the mail (insofar as they interest you).

---

[100] Wiskundig Genootschap. [101] Handwritten remark. [102] Union internationale de
Mathématique. [103] Proofs of [Fraenkel 1923].

It only remains to thank you most cordially for your inspection of the proofs, which was very valuable for me, for your kind support of the use of the library, for the interesting and original lecture course, and finally — and at the same time to thank your revered spouse, the booklet of hers she lent my wife was also very interesting for me — for the nice and stimulating hours in Laren.   Among other things it was very interesting for me to observe the fresh life in intuitionism, which had already been pronounced dead from many sides; within myself these questions are still fermenting.

Please convey at a suitable opportunity my regards to Mr. Weitzenböck, whose inaugural lectures I can't attend anymore, much to my regret. We travel within the next few days.

With best wishes and greetings from house to house,

Your [104]
A. Fraenkel

[Signed autograph – in Brouwer]

1923-08-25

From T. Ehrenfest-Afanassjewa — 25.VIII.1923   Jena

Dear Mr. Brouwer, [Sehr geehrter Herr Brouwer]

Many thanks for your card. Your consent has made me very happy. Now the big question is to organize the matter really as soon and as well as possible. In any case, please let me know, before my departure, the literature that should be considered in your planned rewriting of the book [105] — maybe I still could acquire it quickly. If your own papers are among these, then I would be extraordinarily grateful if you would send the relevant reprints to me in Leyden.

I think I'll be in Leyden from August 29, until the end of the first week of September [106] and then I'll go to Russia via Berlin. So I cannot attend

---

[104] *Ihr ergebenster.*   [105] Apparently a translation into Russian of Brouwer's dissertation (and later papers) was considered.   [106] I.e. Friday, September 7.

the German Naturforschercongress. I could visit you before September 7 or after November 1. I must confess that I am tremendously looking forward to the opportunity to talk with you about those fine things — whether you will enjoy it as much is another question, the more so as I never studied the topics thoroughly.

One request I still have for you: if you don't want to redo the whole set-up of the book, but only isolated places, then please tell me which chapters I can translate immediately: then I can do that already in Moscow and would be perhaps in the position to do the proof correction of one part over there. Your question about Germany I cannot answer in a few words: it goes without saying that I think the situation is terrible and that I wish that it would change soon and that I would *really feel relieved* when that would happen. But I cannot feel so *uniformly* [107] outraged as you seem to be. By the way, I believe that outrage always contains an element of surprise, and being surprised means that one doesn't completely understand the matter. During the war I have too much put myself in the position of the other party — don't forget I'm Russian, and I know too well the contempt and thirst for power of another nation that is unable to immerse itself into the psychology of your own people. I vividly imagine how *we* would feel if the end result was just the opposite of the present one. That is why I can understand now a bit the certainly all too blind rage of the French, without empathizing of course! But after everything I see here, I am convinced that the cultural consciousness and the inner national coherence and also the many cultural practices are on such a high level here, that a destruction of Germany is impossible.

There, dear Mr. Brouwer! Let me hope that you will not completely wash your hands of me after this short extract from my credo.

Please greet your wife most cordially from me.

Your
T. Ehrenfest- Afanassjewa

[Signed autograph – in Brouwer]

---

[107] *'einheitlich'* in the letter.

1923-09-01

To A. Schoenflies — 1.IX.1923   Laren

Dear Mr. Schoenflies [Lieber Herr Schoenflies]

Because, according to the newspapers, there is again a severe food shortage in your area, I am sending you right now a charity package with a few daily needed items. Unfortunately I heard at the post office that packages in Germany are no longer delivered for free, but that the recipient has to pay 300,000 Mark custom duty [108] (which the sender cannot prepay in any form), for which I apologize. I was last year with my wife in Seefeld in Tirol. The view from Mösern into the valley of the Inn there is magnificent! I have received the Jordan curve theorem and looked through it and found it really amusing: for the time being I lack the time for precise checking; my own work is resting completely for three years, because my strength is almost completely occupied with the struggle against our annexation by France, which is so industriously promoted by the Lorentz clique. [109] Nonetheless I certainly hope to come to Marburg [110] and see you there.

With cordial greetings from house to house

your
L.E.J. Brouwer

[Signed postcard – in Brouwer]

1923-10-24

From P.S. Urysohn [111] — 24.X.1923   Moscow
Twerskaja Street, Pimenowski pereulok 8, kb.3, Moskau

Dear Professor! [Hochgeehrter Herr Professor!]

You have requested me in Marburg [112] to communicate to you in writing the objections against your proof in *Crelle's Journal* (Band 142), [113] that

---

[108] The year 1923 was the time of hyper-inflation in Germany, with prices rising about 5 percent *per day*.   [109] Lorentz was advocating a compromise policy, see [Van Dalen 2005], p. 510 ff., [Schroeder-Gudehus 1966].   [110] meeting of the DMV.   [111] Pavel (Paul) Samuilovich Urysohn.   [112] Annual meeting DMV; Urysohn's talk was on 21.IX.1923. [113] [Brouwer 1913d].

I raised in my talk there. Please accept my apologies that I have tarried so long with this letter; I am only since two weeks in Moscow, and after an absence of almost five months I had so much to do that I couldn't get around to writing.

The article concerned is titled 'On the natural dimension concept' [114] and it contains apart from the definition of the dimension concept, the proof of the 'Dimension theorem': *An n-dimensional manifold possesses the homogeneous dimension degree n.* This proof consists of two parts: the reduction to a lemma (p. 149–150) and the proof of the lemma (p. 151–152). The latter is perfectly flawless; but concerning the reduction I have to remark the following.

In the first place — something relatively unimportant — the concept used there, 'the domain set g_1, bordering on the edge $E_1 E_2$ determined by π_2 in τ_1' (p. 150, l. 1) is insufficiently defined. The definition you give in a footnote (p. 150, *)), says nothing about the connectivity situation, which is clearly indispensable for the characterization of g_1. Likewise the concept of the boundary of this domain set is not defined. I will show in a minute that your proof remains inadequate with *any* definition. [115] Hence a more detailed discussion of the possible definitions is superfluous; besides I might remark that in any case a sensible definition of the 'domain set g_1' is not easy to give: for one may not be guided by the analogy with ordinary domains and describe g_1 as *the largest connected*[4] respectively *continuously connected*[5] *subset of* $\tau_1 - \pi_2$[6] bordering $E_1 E_2$, — because the sets[7] defined in this way generally don't have to be domain sets.

Now your proof contains two unfounded statements (p. 150, l. 13–16)

I) $\varepsilon_1, \varepsilon_2, \ldots \varepsilon_n$ converge with ε to zero,

and

II) $\tau_1, \tau_2, \ldots \tau_n$ are contained as subsets in respectively $\pi_1, \pi_2, \ldots, \pi_n$.

I will now show by a simple example that — depending on the definition of the domain set g_1, and its boundary τ_2 — *at least* one of these statements is wrong. It suffices to choose the Euclidean plane as manifold π; let then $E_1 E_2 E_3$ be the line triangle with

---

[4]In the sense of Hausdorff: i.e. g_1 cannot be split into two subsets neither of which contains a boundary point of the other.   [5]Terminology of Mr. Kerékjártó; you call such a set a 'continuum' (loc.cit., p. 147, l. 3).   [6]$\tau_1 - \pi_2$ is the complementary set determined by π_2 in τ_1.   [7]For both definitions lead to different sets.

[114] *Über den natürlichen Dimensionsbegriff.*   [115]Here Urysohn was overly pessimistic. See *Brouwer to Urysohn 14.VI.1924* [Brouwer 1924a, Brouwer 1924d] and [Van Dalen 2005] p. 461.

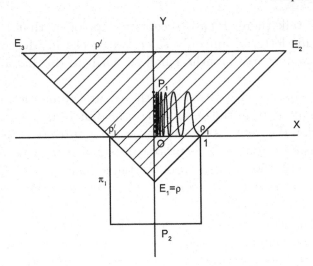

vertices $E_1 = (0, -1), E_2 = (3, 2), E_3 = (-3, 2)$, and let π_1 consist of the following six curves

1) $y = sin^2\frac{\pi}{x}$, $0 < x < 1$ 4) $x = -1$, $-2 \leq y \leq 0$
2) $x = 0$, $0 \leq y \leq 1$ 5) $y = -2$, $-1 \leq x \leq 1$
3) $y = 0$, $-1 \leq x \leq 0$ 6) $x = 1$, $-2 \leq y \leq 0$

and let finally π_2 be the set consisting of both points $P_1 = (0, 1)$ and $P_2 = (0, -2)$.

If one takes the (only natural) definition of g_1, according to which g_1 coincides with part 1) of π_1, then τ_2 is identical with part 2), hence statement II) is incorrect. But when we can define g_1 and its boundary in such a way that statement II) is satisfied, i.e. that τ_2 is a subset of π_2, then τ_2 consists necessarily of the single point P_1. One sees immediately that then ε_2 (p. 150, l. 12) does not go to zero with ε, so that statement I) does not apply.

Hence the proof of the 'dimension theorem' is not correct. Unfortunately I have not succeeded in deciding whether the theorem itself is correct. In any case not only the theorem but also its proof can be made correct by an appropriate *change of definition* of the dimension degree, or more precisely, — of the basic definition of *separation*. Your definition of this concept (p. 147, l. 15–19) must namely be replaced by the following: 'ρ and ρ' are called separated in π by π_1, when $\pi - \pi_1$ can be split into two subsets λ and λ', that contain ρ respectively ρ' and such that neither of them contains a boundary point of the other one.' That this notion of separation differs from yours, one can for instance see from the examples given above. However, the definition of dimension degree thus obtained is, as I have shown,[8] at least for F_σ sets[9] equivalent with the much simpler one that I published last year in the *Comptes Rendus* of the Paris Academy.[10]

The latter definition runs as follows. Let C be any set lying in a compact metric space,[11] and x a point of it. I say that the subset B of C ε-separates *the point x in C*, when the complement set $C - B$ can in such a way be split into two subsets A and D, that

1) none of these two sets contains a boundary point of the other,

2) x belongs to A,

3) the diameter (the upper boundary of the distance of two points) of A is $< \varepsilon$.[12]

Then I define the dimension inductively as follows:

1. The *empty* set has dimension -1.

2. When the point x in C does not have dimension $< n$, but when for every ε it can be ε-separated in C by a set B of a dimension $< n$, then we say that x has the dimension n in C.

3. If the set C contains only points of dimension $\leq n$, and among these also points whose dimension $= n$ is, then we say that C has dimension n.

4. Of sets (and points) that do not have a dimension according to 1—3 we say that their dimension is finite.

---

[8]In Ch. VI of my treatise about the dimensions of sets. The accompanying manuscript is already for several months with the editors of *Fundamenta Mathematicae*; the first part (Introduction and Ch. I – II, maybe also III) will appear in Volume VI of this journal; Ch. VI will appear only in Volume VII or even VIII (the whole treatise is several hundred pages).   [9]I call a set F_σ when it can be considered as lying in a compact metrical space (i.e. ($\mathcal{D}$)-), which means that it is homeomorphic to such a set, and that can be represented as the union of countably many closed sets; every manifold is clearly an F_σ.   [10]Volume 175, p. 440 & 481. The proofs of the theorems that I have stated in these notes without proof, are contained in the above mentioned treatise. [ed. [Urysohn 1922]]   [11]That is, compared to your assumptions, no restriction at all. Indeed, Mr. Chittenden has shown some years ago in the Transactions of the American Mathematical Society that every ($\mathcal{V}$)-set is homeomorphic to a ($\mathcal{D}$)-set, and recently I have proved (I have submitted the proof in July of this year to Mr. Prof. Hilbert for the Mathematische Annalen), that every separable ($\mathcal{D}$)-set (hence *a forteriori* every normal $\mathcal{V}$) is homeomorphic to a subset of a specific compact metric space H_0 (H_0 is the parallelepiped in the infinite dimensional Hilbert space determined by the equations $0 \leq x_n \leq \frac{1}{n}$).   [12]One can also demand — which, as I have proved, doesn't change the final definition of dimension, — that I) B is closed in C and II) also the diameter of the union set $A + B$ is smaller than ε. When one then uses the (modified) concept of separation and if one denotes K the set of points of C whose distance from the point x is $\geq \varepsilon$, then one can also say that K and x are separated in C by B.

The advantage of this definition is not only that it is much better organized than the previous one and that it yields also for non-normal sets a
completely useful result, but it also contains a much sharper definition of the
dimension *in a point* as one can see from the following example: let C be a
plane closed point set that consists of the point x and a countable number of
disks that consecutively touch each other and that converge to x. Then x has
dimension 1 in C, although it is a limit point of points of higher dimension.

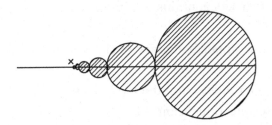

My definition at last permits one to penetrate very far into the properties
of dimension: I refer for example to the theorems stated in my Comptes
Rendus notes (by the way, since then I have found yet other results).

Dearest professor, if you would be interested in the theory thus established, then I would be glad to communicate more details about it. Finally I
permit myself to direct a humble request to you. As I told you already in
Marburg, I have thoroughly studied several of your remarkable topological
articles. Unfortunately only those were accessible for me that were printed
in the German journals. You have published many important articles in the
English language (unfortunately the Dutch language is unknown to me) in
the Amsterdam Academy, publications of which are not available at all in
Moscow. Therefore I venture to bother you, dearest professor, with the request to send me reprints of your Amsterdam articles. [116] My address is as
follows:

Moscow (Russia), Twerskaja Street, Pimenowski pereulok 8, kv. 3

I apologize for the laboriousness of this letter

sincerely yours
Dr. Paul Urysohn.

[Signed autograph – in Alexandrov]

---

[116] *KNAW, Proceedings.*

023-11-29a

To F. Klein — 29.XI.1923[a]   **Laren**

Vertraulich

Dear Mr. Geheimrat, [Hochgeehrter Herr Geheimrat]

I have the honor to send you the enclosed accounting circular of the Mathematische Annalen from Blumenthal, which was sent to me with the request to pass it on.

Last summer I was entrusted by the editorial board of the Annalen with the refereeing of an article by Mohrmann 'On curves of maximal class index'[117] which had been received by Blumenthal. This refereeing cost me a lot of trouble, both regarding the content matter of the submitted work and the personal priority relations: exactly because of that I was very unpleasantly surprised when Blumenthal a few weeks later indicated that Mohrmann wished to retract his submitted article.

The Annalen circular of the last summer mentions both the reception and the withdrawal of the Mohrmann article, and also my name as refereeing editor and the general nature of my objections.

Recently now I got the message from Blumenthal that Mohrmann has again submitted the article in question, and indeed to you. About this I would like to remark that two years ago, on the occasion of an analogous incident, the entire board of editors has jointly and expressly decreed that an editor, once he has been entrusted with the refereeing of an article will remain for his co-editors the one who decides about acceptance, as long as he does not voluntarily part with this duty. Indeed, without this certainty any cooperation between editors is impossible. I am, by the way, the only one who knows the previous history of Mohrmann's submission and also about the mutual priority rights of Mohrmann and Nagy concerned here, and these are based upon the order in which the letters of both these authors have been received by the editorial board (even the authors themselves cannot exactly know this order).[118]

---

[117] *Ueber Kurven vom Maximalklasssenindex.*   [118] The Mohrmann manuscript was probably the cause of Klein's exit from the editorial board of the Mathematische Annalen. The letters *Blumenthal to Ed. Board Math. Ann 16.XI.1928* and *Brouwer to Ed. Board Math. Ann 30.IV.1929* shed more light on Klein's decision to step down. See also [Van Dalen 2005] p. 613, 631.

In the hope that everything is well with you in so far as the prevailing circumstances permit, I am in sympathy with you, greeting you and the other Göttingen colleagues cordially

Yours truly [119]
L.E.J. Brouwer

[Signed autograph –in Klein]

1923-11-29b

To P.S. Urysohn, Summary — 29.XI.1923[b] Zandvoort [120]

Saturday 29.IX.23 from Zandvoort, boarding house John Bückmann, written to Dr. P. Urysohn, Mathematical Seminar of the University, Moscow, concerning the pencilled note (at the separation definition) in the margin of my personal copy of *Über den natürlichen Dimensionsbegriff.* 'This pencilled note, which clears up everything, must date from many years ago; it is very well possible that it has been made after a remark of a colleague (in that case probably Weyl, Gross or Rosenthal). I will try to determine this, and also investigate whether this note was not added to a later publication as Erratum.'

L.E.J. Brouwer

[Signed autograph – in Alexandrov]

Editorial supplement

[*We quote from Freudenthal's comments on Brouwer's pencil remarks, from CWII 1967, p. 549:*]

"There is a hardly visible pencil correction, which in the history of Brouwer's style of writing must be dated before 1923: the word '*abgeschlossen*' [121] in line 18 is deleted and a line with an arrow is drawn

---

[119] *Ihr ganz ergebener.* [120] The following is a private note containing a summary of a letter to Urysohn. [121] closed.

from this word to the margin, where one reads '*zu streichen in Übere-instimmung mit S.150 Fussnote *)*'. [122] In a letter to H. Hahn of 4 August 1929 Brouwer asserted that he corrected the text of his own copy as early as March 1913, which certainly means the pencil rather than the ink correction. This is confirmed in the most unexpected way by a note in the proof sheets of A. Schoenflies 1913 (Brouwer read carefully the proofs of Schoenflies' book and advised the author in the most efficient way (see 1910C and A. Schoenflies 1913, VII Vorwort)). On p. 382 of these proof sheets he elaborated footnote [2]) by adding '*...; ebenso die Untersuchungen Brouwers in Math. Ann. 70, S. 161–165 (an letzter Stelle ist übrigens nach einer Mitteilung Brouwers auf S. 147, Z. 18 das Wort 'abgeschlossen' zu streichen)*.'" [123]

For unknown reasons Schoenflies did not adopt Brouwer's note.

24-01-16

From W. Dubislav –16.I.1924   **Berlin-Friedenau**
                                                              Gosslerstr. 6

Dear Professor, [Sehr geehrter Herr Professor]

It might perhaps interest you that it is extraordinary simple to give examples in which the so-called 'principle of the excluded third', the validity of which you doubt, is demonstrably *not* true. In its usual fashion the principle — the numerous statements in the literature, that partly also differ in content, that are designated as 'principle of the excluded third' had better be ignored — says: 'statement A or statement non-A is correct', and as an aside it may be remarked that the formulation one often meets 'C is D or not' can be easily reduced to this. To show now that the principle is not always true, we consider for example the axiom system drawn up by Hilbert in the '*Grundlagen der Geometrie*' Chapter I (4th edition Leipzig 1913), without axiom group {IV} (parallel axiom) and axiom group {V} (continuity axioms). Now, from the axiom system contracted in this way, let's call it V for short, the following statement, where a is an arbitrary line and A a point outside a: 'There exists in the plane through a and

---

[122] To cross out in agreement with p. 150 footnote *.   [123] ...likewise the investigations of Brouwer in Math. Ann. 70, p. 161–165 (incidentally, in this last place the word 'closed' on p. 147, l. 18 must be crossed out, according to a communication from Brouwer.

A *at most one* line through A that does not intersect a' is not provable.
But also the negation of this statement, namely the statement 'There are
in the plane through a and A *at least* two lines that pass through A and
do not intersect a' is not provable from V. So we have with respect to V
a meaningful declarative statement, let us call it S for which holds that
neither S nor non-S is provable with respect to V. So for S and its negation
— always with respect to V — the 'principle of the excluded third' does not
hold. The statement S is as one says logically independent from V, and as
one can immediately generalize, if one has a statement that is independent
from a totality of statements that together determine a domain of thought,
then with respect to that totality the 'principle of the excluded middle' is not
valid. So the 'principle of the excluded middle is in its general formulation
a logically inadmissible fundamental concept. Q.e.d. [124]

Nonetheless I believe that it is almost always used in mathematics (set
theory included) in a manner that seems admissible to me. Namely one uses
it mostly in indirect proofs, when one has obtained a contradiction from assuming the negation of the statement to be proved, and then concludes that
the statement to be proved is correct. Because according to the 'principle of contradiction', the principle that both a statement and its negation
are true, is wrong; according to the 'principle of the excluded third' one
of both theorems, the statement or its negation, must be correct. If one
however also would want to doubt this application of 'the principle of the
excluded third', then one should consider every indirect proof as an inadmissible justification, which would go too far in my opinion. In other words,
insofar as one uses the 'principle of the excluded third' *only* for statements
of which one knows or can prove (as in the indirect proof by means of the
'principle of contradiction') that they are *not* logically independent theorems with respect to the valid assumptions, I consider its application is fully
legal.

Sincerely yours [125]

Walter Dubislav

[Signed typescript – in Brouwer]

---

[124] Dubislav erroneously identifies the principle of the excluded middle with completeness of theories.   [125] *In vorzüglicher Hochachtung bin ich.*

To P.S. Urysohn — 22.I.1924   Laren

Dear Mr. Urysohn [Sehr geehrter Herr Urysohn]

I received your card of 27.XII.23 all right. With the same mail I dispatch a couple of envelopes with reprints to you and some more will follow.

After my return from Marburg the objection you made there, became immediately clear to me on checking my private copy of the article 'On the natural notion of dimension', [126] where I have on p. 147 an old marginal note at l. 17–20, which says at this place *make it agree with p. 150 at **'. This was the marginal note to which my card from Zandvoort referred.

Coming back now a bit more in detail to the topic, I remark first of all that in my topological articles that appeared in 1908—1914 the expression: 'the domain set g is determined by the closed set α' says exactly the same as 'the domain set g is bounded by α' (c.f. e.g. Mathematische Annalen 69, p. 170, where this is explicitly stated). Consequently the quote on p. 150, "*by π_2 in π_1 determined domain set g_1 bordering on the edge E_1E_2*" on p. 150 at *), can in connection with the text have no other meaning than that of the intersection of a domain set γ_1 that is already available in π_1 determined by π_2, bordering E_1E_2, however not bordering $E_1E_3\ldots E_{n+1}$, with τ_1, so that the existence of the latter domain set γ_1 is postulated by the concept of 'separation of ρ_1 and ρ_1' in π_1 by π_2'. Hence the considerations of the article are actually based upon a separation definition, according to which *ρ and ρ' in π are separated by π_1, only if π_1 determines in π a domain set that contains ρ but not ρ'*. The definition that you indicate in your letter of 24.X.1923 says the same thing in another form. As far as the origin of the oversight on p. 147 is concerned, my notes of that time make it probable that the manuscript of the article originally didn't contain an explicit separation definition, just as in my article that appeared in Annalen 71: 'Proof of the invariance of domain', and that such a definition only much later has been inserted rather thoughtlessly, after a reader of the page proofs pointed out the absence. When not long after the article appeared, the oversight became clear, a quick correction was not forthcoming, because I expected that the article mentioned on p. 151 of the above mentioned paper on the same subject, promised by Lebesgue, would appear soon, and I was convinced that this article would force me to make a rejoinder, in

---

[126] *Ueber den natürlichen Dimensionsbegriff.*

which I could naturally include the necessary rectification as an addendum. When subsequently, the promised article of Lebesgue failed year after year to appear, the whole matter vanished gradually from my mind, and without your interpellation I would maybe never have thought of it again.

Now I have on the occasion of your remarks also studied the published explanations of Lebesgue, that came out with a delay of ten years (and not as agreed in the Bulletin de la Société Mathématique but in Vol. II. of Fundamenta Mathematicae, [127]) and I have seen that these, just as I expected ten years ago, make a contra-publication necessary, for indeed, the proof of Lebesgue of the lemma formulated on p. 150 of 'On the natural notion of dimension', is merely a abbreviated form of my proof of the same theorem. I hope this rejoinder will appear soon. [128] It will at the same time (while mentioning your priority) provide the correction of my old oversight.

I would be very grateful for the promised copies of your Comptes Rendus Notes, and also for more information about your yet unpublished investigations. To be sure, my own researches are since some years of a different orientation, but my interest in topology has remained, and I consider you as one of the few that really can open new perspectives here.

With best greetings

Yours truly
L.E.J. Brouwer [129]

[Signed autograph – in Alexandrov]

1924-03-12

From K. Menger — 12.III.1924   Vienna

Dear Professor, [Hochverehrter Herr Professor]

Thank you very much for the kind dispatch of your article about the natural dimension concept. [130] When I tried in 1921 to define curves and the dimension concept, I was in the first year of my university study and didn't know your article in the Journal für die reine und angewandte Mathematik 142 [131] at all, in which the definition is essentially anticipated. But also

---

[127] [Lebesgue 1921].   [128] Published as [Brouwer 1924a].   [129] *Ihr sehr ergebener.*
[130] [Brouwer 1923a].   [131] [Brouwer 1913A]

later, after I found the publication when studying the relevant literature, I
hoped that I could offer through my results at least a small supplement. For
I have investigated the structure of n-dimensional sets, and I have proved as
a supplement of the theorem: Every open set of R_n is n-dimensional — the
following theorem: Every n-dimensional set of R_n contains an open part. I
hope to be able to send you in the course of this year in printed form the
second part of my article, which I had prepared already long ago.

Please accept, my dearest professor, my expression of my particular admiration and affection.

Sincerely yours

Karl Menger

[Signed autograph – in Brouwer]

•24-03-25a

To G. Mannoury — 25.III.1924^a Laren $^{\langle 132 \rangle}$

For the undersigned significs does not consist so much of practicing
language criticism, but rather of:

1°. tracking down affect elements, into which the cause and effect of
words can be analyzed. By this analysis the affects that relate to
human relations are brought closer to control by conscience.

2°. the creation of a new vocabulary which also for the spiritual life
tendencies of people opens access to their thoughtful exchanges of
ideas and hence to their social organization.

For the realization of the part of the program mentioned under 1°,
cooperation is necessary: for countless affect complexes can not be
analysed unless by the catalytic action of philosophical discussion between unlike-minded.

Also with regard to the creative work meant under 2°, I have believed
for a long time in the great importance of cooperation, here between
like-minded. But I have come more and more to the opinion that this

---

$^{\langle 132 \rangle}$ This letter contains Brouwer's personal statement, which was published as part of
the *Beginselverklaring* (declaration of principles) of the Signific Circle. Published much
later in the '*Signifische Dialogen*', [Brouwer 1937], see also [Schmitz 1990] p. 423.

higher task of significs can only be accomplished by the utmost concentration of the mind of the single individual.

L.E.J. Brouwer

Dear Gerrit,

Above a new version. In the old form it was really too silly. I am glad that your warning has stopped this in time.
Cordial greetings

your
Bertus

[Signed autograph – in Mannoury]

1924-04-06

To K. Menger — 6.IV.1924   **Laren**

Dear Mr. Menger [Sehr geehrter Herr Menger]

Many thanks for your letter of March 12. I am glad that you too have noticed that the definitions we both give for an *n-dimensional continuum* are equivalent, and indeed essentially because for a bounding [set] B of a neighborhood $U(A)$ of a closed set A, finitely many points $P_1, \ldots, P_r$ of A can be given with neighborhoods $U(P_1), \ldots, U(P_r)$, such that B is contained in the union of the bounding sets of the $U(P_\nu)$.

On the other hand we both assign different meanings to the statement: 'the continuum K is n-dimensional *in the point P*', as you certainly will have seen.

Why do you embed the sets M considered by you in metric spaces, instead of considering these sets exclusively as Fréchet normal sets *by themselves*? In the latter case a neighborhood of a point P in M becomes simply a 'domain set' that contains the point P; cf. the definition of domain set in footnote [16]) of the reprint I sent you recently.

As far as the rectification of Crelle volume 142 in footnote [11]) of this reprint is concerned, you probably will have noticed already that the Crelle text can also be put right by deleting the word 'closed' on p. 142 l. 18. But I have preferred a formally thorough change of definition, in the interest of better readability of the new text. [133]

That the rectification of this oversight, which was discovered already in 1913, has been postponed so long, is because in order to come back on the matter, I wanted to wait for the article of Lebesgue mentioned in footnote [19]), and this article was postponed for 10 years.

I am curious to see the proof of your theorem that every open set of R_n is n-dimensional, and even more the results concerning the set theoretic characterization of the topological images of intervals of the R_n that you envisage. In case you find a fast publication of your proof in the interest of your priority, I am quite happy to submit it to the Amsterdam Academy. And for an extensive exposition I would be pleased to put the Mathematische Annalen at your disposal.

Please greet Prof. Hahn from me.

With the best wishes for further success of your investigations, I remain

Yours truly [134]
L.E.J. Brouwer

Am I correctly informed that Prof. W. Gross [135] is no longer alive? And do you know perhaps when he died?
L.E.J.B.

[Signed autograph – in Menger]

24-04-09

To P.S. Urysohn — 9.IV.1924   **Laren**

Dear Mr. Urysohn, [Sehr geehrter Herr Urysohn]

I received your letter of March 20 all right, and also the batch of reprints of yours and Mr. Alexandroff. Unfortunately I must conclude from your

---

[133] [Brouwer 1913d], [Brouwer 1923a].   [134] *Ihr ergebenster.*   [135] Wilhelm Gross, 1886–1918.

letter that of the seven envelopes with reprints, two have been lost. Also a card for which I asked you in the beginning of March to confirm the receipt of my letter of January 22, seems not to have arrived.

The word 'usual' in footnote [11]) of the new version of 'On the natural notion of dimension' was indeed inappropriate, because in mathematical treatises every statement with a subjective or unprovable character must be inadmissible. Consequently I have in turn omitted this qualification in the enclosed communication that appears in Crelle's journal [136]. For the same reason I have also left unmentioned any earlier disclosure of the oversight shortly after the article appeared, by myself and by others, because at the moment I do not have any documents about them in my possession.

Meanwhile I have found in my copy of the book by Hausdorff [137] in the margin of p. 458, § 7 yet another note, according to which it is absolutely necessary to strike the word '*abgeschlossen*' [138] at the place concerned in Crelle 142 p. 147, l. 18, precisely because of the examples Hausdorff gives there. [139]

This deletion produces exactly the separation definition given by Hausdorff on p. 334 of his book. In the new version of 'On the natural notion of dimension', [140] the now published change of definition, which formally goes deeper, decidedly is to be preferred, in view of readability and coherence.

I have read with great interest the theories that you communicate in your last letter. I hope that you will obtain along these lines the axiomatic characterization of the Cartesian n-dimensional spaces among the Cantorian n-dimensional manifolds; I think you are the right man for that.

I will be happy to submit the results of you and those of Mr. Alexandroff that are connected with my article 'On linear inner limiting sets' (which however does not deal with ordinals, but with 'uniform' topological homeomorphy and homogeneity, that preserve their meaning for $n > 1$) to the Amsterdam Academy. I would ask you to write the text either in German or English, and if possible adhere in the formulation to the terminology that I introduced in my article 'Some remarks on the coherence type η' (Amsterdam Proceedings 1913 [141]).

I hope that the reprint of this article that I recently sent you, reaches you and that it does not happen to be in one of the lost envelopes.

---

[136] [Brouwer 1924b].   [137] Unfortunately this book, together with Brouwer's complete library, was sold not long after his death. The whereabouts of the collection has not been discovered.   [138] Closed.   [139] [Brouwer's note on top of page:] *Note not belonging to the letter.* Namely, as connection between $P(0,0)$ and $Q(\pi^{-1}, 0)$ B winds itself inside of B' through boundaries that lie inside B', through which no continuum connecting P and Q can wind.   [140] *Über den natürlichen Dimensionsbegriff.*   [141] [Brouwer 1913c].

Unfortunately, because of many kinds of obstacles, the manuscript of my third communication about 'The theory of the finite continuous groups' is still waiting in a drawer for the 'finishing touch', [142] which however I hope to be able to give it in a not too far future.

I have forwarded your information with respect to the Revue Semestrielle and Matem. Sbornik to the chief editor of the former. I assume that he will shortly write to you in person.

I hope to meet you and Mr. Alexandroff in September at the congress in Innsbruck. [143] We might also meet somewhat earlier, in case you would be in Western Europe during the coming summer. Recently a colleague here (Prof. Van der Hoeve from Leiden) talked about you both: I believe he had been together with you last summer in Norway.

With warmest greetings for you and Mr. Alexandroff

Your
L.E.J. Brouwer

[Signed autograph, draft – in Alexandrov]

24-06-14b

To P.S. Urysohn [144] — 14.VI.1924[b]   Amsterdam

Dear Mr. Urysohn, [Sehr geehrter Herr Urysohn]

Maybe the enclosed variant on the passage in Crelle's Journal 142 [145] between p. 149 l. 2 from below and p. 150 l. 10 from below, by means of which the proof is adapted to the separation definition on p. 147 as now printed (hence without the erasure of the word '*abgeschlossen*' [146] which was needed for the old version of the proof). [147]

(In the accompanying text an 'η-chain' means a finite point sequence in which each two consecutive points have a distance $\leq \eta$.)

---

[142] In the original German text the English expression is used.   [143] Annual meeting of the DMV.   [144] Pavel (Paul) Samuilovich Urysohn.   [145] [Brouwer 1913d].   [146] closed.
[147] This is the notorious *slip of the pen*; by unintentionally adding the adjective 'closed', the definition of 'separation' became too weak. Cf. [Brouwer 1976] p. 541, 547 ff., [Van Dalen 2005] p. 452 ff. Brouwer observes here that even with the unintended separation of the paper a coherent dimension notion arises.

This variant, which I recently found back among my papers from the years 1912 and 1914, has most probably been communicated at that time in correspondence about dimension with Schoenflies, Gross, and others. I will investigate whether maybe the other parties have preserved their correspondence more carefully than I have. My own interests have been diverted for nine years from these subjects, and unfortunately I have always failed as archivist.

Meanwhile, I consider, as before, the separation definition without the word 'closed' more appropriate and productive from the viewpoint of dimension theory.

I am curious to see your article for the Amsterdam communications;[148] likewise the promised communications of Mr. Alexandroff (whose address still is unknown to me).

With warmest greetings

your
L.E.J. Brouwer

[Signed autograph, draft/copy – in Alexandrov; 1 enclosure, not extant.]

1924-06-21

From P.S. Urysohn — 21.VI.1924   **Göttingen**
                                              Hospitalstrasze 1b (bei Assmann)

Dear Professor, [Hochgeehrter Herr Professor]

Last year I have sent at the end of July (i.e. two months *before* the Marburg meeting) a note for the Mathematische Annalen to Mr. Hilbert in which I criticized your dimension notion. I had long since forgotten about this note, when I suddenly received the proofs the day before yesterday. It is not at all clear to me what I should do with it. Maybe you are satisfied with the 'Added in proof'[149] which I have written. Hence I allow myself to send you these proofs and ask you respectfully to inform me whether you agree with the present version, or what changes you deem necessary, or what else?

---

[148] *KNAW, Proceedings.*   [149] See below.

If it is not too much trouble, I would like to ask you to answer me as soon as possible, because the enclosed proof [150] is the one that should go to Mr. Blumenthal; moreover, I stay only for 19 more days in Göttingen.

Finally I must thank you for a dispatch of reprints. With warmest greetings

Sincerely [151]

Paul Urysohn

Added in proof

In the statements above I naturally have based myself on the assumption that one remains within the definition of the notion of dimension in Vol. 142 of Crelle's journal. But since then Mr. Brouwer has published a *rectification*, [13] where he in fact changes the definition of *separation* that is at the basis of the dimension concept. *Thereby* the proof is completely correct and I might emphasize that, as I have learned, the necessity of such a change was already known for a long time to Mr. Brouwer and it remained unintentionally unpublished so far All the same I believe that the above lines may have some use, because Mr. Brouwer did not indicate in his rectification *why* the old definition should be rejected.

Göttingen June 21, 1924.

Register. Deliver by express.

[Signed autograph – in Alexandrov]

Editorial supplement

[*On backside of the envelope in the handwriting of Cor Jongejan:*]

Hello Dad,

This just arrived by mail. The matter is urgent, so I send it on. This evening I sleep here again to check your mail. Tomorrow I sleep on

---

[13]Crelle **153**; the improved text has also appeared in the Proceedings Akad. Amsterdam **26**, p. 795.

[150] In the Brouwer Archive.   [151] *hochachtungsvoll.*

the Overtoom, [152] because little Moek [153] has to leave so early. Bye,
bye. Good luck with your exam period.

Corus [154]

1924-06-24

To P.S. Urysohn — 24.VI.1924   **Bergen aan Zee**

Dear Mr. Urysohn, [Sehr geehrter Herr Urysohn]

Many thanks for sending me the proofs of your forgotten small Annalen
note and for asking my advice about it. I am of the opinion that in both
our interests the publication of this note should absolutely be omitted. For,
publication by scholar A of an oversight that escaped author B is only then
compatible with the dignity of scholars, when either the oversight can be
understood only by an elaborate exposition of new discoveries by A, or
when consultations between the parties concerned has become materially
impossible (e.g. for political reasons or because of the death of B). In any
other case such a publication raises the suspicion that either A has been
carried away by impetuous ambition, and, maybe on purpose, wants to
insult B, or that B refused to acknowledge his oversight to A, alternatively
refuses public acknowledgment, at least to full extent. Fortunately neither of
the mentioned circumstances applies in this case, but rather, in all respects,
the opposite.

For the rest I agree with you that it could be useful when the counterexample you put forward would be brought to the attention of the public. As
a matter of fact, I must after all come back to the matter myself, in order to
show along the lines of my old correspondence with Schoenflies and Gross,
how the proof of the dimension theorem can be put in order also on the basis
of the erroneous separation condition from the year 1913 (more about that
was communicated to you in my letter that I sent to Moscow ten to twelve
days ago). With the publication of this proof I will have a good opportunity to insert the counterexample concerned here (naturally mentioning its
paternity). [155]

---

[152] over the pharmacy.   [153] a private pet name for Lize Brouwer-de Holl; one might
translate it as 'little Mom'.   [154] Nickname for Cor.   [155] See [Brouwer 1924a].

I believe this is the only dignified way of dealing with the matter, and I hope you agree with it. The subsequent suppression of your small Annalen note will cause no problem: as one of the editors of the Annalen I will arrange that with the editorial board and the publisher. Without your message to the contrary the affair will be settled in this manner.

I am looking forward to a, as I hope, reunion before long (you have received my card sent to Göttingen?) and with warmest greetings also to Mr. Alexandroff

Your
L.E.J. Brouwer

On the 27th or 28th of this month I will be in Laren again.

[Signed autograph, copy – in Alexandrov]

·24-06-27b

P.S. Urysohn to W. Sierpiński — 27.VI.1924^b   Göttingen $^{(156)}$

Dear Wacław Constantinovicz, [Hochverehrter Waclaw Constantino-
vicz]

Analyzing my sketch, I have found out that in the introduction of my *Mémoire sur les multiplicités Cantoriennes* $^{(157)}$ I have inserted a remark about the '*Natürliche Dimensionsbegriff*' of Brouwer, in which I have written roughly the following: 'Now the proof of this theorem contains an error that seems incorrigible to me.' Although this remark is justified with respect to the earlier formulation of Brouwer (in his rectification he changes the *definition* of the notion of dimension), it seems to me that after the publication of this rectification it is not appropriate that my remark appears in print. So allow me to beg you urgently to modify the criticism; If possible, to replace it by the by the one given below; if this is perhaps impossible for technical reasons, then at least delete it.

$^{(156)}$In the Menger archive there is a number of translations of letters of Urysohn, Brouwer, and Alexandrov in an unknown handwriting. They were based on documents in the possession of Sierpiński. The originals are presumably not extant. $^{(157)}$Memoir on the Cantorian manifolds.

Here is the text of the desired remark: the memoir was already finished when I learned about the article '*Über den natürlichen Dimensionsbegriff* published by Mr. Brouwer in 1913 in the Journal für die [reine und angewandte] Mathematik (volume 142, p. 146). I hope to come back at another occasion to the definition of Mr. Brouwer and mine.' [158]

Paul Sergiewicz [159] sends his greetings to you and simultaneously sends reprints of his latest article to you, to the editorial board of the Fundamenta and to Mr. Rajchmann.

Sincerely [160]

Paul Urysohn

P.S. Just now the three reprints sent by you have arrived, for which the both of us thank you very much. We stay in Göttingen until July 9, (address Hospitalstrasse 1b with Assmann), and then we go for a few days to Bonn (address poste restante) to Hausdorff, and afterwards probably to Paris.

P.U.

[Handwritten translation – in Menger]

1924-07-09b

To P. Zeeman — 9.VII.1924^b   **Laren**

Dear Colleague [Waarde Collega]

May I ask your assistance for just a moment regarding the enclosed letter? The permission to live outside Amsterdam was at the time one of my conditions to reject the call to Leyden. About this matter there has been correspondence in the summer of 1915 between you as chairman of the faculty and the Mayor of Amsterdam [161], and then you promised me that you would keep the letter of Mayor Tellegen to you in which the pertinent permission was granted. Would you maybe willing to lend it to

---

[158] This text was adopted in the published version, followed by a reference to Menger's work. [159] Alexandrov. [160] *In aufrichtiger Hochachtung Ihr Ergebener.* [161] Tellegen.

me for a short time, so I can use it to plead for my rights with Mayor and Aldermen?

Kindly thanking you in advance and with many greetings

Your
L.E.J. Brouwer

[Signed autograph – in Zeeman]

●24-07-29

From P.S. Urysohn, P.S. Alexandrov — 29.VII.1924   Le Batz

Dear Professor, [Hochgeehrter und lieber Herr Professor]

Only today we finally got around to writing. [162] In Paris we have been walking around every day from 9 o'clock in the morning until 10 o'clock at night [14] — because apart from the city and the museums there was also a police headquarter which made problems for us, and the German consulate, where we asked for a transit visa for the return journey, and so on. After four days we had become so tired, [15] that we decided to postpone the continuation [16] of Paris to the return trip (Urysohn), respectively eternity (Alexandroff). The day before yesterday we arrived here, and it took us a whole day until we could find a quiet place on the coast.

In the same cover you will find our curricula vitae, as well as a letter to you, which should count as the official statement of our wish to come to Amsterdam. [163]

As far as mathematics is concerned we have, naturally, as yet little news. By the way, Urysohn has found a space which not only in the topological sense (like Hilbert space) but also in the metric sense, may be considered the largest metric space with a countable everywhere dense subset. More precisely formulated: there exists a metric space with a countable dense subset which contains for every other metric space with a countable dense subset

---

[14]with the greatest torment: Paris is more horrible than I ever thought.   [15]and Alexandroff has cursed so much and has become so unbearable.   [16]four days

[162]Alexandrov and Urysohn had visited Brouwer in Laren in the middle of July. As Brouwer had to go to Göttingen, the two moved on to France.   [163]Alexandroff and Urysohn – Rockefeller grant.

a subset that is congruent (= admitting a distance preserving mapping) to the latter.— There are actually several 'universal' spaces of this kind, but only one that satisfies certain homogeneity conditions.

We would like to express once more our warmest thanks for the extraordinarily friendly reception, that we found at your place, and we thank also both ladies, whom we caused so much trouble.

Moreover we apologize once more because of the alarm clock. Please write to us whether you didn't forget to take anything with you to Göttingen because of that. In any case, do write something to us about the trip to Göttingen and your stay there; every detail [17] [164] will interest us.

With best greetings to you and both the ladies.

Most cordially yours, [165]
Paul Urysohn
Paul Alexandroff

Our address is (until 25 VIII):
Le Batz (Loire Inférieure), Pension de famille 'Le Val Renaud'

[Signed autograph – in Alexandrov]

1924-08-21

To H. Kneser — 21.VIII.1924   Bad Harzburg
                                                                   Krodotal 4

Dear Mr. Kneser [Lieber Herr Kneser]

You probably will also have received the crushing message that Urysohn has drowned in France while bathing. [166] It is an incredible blow of fate. Alexandroff will probably arrive the day after tomorrow in Göttingen. Should you or Miss Noether learn about the hour of his arrival, would you please

---

[17] I totally reject the responsibility for the use of this word in German, and also for its gender. I call Urysohn nowadays 'Baberuschka', which always makes him mad. Please explain to him that the Russian sense of the word fits him perfectly!
Furthermore Alexandroff develops language- and other theories that differ only from the ones of Denjoy by sign.

[164] Urysohn used in the letter the word 'Detail' with neutral gender.   [165] *Ihre herzlichst ergebene.*   [166] August 17.

inform me immediately, if necessary telegraphically. I, on my part, part will do the same for you.

Your
L.E.J. Brouwer.

[Signed autograph – in Kneser]

24-08-31

To P.S. Alexandrov — 31.VIII.1924   Laren [167]

Dear Alexandroff,

I have received both your letters, and I am all the time in my thoughts with you. Yet I would not pray, in accordance with your statement, that you will not have a long life. In the first place, because we may not pray on account of objective events, but only for the sake of clarification of our consciousness of duty and for the sake of bearing the trials that are imposed on us.

In the second place, because our existence on earth has been granted to us exclusively for purification of our soul of the original sins of fear and desire, and it is only according to the fulfillment of this goal that the life span of the righteous man is measured.

Just for that reason the death of a righteous man has for himself always the character of a satisfaction, a liberation and a redemption, and we must continue to bring him after his death just our love, and not our pity (compassion), in particular not when his death passage has been light.

And for those who are left behind in mourning the following holds: every grief for the heart that suffers it has it its purifying meaning, and in the days of grief it is often easier than in the days of joy, to become aware of the proximity of God, because the grief – to be endured in tranquility – forces to dematerialization.

May this also happen to you.

---

[167] Addressed: Twerskajastr. Pimenowski pereulok 8kb5, Moskau.

For our coming meeting here in the autumn I will also get the necessary things arranged following your indications.

In faithful friendship

your
L.E.J. Brouwer

[Signed autograph – in Alexandrov]

24-09-07

From Mrs. C. Alexandrov — 7.IX.1924   **Smolensk**

Dear Professor, [Sehr verehrter Herr Professor]

In that happy time, when both of my boys [168] were still together, I blessed you for your hospitality and for your touching relation to them. Now in my suffering you have understood with your sensitive heart, that only the certainty that you are with my poor lonely son can sooth me.

The other one, a rarely gifted, happy, innocent child, who had never experienced suffering, was taken by his heavenly Father, to relieve him from all earthly worries, that he would have had sooner or later in order to pay for his cloudless happiness. But the remaining one has gone through many sorrows notwithstanding his youth, and now he is completely broken by this last heavy blow.

Words do not suffice, dearest professor, [169] to express my appreciation for your great compassion and your warm sympathy.

In profound gratitude I shake your hand and wish you the best in life. I send the warmest greetings for your dear family members.

Sincerely yours [170]

Your devoted
C. Alexandroff

---

[168] P.S. Alexandrov and P.S. Urysohn.   [169] *verehrter Herr Professor.*   [170] *Mit Hochachtung verbleibe ich — Ihre ergebene.*

With true fear in my heart I await the reunion with my son. I know I will not be able to console him; he is straying for a long time to come, maybe forever, from the path in life that they went together!

[Signed autograph – in Alexandrov]

––––––––––

24-10-13

To P.S. Alexandrov — 13.X.1924   **Laren**

My dear friend, [Mein Lieber Freund]

I just received your letter of October 5, with your so beneficent, faithful sympathy. My well-being in bed leaves nothing more to be desired and my recovery is making steady progress, only not particularly quickly, and moreover the doctor made it clear that I have to be careful for a long time to come and that I must take care of my health. I have in fact not been able to get a diagnosis with a scientific of a disease name out of him; he only spoke about 'influenza with complications'. With the return of well-being also came giving up copious amounts of sputum, which still persists, but every day in smaller amounts.

From America they further ask me how much your trip from Moscow to Amsterdam will cost. Please inform me about this by returning mail, and make your calculation for a *comfortable* trip.

Sierpiński answers me that the introduction and Chapters I and II of Paul's [171] Mémoire will appear in volume VII of the Fundamenta, that he is willing to include the whole remainder (i.e. Chap. III–VI) in volume VIII, and that this volume VIII probably will appear in the fall of 1925. Maybe he expresses himself a little too optimistically, but anyway I am of the opinion that we should take no steps for the time being with Sierpiński, and at least leave the matter for volume VII as now planned. We can discuss the rest, if necessary, here.

Meanwhile its seems that we unfortunately have to take into account the possibility that Kuratowski already has on his own authority declared the introduction of Paul's Mémoire ready for printing (although on the other hand, such an act without the authorization of Paul's heirs should appear incomprehensible to me) and that hence the footnote 3) which was criticized

––––––––––

[171] P.S. Urysohn.

by me [172] on the ninth proof page cannot be changed anymore. For the rest the corrected proofs don't look ready for printing at all: they still contain many annoying printing errors.

What you write about the depersonalization of your life, are words to my heart. The new path that you will now follow, will actually connect you ever more with eternity, and the awareness of the 're-connection' (= religio) with eternity will bring you ultimately joy and happiness (even though of a quite different, very quiet and pure sacred 'ultraviolet' kind). I hope with all my heart that you have made the right decision in relation to your wife, in fact I assume you did, because you have reached the decision only after consulting with your mother. [173] Unfortunately I have lost the address of your mother, would you please give it me once more?

Preserve your inner peace; my thoughts are with you and I greet you most warmly.

Your Brouwer.

[Signed autograph – in Alexandrov]

1924-10-20b

To P.S. Alexandrov [174] **— 20.X.1924**^{b} **Laren**

My dear friend [Mein Lieber Freund]

I am happy that I can tell you that I am going out again, and that I hope to take up my lectures again next Friday. It seems to me as if I have come back into the light from a dark abyss. Whether I will recover without being permanently affected, the doctor cannot say for certain yet, but he is accordingly [175] on that point rather confident.

I hope to be able to write soon to your mother, I am just waiting for her address. Give the family Urysohn many greetings, and think of me as I do of you.

---

[172] This is the footnote in which Urysohn refers to Brouwer's dimension paper of 1913. It was revised more than once. See [Van Dalen 2005] section 15.5.   [173] Refers to Alexandrov's divorce.   [174] Adressed: Prof.Dr. Paul Alexandroff, Twerskaja str., Staropimenowski pereulok 8kb5, Moskau.   [175] The text is rather enigmatic here; 'nonetheless' would fit better. A slip of the pen?

I look at you and shake your hand.

Your
Brouwer

[Signed autograph, postcard – in Alexandrov]

24-10-21

To H. Kneser — 21.X.1924   **Laren**

Dear Mr. Kneser [Lieber Herr Kneser]

You probably have received the proofs of your article *Ein topologischer Zerlegungssatz.* [176] Unfortunately I didn't succeed in arranging that you will be sent free of charge more than the statutory number of 25 offprints. If you want more, then I advise you to write beforehand a line on the proofs, in which you inquire about the price, and then let me know the answer given to you, together with your view on the appropriateness of that price, so that I can, if necessary, complain about it.

In a few days a student of mine (or rather of Weitzenböck) will come to Göttingen for the winter semester. He is called Van der Waerden; he is very bright and has published something already (in particular about the theory of invariants). I don't know whether at the moment the required formalities for a foreigner who wants to register as a student are difficult; in any case it would be most valuable for Van der Waerden if he would find there some help and guidance. May he perhaps call on you one day to talk things over? Many thanks in advance.

With best greetings

Your
Brouwer

[Signed autograph – in Kneser]

---

[176] a topological decomposition law [Kneser 1924].

1924-11-02

To K. Menger — 2.XI.1924   **Laren**

Dear Mr. Menger, [Sehr geehrter Herr Menger]

Enclosed the revision of your note for the Amsterdam Proceedings, [177] which I ask you to return *to me*, if possible ready for printing. I call your attention to both changes (on p. 2 and p. 5) underlined in pencil on the old proofs, because they originate with me and are of a factual nature: I hope you will agree with them. Furthermore I also enclose for your information the proofs of a note by me on the same subject, with the request to return them.

I welcome your plan to come to here after the end of the winter semester. The Easter holidays in Holland are brief, about three to four weeks in April, whereas the summer vacation starts here earlier than in Germany and Austria.

I would be most grateful if in relation with the slip of the pen of my minor oversight in Crelle 142, [178] you would replace in the manuscript of your article submitted to the Monatshefte: 'On the dimensions of point sets. Part two' [179] on p. 6 in the footnote the words:

'given in a but little known brief article (Crelle Journal 142, p. 146–152) a definition of n-dimensional continua, which after a correction of a clerical error in the (Amsterdam Academy Proceedings XXVI, 1923 [180]) is equivalent with our definition of the n-dimensional continuum.'

by the words

'in a but little known brief article (Journal für [die reine und angewandte] Mathematik 142, p. 146–152; cf. also the correction of a clerical error in there in the Amsterdam Proceedings 26, p. 796), which is equivalent with our definition of an n-dimensional continuum.'

I believe that in this way the reader gets an idea that does more justice to the facts. It is also better to mention the 'Amsterdam Proceedings', rather than of the 'Amsterdam Reports', [181] because the latter usually does not refer to the 'Proceedings', but to the 'Verslagen'.

---

[177] [Menger 1924a].   [178] [Brouwer 1913d].   [179] *Über die Dimension von Punktmengen. Zweiter Teil*, [Menger 1924b].   [180] [Brouwer 1923a].   [181] *Amsterdamer Berichte*.

With best greetings, and in the hope soon to get to know you personally,

Your [182]
L.E.J. Brouwer

[Signed autograph – in Menger]

24-11-06

From C. Carathéodory — 6.XI.1924 [183]

Blumenthal asks me to answer your last letter in which the passage about
Painlevé is contained.

Basically, I completely share your view, but I wonder whether one should
forever recall the nonsense that in all countries has been put together during
the war; because then one would not have had to stop the shooting at all.

Especially where the Riemann volume is concerned, [184] in my opinion
it isn't really necessary to have French mathematicians there too. But if one
so wishes, there is no other way to do this in a decent way, then by turning to Painlevé in the very first place. For Painlevé is among the French
mathematicians the only one who holds a sufficiently secure position to take
part in the Riemann volume, without running the risk that the whole pack
of the narrow minded [185] starts barking at him. Moreover, through Nernst
I know that while he was rector, Painlevé had offered to give a few talks
at the University of Berlin, and that – in spite of the fact that the Foreign
Ministry was interested in the case – this came to nothing because of the
opposition of a few Berlin professors. So you see, that at least according
to this report, Painlevé seems to have forgotten the words that you hold
against him.

[Typescript – in Einstein [186]]

---

[182] *Ihr sehr ergebener.*   [183] This letter was, according to *Carathéodory to Blumenthal
20.I.1925*, not received by Brouwer.   [184] Mathematische Annalen 97, 1927. For the
conflict see [Van Dalen 2005] section 13.3.   [185] *Banausen.*   [186] Fragment in collection
of letters re Riemann volume.

1924-11-13

From K. Menger — 13.XI.1924   **Vienna**

Fuchsthallergasse 2, Wien IX

Dear Professor, [Hochverehrter Herr Professor]

After the settlement of a protracted railway strike that endangered the mail service abroad, sending this letter is my first priority. Dear Professor, I cannot thank you enough for the attention you have paid to my small note, also for your kind letter and for sending me your article which I already had read with great interest in the 'Verslagen' of April 28, 1924. [187] Although one would at first expect that the N- and the MU-dimension [188] would coincide for non-condensed species, this is generally *not* the case.— Basing dimension theory on separation definition b) would certainly be very interesting — both theories [189] would perhaps in a certain respect relate to each other like the theory of simple curves to that of irreducible continua.

In the duplicated [190] text of my article about dimension (II) [191] I have made some small changes before the printing. I have inserted a definition of dimension of sets that are considered in themselves. In particular I have improved in wording the awkward formulation of the footnote on page 6 along the lines that you, dear Professor, suggested, already before receiving your letter.

In the last few days I had to give a talk about research in the foundations of mathematics in a privatissimum of the epistemologist Prof. Schlick. It may have been the first time that an extensive exposition of intuitionism has been given in Vienna. The lecture was followed by a long discussion.

It would be a great joy for me to receive in a few months time instruction from your lectures, dear Professor, about these fundamental questions that

---

[187] [Brouwer 1924c].   [188] N: Natürliche, i.e. Brouwer (separation) Dimension; MU: Menger-Urysohn Dimension.   [189] Brouwer and Urysohn based their definition of dimension on separation; Menger's dimension definition made use of boundaries. All definitions were inductive. The Menger-Urysohn dimension was a *local* one, in the sense that dimension was considered in points; Brouwer's dimension was *global* in the sense that it concerned the dimension of the *whole* space. See [Menger 1928b], [Urysohn 1925, Urysohn 1926]. A new and surprising fact on the relation between Brouwer's dimension and the Menger-Urysohn dimension can be found in [Fedorchuk and Van Mill 2000].   [190] in text 'opalographiert'.   [191] Published as [Menger 1924b].

are very close to my heart. Meanwhile, please receive the expression of my deepest reverence and gratitude.

Your [192]

Karl Menger

[Signed autograph – in Brouwer]

24-12-21

To P.S. Alexandrov — 21.XII.1924   Laren

Dear chap [Lieber Kerl]

Today it furthermore occurred to me that for the fortification of our point of view (indeed, it is a real war, our point of view plays the role of a fortified encampment, in the wall of which a breach has been shot as a consequence of your ill-fated letter to Sierpiński, which we must close with might and men), that it would also be most important for the fortification of our point of view to insert at the end of the first paragraph of page 2 of the outline to Kuratowski (i.e. after the words 'share mine with Mr. Brouwer') more or less the following:

'Moreover, while writing my unfortunate letter to Mr. Sierpiński, I knew already that Mr. Brouwer had asked to read the proofs as well; I was convinced that thereby the correction of the proofs would adequately take into account the exchange of ideas, that took place before the death of my friend, between him and Mr. Brouwer. And because, moreover, I have agreed with Mr. Brouwer that he would not take any important decision without consulting me, one could in fact hardly dispute the necessity that I receive the proofs too.' [193]

Please forgive my insistence in this matter: perhaps I am making a nervous impression on you, but innerly I have the firm and calm conviction of the necessary actions, as well of my own helplessness without your strong support. For, the fact that the people in Warsaw don't bother even to the slightest degree about me, already follows from the fact that after I asked Sierpiński in September, while sending him the mandate of Paul's [194]

---

[192] *Ihr ganz ergebener.*   [193] This passage is in French.   [194] Urysohn.

father, to send the proofs *to you and to me*, Sierpiński, in reaction to your fateful letter, withheld the proofs, not only from you, but also from me, and this without any notification.

One more point is unclear to me: now both Sierpiński and Kuratowski write to you, as if there was from the outset never any plan to send out proofs of Paul's Mémoir, and as if Paul himself had agreed to that. How then was it possible that you received in Batz the proof of the first sheet of the Mémoire!

Should there come a definite refusal from Warsaw, then as last medicamentum heroicum [195] we have still this, that I withdraw the whole printed Mémoir in the name of Paul's heirs, who possess the literary property rights. Then the Fundamenta *may* not publish it, and the editorial board will, with a probability of 95 percent, back off from the ensuing complications, and at last conform to our wishes. Should the editors even then not give in, then we get the manuscript back, which will be printed again within a few weeks in Amsterdam, and in much better form than in the Fundamenta with its bad paper and the many printing errors. In that case the management of the Fundamenta will have a damage claim because of the wasted typesetting work, but I'll gladly bear that.

In connection with your pass and your residence permit Pannekoek has now written to Rutgers (a Dutch mountain engineer who has a high position in the service of the Russian government) and I myself to Varjas (professor of the red professorate, Ostoschenka 53, Moscow). Moreover, within a few days a letter will be sent from my faculty to yours.

Now, my dear boy, very soon more. If only you were just here! But for now, we meanwhile stand, distanced as we are, calmly and firmly side by side, in unflinching passive resistance!

With the warmest greetings, also from my family

Most cordially yours [196]
L.E.J. Brouwer

[Signed autograph – in Alexandrov]

---

[195] Kill or cure medicine.   [196] *Herzlichst Ihr.*

25-02-11

From K. Menger — 11.II.1925   **Vienna**

                                                          Fuchsthallergasse 2, Wien IX

Dear Professor, [Hochverehrter Herr Professor]

Please accept my sincere thanks for your kind card, which made me
very happy. Above all because I see from it that your health has improved
and that I may hope to attend in the spring your lectures, dear Professor.
And then, because of your great kindness to submit my article about curves
to the Amsterdam Academy and to consider the extensive article that I
enclose with this letter for the Mathematische Annalen. [197] I cannot thank
you enough, dear Professor, for the extraordinary support that you give to
my work.

Recently I have studied your articles on the foundations of mathematics
again, first of all 'Mathematics, Truth, Reality'. [198] Although I still need
ample instruction concerning your positive construction of mathematics,—
I feel the urge to tell you, dearest Professor, that your criticism of pure
existence statements in arithmetic has now convinced me. Theorems of
that kind are empty forms, which only can acquire a meaningful content by
constructive realization. That such a constructive realization would always
be possible,— for that no reason has been given until now, and when one
bases oneself on constructive foundations, may perhaps not be given at all.
One can at most *believe* in the possibility of such a completion, but then
the rigor of constructive argument has come to an end. Since all of this has
become clear to me, I look with deep admiration at your work, by which
you take hold of age-old prejudices by the root, and pursue them to their
far-reaching consequences.

Dear Professor, accept the expression of my greatest admiration and sincere thanks.

Sincerely yours [199]
Karl Menger

[Signed autograph – in Brouwer]

---

[197] [Menger 1925a, Menger 1925b].   [198] *Wiskunde,   Waarheid,   Werkelijkheid,*
[Brouwer 1919e].   [199] *Ihr ganz ergebener.*

1925-06-22b

From B. Kagan — 22.VI.1925[b]   **Odessa, Ukraina**
                                           Tschernomorskaja, 20

Dear colleague, [Sehr geehrter Herr Kollege]

With great joy I have obtained a couple of days ago a series of your
articles and the longer treatise on the foundations of science, which you
sent to me through my friends P. and T. Ehrenfest. Unfortunately they
have been written in a language hat I have no command of. But I hope to
conquer this obstacle, many a page I have mastered already. Soon the holidays will start, then I will have time enough to master the Dutch language
through your works. Probably only the first pages will offer serious problems. In any case, I will not give up this enterprise because the questions
to which your works are dedicated interest me highly; they were cultivated
in our school in Odessa for quite some time, as I believe not quite without success. The border areas between mathematics and logic pose very
great difficulties to a strict scientific treatment, which are mainly rooted in
logic.

From a number of references in the literature I have found that you have
published in 1920 a treatise about the law of the excluded third. This question was posed already several years ago by Professor S.O. Schatunowsky
here and, insofar as I can judge from the scant indications in your article
about the 'Set Theory' of Schönflies and Hahn, Schatunowsky's ideas in
essence hardly differ from yours. Prof. Schatunowsky has published a substantial treatise in Russian, 'Algebra as theory of congruences on functional
modules,[200] which has mainly (though not exclusively) the aim to develop
algebra while completely avoiding the law of the excluded middle, so also
while avoiding the theory of irrational numbers based on it. We had planned
to have this work published also in German, but the war and further events
have prevented that. So we were looking forward to the above mentioned
article with special interest, and we regret very much that we didn't get it.
Don't you have a copy of it? We would be most happy to have this article
in our library.

Both in my name and also in that of my colleagues I thank you most
warmly for the articles sent to us, and we politely request to make also your

---

[200] [Shatunovsky 1920].

further articles in this way accessible to us.

    Sincerely [201]

Ben. [202] Kagan

Professor in Odessa.

[Signed autograph – in Brouwer]

25-07-03

From K. Menger — 3.VII.1925   Semmering (Niederösterreich)

Kurhaus Semmering

Dear Professor, [Hochverehrter Herr Professor]

I thank you most warmly for your kind words of condolence for the terrible stroke of fate that has befallen me.

I cannot express in words what I have lost in my dear mother, her good-heartedness was boundless. And to the sorrow that she has been taken away from me is added the indescribably tormenting thought that she, who has since I was born, done and sacrificed so much for me, died right now, when finally a more peaceful evening of life had begun for her, to which she was looking forward to with great pleasure, still being able to enjoy it.

Deeply interested, she followed from a distance everything that concerned me, with gratitude in particular towards you, dearest Professor, [203] for all the favor and support you showed me. These tidings were her last joy.

An emptiness that cannot be filled has been struck in my life.

My mother was never ill, except for colds and in the last years occasionally a lumbar pain, which she thought was rheumatic. Now we know that this must have been the unobtrusive symptoms of an advancing nephritis. Because after a seemingly slight indisposition of two days she succumbed to a sudden kidney attack. The slight mental confusion that commonly in the last hours goes with this treacherous disease let her pass away without any inkling of her condition and without pain.

---

[201] *Hochachtungsvoll ergebensts.* [202] Benjamin Fedorovich [203] *verehrter und gütiger Herr Professor.*

Only after this terrible event the telegram was sent to me; I obtained it the same night in Heidelberg, thanks to the prudent forwarding from Laren, so that I could arrive the next day in Vienna. If the only relatives that I have at all, a sister of my dear mother and her husband, had not taken care of me, taken me in and sacrificed themselves to nurse me with the help of friends,— then I wouldn't know how I could have survived these days without going mad. Even so I laid down for a week, ill and half out of my mind. As of today I am in the Semmering, [204] where I must regain my strength through a rest-cure of several weeks in the open.

During this time I often thought of the poor Urysohn and I wished that I had perished in his place. Only the thought that I should not destroy what my beloved mother had built up with so much effort in her life, now gives me the will to regain my health, if possible, and then to achieve something.

I stop for now, dearest Professor, to write again to you as soon as I have gathered more strength. Meanwhile, rest assured of my sincere veneration and gratitude.

Yours devoted [205]
Karl Menger

P.S. I had written the enclosed letter [206] that evening, unaware yet of the events. In the confusion it got into my luggage, where I found it only today.

[Signed autograph – in Brouwer]

1925-07-08

To K. Menger — 8.VII.1925   **Laren**

My dear Mr. Menger, [Mein lieber Herr Menger]

I thank you for your letter of the third of this month, which gives me in any case the relief that you have withstood the stroke of fate which has struck you so suddenly. For that reason I have worried very much, because during your stay here I have felt strongly to what great extent the aura of your mother irradiated your life. So I surmised how great your loss was, and I expected the crisis that would be unleashed in you by the sudden emptiness

---

[204] a sanatorium.   [205] *Ihr ganz ergebener.*   [206] Not extant.

and the sudden necessity to assume a different spiritual way of inhaling. But after you have weathered the first crisis, I am certain that you will find the necessary concentration and religious dedication to work your way through, and that the certainty about the wish in that direction of the dear departed, and also the during, serene memory of her will help you with that.

Because I don't know whether your situation has possibly now also worsened in pecuniary respect, I have preferred to propose you already now for the assistant's position, mainly because one also can't be sure whether and when the Rockefeller stipend will be awarded. (I just received a letter from Paris in which in the first place recommendation letters are required from your teachers in Vienna, not including prof. Hahn, and secondly reprints of your publications until now. I would like to ask you to send the reprints directly to Dr. Trowbridge, Agent for Europe of the Rockefeller Foundation, 22 rue de l'Elisée, Paris 8^c; for the recommendation letters Weitzenböck will turn to Wirtinger). I have managed to get a salary of 3000 guilders for the assistant's position, to which can be added a personal extra allowance of 500 guilders, if necessary. Of course you should not come earlier than your health allows; but if you can be here on the first of October, your salary will start on September 16.

Within a few days I will travel with my wife to Switzerland; but until further notice my postal address will remain in Laren. I would appreciate to be kept informed about your well-being; please rest assured that my best wishes accompany you.

With cordial greeting, also from my family [207]

Your
L.E.J. Brouwer

[Signed autograph – in Menger]

25-12-15

To W. von Dyck — 15.XII.1925   **Laren**

Dear Colleague, [Hochgeehrter Herr Kollege]

At the same time I send you 50 copies of the enclosed document about the Conseil Internationale de Recherches, that I put together earlier. In

---

[207] *Mit herzlichem Gruss, auch von den meinigen.*

fact this is part of the Karo brochure, [208] but it may create a stronger impression if read by itself.

I would like to ask you to make arrangements that every member of the science section of the Bavarian Academy of Sciences gets a copy. Because soon the union of the German Academies of Sciences will be invited to join the Conseil internationale de recherches, which was founded only to malign and boycott Germany. Maybe some will say then: 'Who accepts the League of Nations, can also accept the C.I.R.' [209] But that would be wrong, firstly because the material necessity that pushes one to the former, does not exist in case of the latter, and secondly because the League of Nations is in the end a humanitarian American idea, while the C.I.R. is only a product of the French wish for destruction, as the enclosed composition may show unambiguously.

Sincerely yours

your [210]
L.E.J. Brouwer

[Signed typescript, copy – in Brouwer]

1925-12-21a

To H. Hopf — 21.XII.1925[a]   Laren

Dear Mr. Hopf, [Sehr geehrter Herr Hopf]

I have read with great interest the proof sheets of your article about vector fields on n-dimensional manifolds [211] (just as, by the way, the ones of your preceding Annalen article). As far as the quotation in § 42 of the Hadamard note [212] mentioned by you, is concerned, I agree with you that this is not correct. The explanation is that the contents of both my article *Ueber Abbildungen von Mannigfaltigkeiten* [213] and the note by Hadamard in the book by Tannery have been discussed by Hadamard and me around Christmas 1909 in Paris. On that occasion I have—referring to a couple of articles of mine that were in part in print, in part waiting for the final

---

[208] [Karo 1926].   [209] Conseil Internationale de Recherches.   [210] *Mit hochachtungsvollem Gruss – Ihr sehr ergebener.* [211] [Hopf 1926]. [212] [Hadamard 1910]. [213] On mappings of manifolds.

editing—stated, among other things, also the theorem, proved now by you for the first time, and I was holding out the perspective of a publication of its proof in my article 'On mappings of manifolds', that was at the time available in preliminary version, and originally intended for publication in the Amsterdam proceedings [214]

To this circumstance on the one hand the quotation in § 42 of the Hadamard note, and on the other hand the description of the theorem in § 40 as 'théorème de Brouwer', are to be attributed. It was an omission on my part that later I didn't tell Hadamard in time that I would submit the article not to the Amsterdam Proceedings but to the Mathematische Annalen, and also that the implementation of the proof of this theorem finally became so complicated that I had to abandon its publication for the time being.

So because of the above I would like to ask you to make the following changes (the present formulation would among others imply that my article 'On mappings of manifolds' was based upon the already present note of Hadamard and that the latter note was written independently from me): [215]

l. 11 *'bereits kurz vor'* to be replaced by *'ungefähr gleichzeitig mit'*
l. 12 *'von Hadamard'* to be replaced by *'von Hadamard ohne Beweis'*
l. 18–22 *'genügt; Hadamard will ...... Beweis befindet.'* to be replaced by *'genügt³). Wie mir Herr Brouwer mitteilt, sind übrigens die Brouwersche und die Hadamardsche Arbeit unter Gedankenaustausch zwischen den beiden Verfassern entstanden.* [216]

I was sorry that I could not get you as assistant in Amsterdam. For a single semester it would serve no purpose to come, I absolutely need someone who can stay long enough to immerse himself thoroughly in the local activity.

In the hope that I can get to know you soon personally, I remain with best greetings

Your [217]
L.E.J. Brouwer

[Signed typescript – in Hopf]

---

[214] *Proceedings KNAW.*   [215] Hopf adopted Brouwer's suggestions in his manuscript.
[216] As Mr. Brouwer informs me, the articles of Brouwer and Hadamard have grown out of an exchange of thoughts between the two authors.   [217] *Ihr sehr ergebener.*

1926-04-10a

From H. Hahn — 10.IV.1926[a]   **Vienna**

Hahn to Brouwer 10.4.1926[218]

Dear Colleague, [Lieber Herr Kollege]

I will tell you with great pleasure what I know about the genesis of the
first articles of Menger. I am in the position to do that as well in my quality
as university professor because Mr. Menger sought my advice repeatedly
when he was writing his Ph.D. thesis, as in my quality as publisher of the
Monatshefte für Mathematik und Physik, because I attached importance to
the publication of the first results of Menger in this journal.

I conducted a seminar in the summer semester of 1921 on some problems in the theory of point sets. I opened this seminar in the first days
of May with a talk in which I pointed out that a fully satisfactory definition of the curve concept didn't exist yet. Quite soon afterwards Menger,
whom I had not known until then, came to see me, to find out what I
thought about a definition of this concept which he had thought out, stimulated by my talk. I saw immediately that Mr. Menger was on the right
track, which had been before my mind's eye since 1914 as the one that
should lead to a natural definition of dimension, without however pursuing these quite vague thoughts at that time. I was especially glad that
now a young man all by himself followed precisely that direction. As is
unavoidable with a young student who is in the stage of familiarizing himself with some field, Menger's definition of a curve at first had an essential defect (but the fundamental idea was already the final one); I pointed
out this defect and challenged him to deal with it by thinking a bit more.
A first written sketch from that time is still available. Mr. Menger succeeded very quickly in redressing the defect. Also, in February 1922 he
had already recognized with complete clarity that the path he took would
give a recursive definition of the concept 'n-dimensional'. This definition
is described in extenso in a letter to me of February 15, 1922, which is
in my possession. In fact Mr. Menger must have possessed the essential
parts of this definition even earlier, because in the letter it says: 'I had
ended the small article which you, Professor, have been so very kind to
read, with a definition of the n-dimensional set, which should have been,

---

[218] In pencil in Brouwer's handwriting.

as I now believe, as follows etc.' But I can't recall this earlier version anymore.

I received from Mr. Menger a completely revision in November 1922. Now everything was completely correct, only the importance of the covering theorem that now bears Menger's name wasn't recognized, which is not essential for the question now at hand.

Summarising, I observe: Mr. Menger was stimulated by my seminar talk in May 1921 to search for a satisfactory definition of the curve concept. In next to no time he had found the right way. Pursuing this route he had found in February 1922 a recursive definition of the concept 'n-dimensional'. A final written exposition was in my hands in November 1922. That also elsewhere work had been done on these concepts nobody here in Vienna knew.

I hope that with this I have clarified everything that needs to be known. With best greetings

Your [219]
H. Hahn.

[Signed autograph – in Brouwer]

26-04-10b

From K. Menger — 10.IV.1926[b]   **Vienna**
                                                    Fuchsthallergasse 2, Wien

Dear Professor, [Hochverehrter Herr Professor]

I write to you only today, because the priority matter is now completely settled: not only all evidence mentioned by me, but also other material has been found that I myself had already forgotten. I have put all the material in a safe and I will hand over personally all the originals to you. Today I only mention shortly the *officially* certified documents:

1) A manuscript submitted in *June* 1921 to the Monatshefte, containing a definition of the constructs that I later named 'regular curve', furthermore a definition of end- & branching points.

---

[219] *Ihr ergebener.*

2) The letter at the Academy containing the full curve definition, the definition of surface, the definition of the n-dimensional continuum, of end points and branching points, and a few important theorems about curves.

3) A letter delivered on February 15, 1922 to the Monatshefte, which has been placed in safekeeping by the editors, containing *literally* my general definition of dimension (including the empty set as -1-dimensional) and other matters.

4) A manuscript entered at the editorial board of the Monatshefte in November 1922, containing numerous theorems *together with full proofs* (among others the theorem that the union of finitely many closed n-dimensional sets is n-dimensional, with a proof, & implicitly the proof of this theorem for the union of countably many sets).

So much for your preliminary orientation. About the tension that the collecting of these documents caused me, I'd rather remain silent: if I had not kept in mind that I had to put the documents into your hands for all that you have written about the theory, and that you have done for me,— then I would not have been able to bear all I had to go through!

Now my nerves have gone completely to pieces. Yesterday I visited an excellent doctor who says that my nerves and my general condition are in a terrible state and who told me that I should spend every day I can possibly make free in absolute rest somewhere around Vienna. At the same time he advised me to be as careful as possible for some time, if I don't want to risk that my ability to work will soon be permanently lost.—

If you, dear Professor, could bring yourself to drop me a line to say that you have received my letter all right, and when you will be in Amsterdam, then I would be very grateful. I would be very happy to hear that you are having a nice vacation and recuperate well.

Sincerely yours

Your grateful [220]
Karl Menger

[Signed autograph – in Brouwer]

----

[220] *Empfangen Sie inzwischen den Ausdruck meiner verehrungsvollen Ergebenheit – Ihr dankbarer*

26-05-11d

To A. Heyting — 11.V.1926[d]   **Laren**

Dear Mr. Heyting [Waarde Heer Heyting]

I have glanced through your manuscript, to my great satisfaction, although I have by no means checked the details (for which I hope to find time later), but so much is clear to me that your work is ready for international publication. So I would like to suggest to you to write a German (or, if you prefer, a French) treatise, which contains both your dissertation and these last results, and which from the outset aims at deducing the non-Pascalian number geometry from the non-Pascalian axioms, while the 'Pascalian, non-Archimedean' and the 'Archimedean' geometries are dealt with as specializations at the end in an appendix. We can discuss in more detail the manner of publication of this German treatise when it is finished or almost finished; indeed, this will depend on the size and disposition of the work. Maybe the article is suitable for one single *Treatise*[221] of the Academy of Sciences; maybe a series in a professional journal will be preferable.[222]

With friendly greetings,

Your
L.E.J. Brouwer

[Signed autograph, postcard – in Heyting]

––––––––––––

26-07-23

From M. Planck — 23.VII.1926   **Berlin**
                                                                        Berlin-Grunewald

Dear Colleague, [Hochverehrter Hr. College]

In the matter about which you were so kind as to inform us recently in the Academy, we meanwhile have received new information, which I think

––––––––––––––––––––––––––––––––––––––

[221] Verhandeling.   [222] The papers were published in the Mathematische Annalen, [Heyting 1927a, Heyting 1927b].

I should communicate to you, before I wait until I receive the letter you promised me.

Mr. Schuster-Manchester has formally communicated to the secretariat of our Academy [223] that, in the statutes of the Conseil des Recherches the passage that referred to the London declarations has been *struck out*. If this is really true — and from the whole nature of the letter we actually have no cause to doubt — then there are two statements that are diametrically opposed, and we would be sincerely indebted to you, if you would be in the position to clarify the matter. Because any further step that we can make depends essentially on what are the facts at hand.

With collegial greeting

Yours sincerely [224]
M. Planck

[Signed autograph – in Brouwer]

1926-08-08

From M. Planck — 8.VIII.1926   **Berlin**
                                          Preussische Akademie der Wissenschaften
                                                           Unter den Linden 38

Dear Colleague, [Hochverehrter Herr College]

I have received your kind note of the 31st of last month, and as agreed I have informed the secretariat of our Academy [225] insofar as it is represented here during the vacation. It is a mystery to us how the Royal Society acquired the 'certainty' that Germany would unconditionally comply with an invitation to join the Conseil, [226] and I will take the trouble to find out what is the source of this myth.

We don't think it useful to direct a formal request to the government, because we cannot at the same time produce tangible evidence, and as a consequence we can be certain that we will not get an adequate answer. However, the main thing is that this astonishing statement of the representative of the Royal Society doesn't have the least significance for us, and that

---

[223] *Preußische Akademie der Wissenschaften.*   [224] *Ihr aufrichtig ergebener.*
[225] *Preußische Akademie der Wissenschaften.*   [226] Conseil Int. des Recherches.

we have an entirely free hand with respect to our position on the question
of Germany's entering the Conseil.

   With collegial greetings

   Sincerely yours [227]
   M. Planck

[Signed autograph – in Brouwer]

26-08-19

From K. Menger — 19.VIII.1926 Vienna

Fuchsthallergasse 2, Wien

Dear Professor, [Hochverehrter Herr Professor]

After you saw me off at my last visit with the request that I call on you
again in a couple of days (in particular for the formulation of my dedication
of my *Bericht*), I tried to do so four times, during the next two days, but
each time I found that nobody was at home, and the housepainters confirmed
this. The following day I woke up with a violent influenza, which tied me for
four days to my room. My first steps as soon as I could leave the house, were
to you, where I heard that you had left that same morning for an indefinite
(and in any case, a longer) time. Miss Jongejan added that you had neither
written to me, nor left a message for me, because you thought I had left the
country without informing you. — I must tell you, professor, that it is the
first time in my life I had to hear such an unjustified attribution of lack of
character, education & manners.—
   In view of my request concerning the dedication of the *Bericht*, I permit
myself to submit it in writing, as I have now gone to my country. I was
going to write:

<table>
<tr><td></td><td>Herrn L.E.J. Brouwer,</td><td></td></tr>
<tr><td>entweder:</td><td>dem grossen Förderer der Topologie</td><td>[228]</td></tr>
<tr><td>oder:</td><td>dem bahnbrechenden Bearbeiter der Topologie</td><td></td></tr>
<tr><td></td><td>zugeeignet.</td><td></td></tr>
</table>

---

[227] *Mit der Versicherung ausgezeichneter Hochachtung und collegialen Grüssen – Ihr
aufrichtig ergebener.*   [228] Dedicated to L.E.J. Brouwer, either: the great promoter of
topology / or: the pioneering developer of topology.

I pray to you, revered professor, to let me know as soon as possible about this matter.

It remains for me to thank you sincerely that I had the honor to be your assistant for a year, and that you have made it kindly and magnanimously possible for me to prepare and publish a series of articles. When I join to my gratitude a plea that you do not effectuate an extension of my assistantship, then this is a decision that was hard to take, but carefully considered, of which I am certain that it also conforms to your own wishes.

I hope, dear professor, that you will soon come to Vienna and visit me, and that I may guide you through Vienna. I hope also that you will like it here. I just want to ask you, in order that I will indeed be in Vienna, that you send me a telegram two days before you arrive, and that you tell me in time what hour of your arrival, so that I can meet you at the station.

I assure you, dearest professor, of my permanent gratitude and unshakeable veneration

  Karl Menger

[Signed autograph – in Brouwer]

1926-08-20

From H. Scholz — 20.VIII.1926   **Baarn**
                                                   Huize Ekely
                                           (p.A. Herrn Dr. W.H. Patyn)

Dear Professor, [Hochverehrter Herr Professor]

You may as well hold me for one of the most ungrateful people that you have ever met; but I just had the misfortune, to be so badly pursued by misfortune, that it had really been impossible for me to discuss with you the theorem of the equivalence of the absurdity of the absurdity of the absurdity with the simple absurdity, which has manifested itself in the hardest conceivable form as truth to me, now also in theoretical form.

But that would have been necessary and some other things as well, as is indicated on the enclosed sheet.

I soon understood that now a renewed personal discussion could be a bit more useful for me, and that it can be organized in such a manner that no excessive claims on your time are made.

So now I choose the shortest way, namely that I put the enclosed sheets as reference documents for such a consultation into your hands [229] and that at the same time I ask you whether you could be so kind as to inform by telephone Mr. Patyn (149 Baarn) if and when I may visit you once more next week Tuesday, Wednesday or Friday about this matter.

For you may not conclude from my silence, that I didn't struggle the whole year with these questions that I was allowed discuss with you in the last summer.

Otherwise I couldn't have given a lecture this summer on the axiom system of classical logic and its correction by Brouwer.

At this occasion I have not only found out how much I still lack, but I believe also that I have brought the problem of consistency into a new light, for which I am indebted to your constitution of concepts as first stimulus.

The problem is this: Can we, in an ultimate reduction, prove at all the consistency of some mathematical concept in any other way than by constructing at least one object that falls under this concept?

Hilbert's consistency proof by means of the inference from n to $n + 1$ stands or falls by such an existence proof is essential, because its consistency can only be shown when there is at least one class of entities to which it can be applied.

I would like to elaborate this a bit more extensively with you.

Because what was shattered last winter, should be accomplished this winter.

In January or February, I will speak about the crisis in the foundations of logic for the Berlin Kant society.

But in any case you must finally know now, that I have a better memory than you thought, and that the severe personal inhibitions that also prevented my access to you don't prove at all that I didn't remember you in the most sincere gratitude.

Sincerely yours [230]

Your

Heinrich Scholz

---

[229] Enclosures: 1. Classification of consistency [We have translated *Widerspruchsfrei-heit* (in Scholz's text systematically abbreviated as 'WF') by 'consistency', where 'freedom of contradiction' would be somewhat artificially archaic.] consistency - propositions, 2. On the place of Hilbert's concept of consistency, 3. Consistency etc. 4. Problematic and unspecified concepts   [230] *In grösster Hochschätzung – Der Ihrige.*

From August 28 on my address will be again:
    c/o Mr. Justus Meyer
    30 Zandvoortsche Laan
    Zandvoort.

Postscriptum: (1) Mr. Patyn will drive me in his car to you, so that we
                don't have to reckon with the trains.
                (2) Only after many doubts I decided to send you the en-
                closed material. Please consider it merely as a preliminary
                study, and allow me to ask it back so I can elaborate it fur-
                ther.

[Signed autograph – in Brouwer]

1926-12-13

To H. Hopf — 13.XII.1926   Laren

Dear Mr. Hopf [Lieber Herr Hopf]

I believe that it is best for Miss Gawehn (also in the interest of her
possible later scientific career) that she first takes the state examination,
and then coming fall applies for a Rockefeller grant to study for a while
in Amsterdam. If she qualifies herself well during these studies with me, I
would be happy to consider her *subsequently* for an assistantship. At the
moment she would not be of use for me notwithstanding her evident talent;
she has not enough command of the subject matters and also too easily
makes errors. (I base this all on her manuscript that she submitted months
ago to the Annalen [231] which I see gradually getting ready for printing, and
about which Menger as my assistant is corresponding with her.) In case I, as
I hope, soon come to Berlin for a few weeks, I will also find an opportunity
to speak with Miss Gawehn about her plans for the future. [232]

Would you be so kind to read through the continuation of the investigation of Wilson on the mapping degree, of which you have refereed the first
part during the last summer? I permit myself to send the manuscript concerned with the same mail, together with the page proofs of the first part,
which you know already. [233] Many thanks in advance for your efforts.

------

[231] [Gawehn 1928].   [232] For Gawehn see [Van Dalen 2005] p. 567.   [233] [Wilson 1928].

I am very eager to see your own further publications and I am very much looking forward to our meeting again, hopefully before long, in Berlin.
Cordial greetings!

Your
Brouwer

[Signed autograph – in Hopf]

26-12-21

To A. Fraenkel [234] — 21.XII.1926   Laren

Dear Mr. Fraenkel, [Lieber Herr Fraenkel]

I cannot tell you how dumbfounded I was, when hardly three weeks after I received the first proofs, [235] you gave me to understand that the time for taking into account possible suggestions for changes had already expired. What kind of wizard you must have taken me for that you required me, in the middle of the semester and with all my time as good as completely occupied with other things, to study a book of more than 100 pages, and do it so thoroughly that I could bear the responsibility for suggestions to change something. Even today I haven't yet finished my judgments concerning details, I will indeed still need also the Christmas week for that. With the inexplicable hurry, which in my opinion is damaging for all parties (author, publisher, public) the only way out for me is, that I incorporate all my marginal remarks into a review of your book, in which I will however have to put right quite a lot (especially as far as intuitionism is concerned), but it is maybe just as well that I have a reason to deal with the erroneous information about intuitionism which is given to the public from so many sides. In order that meanwhile my review can remain as free of personal matters as possible, I would like to suggest to you three small changes, which certainly can still be corrected on the proof sheets: 1) delete the (indeed completely unfounded) insinuation in footnote [12]) (sheet 18); 2) in the text of sheet 20, lines 21 from below, 15 from below and 7 from below,

---

[234] Addressed: Breiter Weg 7, Marburg (Lahn).   [235] proofs of Fraenkel's *Zehn Vorlesungen über die Grundlegung der Mengelehre*. (Ten lectures on the foundations of set theory).

speak of Brouwer rather than of intuitionists in general; 3) include in the literature references to all my intuitionist articles (among which actually the only publications about intuitionism in existence that 'don't just talk but do something' — apart from Heyting's dissertation — are to be found.). [236]

All the best greetings and holiday wishes from house to house

Your Brouwer

P.S. A package of reprints is sent today to you. It is obvious that under the present circumstances I *cannot* allow that anywhere in your book or preface the fact can be mentioned that I have seen your proof sheets.
Your B.

[Signed autograph, postcard – in Fraenkel]

1927-01-12

To A. Fraenkel — 12.I.1927   **Laren** [237]

Dear Mr. Fraenkel, [Lieber Herr Fraenkel]

That the main theorem of Cantor evidently holds for completely deconstructible point sets, but that it is 'false' for general point sets, has nothing to do with a 'gradual refining' of the fundamental concepts, but only with the fact that the intuitive basic construction of mathematics (which nowhere exceeds the countable, where it occurs with my predecessors) was explained by me first (1907) as *completely deconstructible finite spread*, [238] next as *completely deconstructible (but not necessarily finite) set*, and finally as *spread without further qualifications*, but which was always in the phase of its introduction called "spread", for short. [239] One cannot keep intro-

---

[236] For a discussion of Fraenkel's views on intuitionism see [Van Dalen 2000], [Van Dalen 1999] section 10.5.   [237] Addressed: Breiter Weg 7, Marburg (Lahn). [238] What Brouwer called *Menge* and *finite Menge* is now known as 'spread' and 'fan'. The notion of 'deconstructible' is essentially taken from the transfinite proof of the Cantor-Bendixson theorem. See also [Van Dalen 1999], section 10.5.   [239] '1907' seems surprising; the notion of *afbreken* occurs in [Brouwer 1917a, Brouwer 1917b], and *abbrechen* and *abbrechbar* occurs in [Brouwer 1918a]. In the dissertation one can however, reading between the lines, recognize the notion of "breaking off" (p. 64 ff.). From the present letter one may conclude that Brouwer had recognized that his implicit notion of 'fan' required extra conditions.

ducing new terminology all the time; therefore I have denoted my intuitive basic construction by 'spread' again, each time when it needed an extension; even a few months ago such an extension became necessary as one can read in my article 'Intuitionistic introduction of the dimension concept' [240] After this extension too, certain so far 'self evident' theorems will turn out to be 'false'. Nonetheless, admonishments from your side, as in the mentioned footnote, do not have the least justification. Should you want to stick to this humiliating and hollow insinuation, even after my urgent request and my urgent advice to delete it, then the competent reader (I too, claim to qualify as such) can only view that as a declaration of war to me; I am asking myself in vain what grounds I could have given you. Excuse me that I write sharply and clearly; but I will have to do that subsequently in public too, and then it should not be said that I didn't call your attention to the implications of the statement, and didn't warn you.

  With friendly greetings

  Your Brouwer

[Signed autograph, postcard – in Fraenkel]

27-01-28

To A. Fraenkel — 28.I.1927   **Berlin-Halensee**
                                       Joachim Friedrichstr. 25[II] [241]

1 Enclosure.

  Dear Mr. Fraenkel, [Lieber Herr Fraenkel] [242]

  You are really mistaken, and you really hurt me again, if you attribute my latest card to an 'irritability independent of you'; please keep in mind that in your letter of December 31, 1926, you quote my position in 1913 and 1919 on Cantor's main theorem as an example of the phenomenon that in connection with the gradual sharpening of fundamental concepts the term 'self evident' easily gives rise to errors, a claim that after my exposition

---

[240] *Intuitionistische Einführung des Dimensionsbegriffes* [Brouwer 1926]. [241] 'Adresse bis Mitte März'   [242] For more information on the topics of this letter, see [Van Dalen 2000].

given on my last card must appear to you too as both unfounded and insulting.

How little one can speak of a "declaration of war" on my side, and how strongly, on the contrary, I strive with all my strength to avoid a public fight between the two of us, you can see from the fact that I have succeeded in getting a statement from Teubner that he is willing even now to incorporate substantial changes into your book before the printing. [243] And so I would like to implore you not to continue the expropriation that the German mathematical literature has practiced on me, by making me share what is exclusively my personal intellectual property with Poincaré, Kronecker and Weyl. (By the way, to a certain degree I am to be blamed for that myself, because I have now and then, in a for the superficial reader easily misleading manner, brought myself and my predecessors, with whom I merely share the struggle against formalism, under the common denominator of "Intuitionist".)

For your information I enclose (with the request to please send it back some time, because it is my last copy) the German translation of a section of an article which I will publish in the Revue de Métaphysique et de Morale [244] and in line with that I propose the following changes for your book which are minimally required by justice:

α   To edit the second paragraph of §6 of the 3/4-th lecture: [245] "[in] this intuitionism two phases can be distinguished, of which the first one is only a phenomenon of reaction [...old text ...] of the last quarter [...old text ...] by Cantor; at the beginning of this century [...old text ...] adopted a far milder position.

The second, much more radical phase, which does not just concern the *founding* of mathematics, but which reshapes *the complete doctrine of mathematics*, was inaugurated by Brouwer, who was joined by Weyl as an adherent. According to a formulation of Brouwer this neo-intuitionism [246] is based on the two following principles:

1. *The independence of mathematics* [...old text ...] will be capable.

2. *The constructive definition of set* [spread] [...old text ...] without using the Bolzano-Weierstrass theorem".

---

[243] '*Zehn Vorlesungen*', [Fraenkel 1927].   [244] Paper not published. No manuscript extant.   [245] cf. [Fraenkel 1927] p. 34, 35.   [246] neo-intuitionism

(These two principles are on the one hand exclusively mine, [18] on the other hand they implicitly embody in a completely rigorous way the whole future rebuilding of mathematics.)

β Lines 15-21 of section §9 of the 3/4-th lectures [247] are to be revised for example as:

"[... old text ...] of a real function which is continuous in a closed interval; the deficiency of this proof is matched in intuitionism (cf. Brouwer 5) by the curious (in fact in no way obvious, but rather deep) fact, that each function which is defined everywhere on a continuum, [248] is uniformly continuous".

(In the summer of 1919 I have once in personal conversations with Weyl in the Engadin, as a result of which he was converted to my views, in connection with the definition of the continuous function in §1 of my *Begründung der Mengenlehre unabhängig vom logischen Satz vom ausgeschlossenen Dritten* stated and motivated the conjecture that these functions are the only ones existing on the full continuum (cf. in this connection p. 62 of my paper *Über Definitionsbereiche von Funktionen*, which has just appeared in the Riemann volume of the *Annalen*). The legend which has since then been circulated about Weyl, "that it is obvious in Brouwerian analysis that there cannot exist on the continuum any but uniformly continuous functions", can only be based on this ([as many?] other ones, half understood by Weyl) conjecture, stated by me).

γ Extend line 16 of the first paragraph of §10 of 3/4-th lecture [249] as follows:

"in an inductive (or recurrent) way. Over and above this, Brouwer has subsequently made the step (already mentioned in §6), that he unfolds the ur-intuition further to the general spread construction, and in this manner extends the intuitionistic founding of (discrete and denumerable) arithmetic to (continuous and non-denumerable) analysis. From this ur-intuition, stressed with special emphasis . . . ."

δ To complete the part of the References which concerns me, at least as follows:

---

[18] so that it is a crude injustice towards me to claim that "these considerations of the new adherents to intuitionism have emerged, at totally different places, independent of each other, in a remarkable agreement".

[247] cf. [Fraenkel 1927] p. 48.   [248] i.e. a connected compact set.   [249] cf. [Fraenkel 1927] p. 50.

" 1. Begründung der Mengenlehre unabhängig vom logischen Satz
     vom ausgeschlossenen Dritten I–II. Begründung der Funktionen-
     lehre unabhängig vom logischen Satz vom ausgeschlossenen Drit-
     ten. I. *Amsterdamer Verhandelingen*, **12** no. 5, 7, 13, no. 2
     (1918–1923).

  2. Intuitionistische Mengenlehre. *Jahresbericht der Deutschen Math-
     ematiker Vereinigung*, **28** (1919), p. 203–208.

  3. Über die Bedeutung des Satzes vom ausgeschlossenen Dritten in
     der Mathematik, insbesondere in der Funktionentheorie. *Journal
     f.d. reine u. angewandte Mathematik*, **154** (1925), p. 1–7.

  4. Zur Begründung der intuitionistischen Mathematik I–III. *Math-
     ematische Annalen*, **93** (1925), S. 244–257; **95** (1926)
     S.453–472; **96** (1926), p. 451–488.

  5. Über Definitionsbereiche von Funktionen, *Mathematische An-
     nalen*, **97** (1926), p. 60–75."

(The citing of the three Amsterdam essays would in any case be more necessary than that of the three *Annalen* papers, which altogether only bring a
technical elaboration – without any philosophical addition whatsoever – of
the first (least important one) of the three mentioned *Verhandelingen*. And
the citation of my paper which appeared in the Riemann volume, which is
of central importance for the continuity question for full functions (cf. above
under β) and in general for the continuum problem, seems to me of the utmost urgency, where for the rest you mention indeed every philosophy [. . .]
textbooks on set theory).

   In the last paragraph of §8 of the 9/10-th lectures, line 17 from the
   bottom, mention instead of "the opinion of the radical intuitionist",
   "the opinion of Brouwer" (this opinion is, even if it has since then
   been repeated after me by others, nonetheless to no lesser degree my
   intellectual property).

According to a statement of Schopenhauer, there will be practiced against
each innovator, by the automatically appearing opposition, at first the strategy of (factual) ignoring [250], and after the failure of this strategy, that of
priority theft. Should this also bear on my case, then I am convinced that
you do not belong to my enemies, that on the contrary you harbor the wish—
and after learning the above—will cooperate to make the above-mentioned
strategy against me as little successful as possible.

---

[250] *totschweigen.*

Finally I beg you to believe that the purely objective content of this letter is accompanied only by benevolent and friendly feelings towards you.

With best greetings [251]

Your Brouwer

[Signed autograph – in Fraenkel]

●27-02-03

To P.S. Alexandrov — 3.II.1927   Vienna

Wien

Greetings and a handshake. (The card will, by the way, not be sent today, because the postal drivers are on strike, so the mailboxes will not be emptied). I am here for a few days for discussions and for a visit to Dutch friends. Tomorrow I will dine with Wirtinger, Ehrenhaft, Hahn, Vietoris and Loewy. In Berlin the colleagues are very nice to me and my lectures are attended very well. [252] That Blumenthal sent the Kuratowski paper to you, while bypassing me and also without informing me in advance, was against the rules of the editorial board, and it was unfriendly, offensive and inappropriate (maybe offensive on purpose because of the many conflicts between him and me; he still is regularly changing my articles after they have been declared ready to print; in the Riemann volume [253] he has introduced again a gross error). When he does something like that again, please answer him that you can accept these refereeing requests only from me, because for the outside world I am the editor in charge of topology.

Greetings to your family members. [254]

Your Brouwer.

---

[251] *Mit den besten Grüssen.*   [252] The Berlin lectures on Intuitionism.   [253] The commemorative volume for Riemann's birth 100 years ago; see also [Van Dalen 2005] section 13.3.   [254] *Grüsse an Ihre Hausgenossen.*

Brouwer is celebrated a lot. He will have to buy a dinner jacket! He drags me everywhere with him.

Warmest Greetings   Your Corrie Jongejan.

[Signed autograph, picture postcard – in Alexandrov]

1927-03-08

To H. Hopf — 8.III.1927   **Berlin-Halensee**
                                                             Joachim Friedrichstr. 25^{II}

Dear Mr. Hopf, [Lieber Herr Hopf]

It just occurred to me that I owe you and Feigl an addendum to what I said at the end of my talk about fixed point theorems; and before I leave, I want to settle that debt. When I remarked that the classical fixed point theorems *cannot* be saved intuitionistically *as fixed point theorems*, I didn't mean at all that intuitionistically these theorems don't admit an interpretation that is still valid there. [255] On the contrary: the classical theorem that the transformation τ of a compact space R (which we will suppose to be a metric space) has a fixed point, has a meaning which remains intuitionistically correct, namely that for every $\varepsilon \diamondsuit 0$ [256] a point P of R can be determined, that is less than ε removed from its image point. And the classical theorem that the transformation τ possesses n mutually distinct fixed points, has the intuitionistically correct meaning that there exists an $a \diamondsuit 0$ with the property that for every $\varepsilon \diamondsuit 0$, there can be determined n points $P_1, P_2, \ldots P_n$ of R, which all are less than ε removed from their image points and of which every two have a mutual distance $\geq a$. But these theorems are not fixed point theorems anymore, because one doesn't have means to indicate a fixed point, i.e. to approximate it.

Please show this card to Feigl too. It is intended for you both.

Cordial greetings!

Your Brouwer

[Signed autograph, postcard – in Hopf]

---

[255] Cf. [Brouwer 1992] p. 56.   [256] $\diamondsuit$ is Brouwer's notation for the natural order relation: $a \diamondsuit b$ if the difference of a and b is greater than a suitable 2^{-k}.

)27-04-09

To H. Hopf — 9.IV.1927   Laren [257]

Dear Mr. Hopf, [Lieber Herr Hopf]

Many thanks for your letter of March 20. I have written immediately to Dr. Trowbridge [258] in Paris (but I had spoken already about your case with Dr. Tisdale in the fall), and I have received from him an answer that appears to be very favorable.

I was only today in the position to send three copies of my article about domains of functions [259] to Miss Gawehn, [260] one for herself, one for you and one for Feigl. You would do me great favor when you would keep an eye on Miss Gawehn, and try do something so that her philosophical article would be ready to print and printed as soon as possible.

Please give many greetings from me to Mrs. and Prof. Courant, and recover completely.

Cordial greetings from your

Brouwer

If you have time, then go to Arosa to Miss Alice Beyreiss (teacher, lives in Chalet Valbelle, somewhat above Sporthotel Merkur), and bring her my greetings. You would do me a pleasure. Your B.

[Signed autograph – in Hopf]

27-09-07

From H. Scholz — 7.IX.1927   Baarn
                                                                Huize Ekely

Dear Professor [Hochverehrter Herr Professor]

It is an old experience that one only knows what one lacks, when one has learnt something new.

[257] Addressed: Hotel Pratschli. Arosa; forwarded to 'Prof.Dr. Courant, Universität Göttingen.'   [258] of the Rockefeller Foundation.   [259] [Brouwer 1927].   [260] See [Van Dalen 2005] p. 567.

In this sense I would like to ask you kindly to make a few remarks about
the enclosed page that would redeem me.

Because by what you said to me today, my interpretation until now

I can neither show: $r = 0$

     nor show: $r < 0$

     nor show: $r > 0$

is completely thrown into confusion.

I thank you once more most cordially for the two *wonderful* hours of this
afternoon and I remain

in the greatest veneration

Yours [261]

Heinrich Scholz

10.–15. September: 30 Zandvoortschelaan, Zandvoort.

[Signed autograph – in Brouwer]

1927-11-08

From L. Herzberg — 8.XI.1927   **Berlin-Tempelhof**

Dear Professor, [Sehr geehrter Herr Professor]

The *Berliner Tageblatt* has the plan to acquaint its readers with the main
thoughts in the modern dispute about the foundations of mathematics, and
for this purpose it wants to make one page of the newspaper available. The
theme might be perhaps: 'What about the validity of the theorem of the
excluded middle'?'

The *Berliner Tageblatt* would be grateful to you, dear professor, if you
could write something on this theme in an article from the *intuitionistic
point of view* of about three or four typewritten pages. If you would decline
to produce *yourself* a popular article for a mostly lay public, then it would
be very kind if you could send me a few statements about this theme in a
letter, especially also about the consequences for people's world view, which

---

[261] *in grösster Verehrung – der Ihrige.*

might follow from intuitionistic mathematics. Then I would convert these into a newspaper article and submit this to you to sign.

In case you agree we will ask professor Hilbert in Göttingen to treat the same theme from the formalistic point of view. If you would agree with our request, we would be much obliged.

Sincerely yours [262]
(signed) Dr. Lily Herzberg.

[Typescript copy –in Brouwer [263]]

27-11-16

From H. Scholz — 16.XI.1927 [264]   Kiel

A) ATTEMPT OF A CONSTRUCTION OF THE BROUWER THOUGHT CONCEPTS.

(1) Thinking is

   a) constructing of relations,
      for short: constructing

   b) deducing new relations from relations already constructed,
      for short: deducing.

(2) What constructing and deducing is cannot be defined, but can only be learned by demonstration and imitation.

(3) Thinking is basically a 'soundless' process, i.e. a process which is fundamentally independent of all (symbolic) means of representation by means of speech or without speech, by which we preserve the results for ourselves and for others

(4) So, thinking is

   a) a soundless constructing of relations

(5) More precisely of relations between 'objects' and 'concepts'.

---

[262] *Mit vorzüglicher Hochachtung — Sehr ergebenst.*   [263] The copy was most likely made at Brouwer's request; there is no original letter in the archive.   [264] Date - postmark. This is more a (drafted) manuscript than a letter. There are some lines of correspondence inserted in the text.

(6) More precisely of relations between objects and concepts, about which one can t h i n k.

(7) About an object or concept one can only think when both can be clearly grasped.

(8) This is only the case when they can be 'constructed'.

(9) A concept can only be constructed when the objects it encompasses can be constructed.

(10) Objects cannot be created from nothing, hence also not by 'purely' thinking; because then this must create them from nothing. Thinking is itself only a construction tool, but it is not able to produce the construction material by itself.

(11) This material (and certain intuitions, that are unconditionally necessary for the evaluation of the material) is produced by the sense of time, and o n l y the consciousness of time. (More under 'mathematics'.)

(12) The objects thus generated are called, with reference to their number nature, mathematical objects, correspondingly the concepts built from them are called mathematical concepts.

(13) Hence thinking is

   a) A constructing of relations between mathematical objects and concepts.

(14) Thinking is

   b) Deducing new relations from already constructed relations; but certainly not according to abstract reasoning schemes given in advance, but so that the 'deduced' or 'deducible' relations must follow instinctively and evidently from the intuition of the already constructed r elations, and only from this intuition.

   Example: I have proved:
   a) Every x from K is an x from K',
   b) Every x from K' is an x from K''.
   Then it is evident that I have proved:
   c) Every x from K is an x from K''.

(15) Summary:
   Hence thinking is

   a) in its constructive function:
   operating on mathematical objects and concepts.

b) in its deductive function:

interpreting results of constructions based on intuitionistic consideration of one's own constructions.

So all deductive thinking is thinking that is based on the intuitionistic consideration of mathematical constructions, not based on pre-existing abstract reasoning schemes.

(16) Operating with non-mathematical objects and concepts is only thinking in so far as these non-mathematical objects and concepts can be reduced to mathematical objects and concepts, i.e. can be replaced by these. Whether and into what extent this is possible can only be decided on a case by case basis.

B) THE BROUWERIAN CONCEPT OF LOGIC

(1) Logic, as theory of the forms of valid thinking, is

a) not a system of aprioristic deduction schemes, when 'aprioristic' roughly means 'independent of intuitionistic consideration of mathematical constructions'.

Follows from the nature of reasoning, characterized under A15.

b) and certainly not a system of universally valid aprioristic deduction schemes when 'universally valid' amounts to 'directly applicable to all classes of objects and concepts'.

Already not because then one would burden oneself with the absurd set of all things.

Especially because of A16.

c) and certainly not a system of arbitrary aprioristic, universally valid deduction schemes, if 'arbitrary' = 'only satisfying the postulate of freedom of contradictions'.

Because such a thinking has with the thinking characterized under A simply nothing but the name in common.

(2) Logic, as theory of forms of valid thinking, is the system of those and only those schemes that I obtain, when I

a) have somehow symbolically represented both the 'constructions of constructive thinking' [265] that are soundless by themselves and the likewise soundless deductions of deductive thinking from these constructions,

---

[265] The end of quote mark seems to be missing here in the original.

b) study the invariants of this symbolically represented thinking.

Hence the theory of the valid invariants of symbolically represented thinking.

Or the system of fundamental reasoning schemes that are abstracted from the symbolically represented accomplishment of thinking and hence, strictly speaking thus only for the fundamental deduction schemes that are basic for these symbolized accomplishments but not for the proper (soundless) thinking.

But now all 'thinking' is a mathematical operating.

Consequently the symbolized thinking is a symbolic mathematical operating.

Or shorter: mathematics in a verbal (symbolic) representation.

Consequently logic is the theory of valid forms of a verbal representation of mathematics. (And not the theory of valid forms of mathematical construction as such!)

B') THE BROUWERIAN CONCEPT OF LOGIC

(1) The assumptions of logic:
  a) Thinking,
  b) the verbal (symbolically represented)
     expression of what is thought,   Executed   in
     from which thinking as such is   more   detail
     fundamentally independent.   under B)
(2) The object of logic:
     the forms of symbolized thinking.

(3) The task of logic:
    the analysis and synthesis of forms of symbolically represented thinking.
    Analysis = formation the system of the original forms of thinking,
    Synthesis = formation the system of the 'deducible' forms of symbolized thinking, where deducibility is determined by well determined formal constellations and substitution rules.

    Summarizing: the theory of forms of symbolically represented thinking.

    But now thinking is defined as a mathematical operating.

    Consequently the verbally expressed (symbolically represented) thinking as a verbally expressed (symbolically represented) mathematical operating.

Therefore logic is the theory of forms of verbally expressed mathematical operating, or the theory of forms of verbal expression that accompany mathematica l operations (but in such a manner that they don't fundamentally depend on these forms) or shorter: the theory of mathematical language.

(4) Consequences

  a) Logic is not a necessary condition for the construction of mathematics; for it is only the theory of the mathematical language (which is as such basically irrelevant for mathematics).

  b) Mathematics is a necessary condition for the construction of logic; because it produces the material for logic, the verbal (symbolical) formulation of which is the object of logic.

  c) From a rational logic must be demanded:

   1. that it restricts itself strictly to formulas that admit an mathematical interpretation at all;

   2. that it applies these formulas basically only to the extent that they, after a mathematical interpretation has been achieved, can be confirmed by the thinking mathematician.

Already the classical logic has most severely violated 2. Symbolic mathematical logic [logistics] has relieved itself from 1, and consequently compromised itself even more severely than classical logic.

C) THE THREE MAIN FAILURES OF FORMAL LOGIC. [266]

(1) the misuse of the Tertium non Datur:

  consisting of
  a) the illegitimate application to arbitrary properties of a given individual,

  b) in the use of it in the form: either all x from K are also x from K', or there is at least one x from K, which is not x from K'' for transfinite classes.

(2) the abuse of the notion of class, resp. property.

  consisting of the use of the above for the creation of non-constructible sets, and in particular totally unrestricted. or, as this unrestricted use has led to logical 'catastrophes, under the determined conditions of the sharpened axiom of separation of Zermelo.

---

[266] Section ends here.

(3) the misuse of the notion of consistency

consisting in the identification of the mathematically totally incon-
sequential 'logical' phenomenon of consistency with constructibility
(crucial difference with *Poincaré*)

based on the arbitrary introduction of non-mathematical objects, the
existence of which is identified with the consistency of the properties
that define them, and which the domain of mathematical objects is
allegedly made part of.

D) On the Brouwerian interpretation of logic

For the precise understanding of Brouwer's notion of logic it is of the
greatest importance, that one grasps clearly, what it means that the logic in
the sense of Brouwer is the theory of valid forms of s y m b o l i s e d thinking.

This means that the laws of logic are the laws of s y m b o l i s e d thinking, and not the laws of thought in general. This in particular, because
they can only be formulated at all for symbolized thinking. I cannot even
formulate the excluded contradiction, if I do not have p and $non - p$, resp. p
and $abs\ p$. And I have p and $abs p$ only in the domain of symbolized thinking.

Thinking as such is, strictly considered, just as little contradictory or
consistent, yes, even just as little true or false (absurd), as the building of a
house, or the experimenting of an experimental scientist.

Contradiction, consistency, truth and falsity (absurdity) are therefore
not properties of thinking in general, but properties of symbolized thinking.

Thinking as such can rather, like all constructing, either be carried out
(is crowned with some success), or cannot be carried out (ends in failure).

Thinking ends then and only then in failure if the objects, with which
it is operating, disintegrate in the course the operation, but then and only
then if a distinction $(0 \neq 0)$ is intrinsically forced upon the operations that
have been tried.

The supreme basic law of Brouwerian thinking could thus be formulated
as: each object of thought 'disintegrates' when processed, under the influence of thinking, if by means of this processing a distinction from itself is
forced upon it. That is to say: if at least one property can be constructed,
that is both given to it through the processing by virtue of thinking, and
withdrawn.

Thus we are back to the Aristotelian formulation: 'It is excluded that a
(not disintegrating) object has the property E, and also not has the prop-

erty'; but we interpret it now with Brouwer ontologically, and basically with so little concern for any 'logical' interpretation, as has, at least in the domain of mathematics, not happened since Aristotle.

E) THE BROUWERIAN NOTION OF MATHEMATICS

(1) I operate – I construct relations between "objects".

(2) I operate with operations – I derive from already constructed relations new relations.

(3) I cannot define what it means to construct, but only demonstrate, and learn through imitation.

(4) Constructing, and operating as well, is a fundamentally speechless act. That is to say: it is basically independent of all symbolizing, communication, linguistic means of expression, by means of which we preserve the results of construction for ourselves and for others.

(5) Mathematics is not a game of formulas, of the results of which only consistency is required, but an operating with objects.

I add: and with operations and objects; for if I cannot deduce the successful embedding of each x from K in K'' from the successful embedding of each x from K in K' and each x from K' in K'', then I can not build up a mathematics.

I can thus also say: mathematics is the totality of all results, that I obtain by constructing relations between objects and from the constructed relations derive new relations.

NB. This deriving does not mean a concluding in the logical sense of the word, where it means:

If I have the formula $F = (p\ q)$,
then I can write the formula F':
$$(p\ q)\ (q\ p);$$

for this is already a statement on sign-complexes, through which we symbolize mathematical constructions, has thus nothing to do directly with mathematics as such. Instead, deriving means here an immediately intelligible drawing of conclusions based on the nature of the executed construction.

(Cf. the syllogistic interpretation of the "Cogito, ergo sum" by Descartes, and Descartes' position on logic at all!)

(6) Mathematics can only operate with sharply graspable, i.e. with constructible objects.

(7) Where does mathematics gets the material from which it can generate its objects?

Not from logic:

for logic operates either with signs for objects; then it already presumes the objects.

Or it operates with 'meaningless' signs: then, in any case, it does not yield material for generating objects.

Finally one can ask, in how far it can do without numbers, which should be created first.

Not from observation either (see below).

There only remains as material-providing principle a field of sources of unfailing intuitive certainties.

Spatial consciousness can not provide this field, for

1. it is so intrinsically vague, that it becomes comprehensible, when it is understood as the expression of a Riemannian manifold.

2. the delicate question, not yet existing for Kant, indeed incorrectly declared to be impossible, arises, which spatial consciousness we should accept as fundamental; for, to each Riemannian manifold (with its own measure of curvature) corresponds then a specific spatial consciousness.

Thus only the consciousness of time remains.

This provides us with

1. distinct 'now'-points, i.e. points that are separated by means of time; that is, discrete objects, or rather at once natural sequences of such objects.

2. It provides us with these points in arbitrary number, i.e. more precise, with the consciousness that the sequences of these points will never stop.

   Comment: Observation can never achieve this; therefore it cannot be the foundation of mathematics either.

3. the equally unfailing certainty that between any two 'now'-points there can always be interpolated a third.

From 1. and the ability to collect discrete objects, and to create, beginning with one, through repeated addition of a new thing, ever new units, we obtain the natural numbers.

NB. It is not clearly seen, whether first ordinal- or cardinal numbers!

From 2. we obtain the constitutive consciousness of the unbounded continuation of this sequence.

From 3. we obtain the basis for a constructive composition of the continuum.

F) On the theory of indirect proofs

Euclid 1.1.6: Every triangle with equal base angles is isosceles. =
If x is a triangle satisfying the condition: $\beta = \gamma$, then x is a triangle that satisfies the condition: $b = c$.

Proof:

x is a triangle that satisfies the condition B: $\beta = \gamma, b \gtrless c$ (1)
$\rightarrow_x$ x is a triangle satisfying the condition B': I can construct for x a x' with

$b' = c$ and construct the sides $b', c'a$ such that x'
is fully contained in x (resp. such that x is fully contained in x'.
$\rightarrow_x x \gtrless x'$
Now, however, I can show: $x = x'$.

Therefore (1) is false (absurd).
Therefore there is no triangle, satisfying the condition: $b = c$.
Therefore every triangle with equal base angles is isosceles.

In this form Euclid's proof seems me to be also intuitionistically completely correct.
But it is only so, if one acknowledges the implication
    $abs(p\ abs\ q) \rightarrow (p \rightarrow q)$ [267]
For without the acceptance of this basic implication, an indirect proof of an implication is not possible at all.
    Then we would get: if I can show: the assumption: 'there is at least a triangle x (= I can at least construct an x), for which $\beta = \gamma$ is true, and $b = c$ absurd, is itself absurd; then I have shown: if x is a triangle with equal base angles, then x is a isosceles triangle.

I repeat: If this conclusion is *not* admissible, then I can not see any possibility at all to show the implication indirectly; and in particular: the proof given by himself [268] runs, when precisely analyzed, exactly according to this schema.

    I would like to go one step further and claim that the converse:

$$(p \rightarrow q) \rightarrow abs(p\ abs\ q)$$

---

[267] I.e. $\neg(p \wedge \neg q) \rightarrow (p \rightarrow q)$.   [268] possibly 'yourself'?

is also completely intuitionistically correct too.

Then we have the equivalence:

$$(p \rightarrow q) = abs(p \; abs \; q)$$

In words: $(p \rightarrow q) =$ it is absurd that there is an x ($=$ that I can construct an x), for which p is true, and q absurd.

This equivalence is all the more legitimate, as certainly also in intuitionism a 'there is *no* x, which... ($=$ there is *no* x constructible, which ...) is just as little an existential statement, like any implication.
[handwritten] Thus there is both on the left hand side and on the right hand side of the equation a *non*-existential statement.

If, however, one admits the above equivalence, then the following deep apories:

$abs(p \; q)$	$= abs(p \; q)$
$= p \rightarrow abs \; q$	$= q \rightarrow abs \; p$

I should then have to proceed accordingly:

$abs(p \; abs \; q)$	$= abs(abs \; q \; p)$
$= p \rightarrow abs^2 \; q$	$= abs \; q \rightarrow abs \; p$

I would thus only be able to get $p \rightarrow q$ along this line because in this case, by way of exception, I start with classical logic: $abs^2 q = abs \; q$!

If, in order to avoid this, one does not admit the equivalences, then
1. I do not see how $p \rightarrow q$ can be shown indirectly at all,
2. it remains unclear what the relation between $p \rightarrow$ and $abs(p \; abs \; q)$ is.

In order to make the consequence of these apories quite clear, I add the following confrontation with the table of Wavre:

Scholz	Wavre
$p \rightarrow q = p \rightarrow abs^2 q = abs \; q \rightarrow abs \; p$	$p \rightarrow q \neq p \rightarrow abs^2 q \, p \rightarrow q = abs \; q \rightarrow abs \; p$
$p \rightarrow abs \; q = q \rightarrow abs \; p$	$p \rightarrow abs \; q = abs^2 q \rightarrow abs^2 p$
$abs \; p \rightarrow abs \; q = q \rightarrow abs^2 p = q \rightarrow p$	$abs \; p \rightarrow abs \; q = abs^2 q \rightarrow abs^2 p$

Finally I remark that the for the intuitionistic proof of

$$abs^3 = abs \; p$$

required equivalence

$$p \to q = absq \to abs\,p$$

from my point of view, can only be justified by the evident

$$abs(p\ abs\ q) = abs(abs\ q\ p)$$

which, however, requires (see above) that one decides at this point to accept $abs\,p = abs^2\,p$.

Otherwise I should have to beg for a precise intuitionistic justification of this equivalence.

[Typescript – in Brouwer]

28-01-17

From K. Menger — 17.I.1928 [269]

Dear professor, [Hooggeachte professor]

Enclosed a typescript of '*Allgemeine Räume & Cartesische Räume III*.[270] Maybe it will give you some pleasure. A detailed exposition of the entire proof will of course greatly exceed the space for a Note.

Coming back to the Encyclopedia article,[271] I must confess that perhaps I wouldn't really like to read the page proofs, and that even if they would be sent to me, I must *reject* any thanks for advice I haven't given. For, I meanwhile met Vietoris in the Vienna seminar with the proofs, and he refused explicitly to even *show* them to me just for a moment, and he declared that it was the wish of Tietze and himself that among the German scholars only Rosenthal and Kneser receive the proofs, and he added to this verbatim (it is incredible!) that both authors [272] hadn't shown the proofs *to me* already last autumn!! Well, in case the authors of the Encyclopedia article expect more help from the two gentlemen than from me, they are welcome to believe that (it doesn't reflect, I think, on *my* intelligence). It is clear that under these circumstances, and also in view of the fact that Vietoris in the conversation appeared to be totally ignorant of fundamental

---

[269] Original in Dutch.   [270] General spaces and Cartesian spaces III, [Menger 1929].
[271] Cf. *Brouwer to Menger 3.I.1928*; the topic is the contribution of Vietoris and Tietze to the Encyclopedia.   [272] '*uitgevers*' in original.

dimension theoretical theorems (published in 1926), I am afraid that I will have serious objections to this article. I naturally suppose that you have arranged for the acknowledgments to you in the preface formulated in such a manner, that it will still be possible for me, notwithstanding my highest esteem for you, to express my objections to the fruit of the Tietze-Vietoris labor!—

In my tax affair, I'm sorry to have to bother you again. The letter which you were so kind to send me, I cannot post [273] because I have not received the tax assessment for 1927/28, and the form of 1926/27 was sent by the Laren/Blaricum tax office, to which I reported my moving, respectively departure, in the summer of 1926. Maybe you can inform me through a word from Hurewicz where I should direct my letter of my checking out.

From what I heard, you will receive one of these days an extensive letter from Ehrenhaft-Hahn-etc.

With respectful greetings, good bye

Yours sincerely [274]
Karl Menger

[Signed autograph – in Brouwer]

1928-01-20

From L. Bieberbach — 20.I.1928   **Berlin-Schmargendorf**
                                                            Marienbader Strasse 9

Dear Brouwer, [Lieber Brouwer]

First I would like to thank you cordially for your kind report on Mr. Süss; consequently I have proposed to approve his research grant.

Concerning the proposal of your article [275] for the academy: there is in some cases a difficulty because of § 17 of the academy regulations, which read as follows in paragraph 1:

---

[273] Menger refers here to a tax form.   [274] *Met waardeeringsvolle groeten en tot ziens – Uw dienstwillige.*   [275] *Intuitionistische Betrachtungen über den Formalismus,* [Brouwer 1928a, Brouwer 1928b, Brouwer 1928c].

> 'A scientific communication intended for the publications of the academy may in no case before it is published there, be published elsewhere in the German language, whether as abstract or in more extended form. If the editing secretary becomes acquainted with a publication that violates this rule, before it is published by the academy, then he must cancel the communication.'

Under these circumstances I ask you to inform me that your article in the 'Amsterdamer Berichten' [276] will *not* appear in the *German* language; then I think I may assume that the academy will consent to inclusion in the *Sitzungsberichte*. Unfortunately there is no possibility to deviate from this session regulation.

Finally, in the matter of your statements about the Conseil de Recherches, I see no possibility to include them in the Jahresbericht, [277] because it would create a novum when we would accept political statements in the Jahresbericht; thus it has been avoided until now, because of the political aspects, to mention the planned congress in Bologna. It seems to me that the proper place for your statements would be perhaps the *Hochschulnachrichten*. Personally I agree with you and I will not go to Bologna. [278]

The works of Weierstrass do not belong to the ones that members of the DMV can get for a reduced price. But if you tell me which volumes you want to obtain, and whether you want them bound or unbound, then because of our personal relations I will try to get a cheaper copy in some other way.

I have received now the proof sheets of his first communication, corrected by Mr. Menger. I assume that you would prefer to look at them when the handwritten corrections of Mr. Menger are in print.

With cordial greetings

Bieberbach

For the academy a short abstract is required. Do you think the enclosed one is all right for you? [279]

[Signed typescript – in Brouwer]

---

[276] KNAW Proceedings.   [277] JDMV.   [278] Sentence added in handwriting.
[279] Added in handwriting.

1928-01-23

To L. Bieberbach — 23.I.1928   Laren

Dear Bieberbach, [Lieber Bieberbach]

The translation of my article in another language than German is so difficult that I must for once, without exception, abandon writing a Dutch text for the Amsterdam communications and restrict myself to the publication of the German text in the Proceedings. [280] However I can promise to take care that the publication in Amsterdam in the Proceedings of the already submitted article will happen at least a month after the appearance in the communications of Berlin. It seems that by this promise the rules of the Berlin academy statutes are satisfied. Please tell me whether this solution is also satisfactory to you. In the opposite case I would also agree to a publication in the Jahresbericht, [281] but only when this publication by way of exception could be effected *immediately*.

As far as my statements about the Conseil des Recherches [282] are concerned, they are only in form, but not in actual content, more political than the invitation to the Bologna Congress (precisely this is explained to each reader by my arguments which expose the hidden meaning of the invitation). So when you cannot print my arguments in the Jahresbericht, then I will have it printed as a pamphlet, and I will ask you to send it together with the Jahresbericht as a separate supplement, just as it was done with the invitation for the Congress. [283] Please let me know whether you or Teubner agree with this proposal. It would be especially pleased if a few German mathematicians would cosign the pamphlet.

In the matter of the Menger proof sheets, the copy with the handwritten corrections would be most welcome, because I know the original text which was written in agreement with me, and I would like to get a quick survey of the subsequent changes. [284]

In case there would be a publication of an article in the Berlin communications, the abstract you wrote, which I return hereby, is completely adequate.

---

[280] [Brouwer 1928a, Brouwer 1928b, Brouwer 1928c].   [281] JDMV.   [282] Conseil Internationale d. Recherches.   [283] Bologna congress.   [284] Refers probably to [Menger 1928a], which deals with spreads from a classical point of view.

With cordial greetings

Your
Prof.Dr. L.E.J. Brouwer

[Carbon copy of typescript – in Brouwer]

28-02-16b

To H. Weyl — 16.II.1928[b]   **Laren**

Dear Weyl, [Lieber Weyl]

I was really pleased with the card you sent together with Révèsz and Geiger, from Arosa. Today I repay you with a more businesslike sign of life. For, in Utrecht there is an important mathematical list of candidates of the faculty: Barrau, Beth, Schaake (all three insignificant). [285]

My (alphabetic) list: Heyting, Hurewicz, Van der Waerden (in that order an intuitionist, a topologist and an algebraist). Heyting and Van der Waerden are Dutch, Hurewicz (my assistant) is in fact of Polish nationality and educated in Moscow and Vienna, but has settled already for a long time in Holland. To document my list for the minister, I need foreign testimonials. For Heyting (until now my only truly gifted intuitionistic student), only you qualify as a suitable author of a testimonial. Such a testimonial should on the one hand in general terms stress the importance of intuitionistic investigations at the present stage of development of mathematics (this is namely not at all believed outside of Amsterdam in Holland), and on the other hand it should qualify Heyting's articles (which I send you simultaneously) as pioneering.

The matter is extra difficult for me, because all three of my candidates are still young (well under 30) and the candidates of the faculty respectively 55, around 45, and around 40 years.

Many thanks in advance, and please rest assured that also without a sign of life your existence is something that is essential for me.

The first half of March I give talks in Vienna. What is that man Scherrer doing who sent me some time ago letters and articles on topol-

---

[285] H.J.E. Beth, the father of E.W. Beth.

ogy? Greetings to your wife and also to A[?]la and Mrs. Geiger when you see them.

In true friendship [286]

Your Brouwer

[Signed autograph – in Weyl]

1928-03-24

From A. Sommerfeld — 24.III.1928   **Munich**
                                              Leopoldstrasse 87, München

Dear Mr. Brouwer, [Lieber Herr Brouwer.]

Faber didn't write to you because he thought you would have heard from Bieberbach everything what he had to say to you about his negotiations with Bologna: he didn't know more than what had been discussed in the committee of the Mathematics Society. [287]

Today a new invitation to Bologna arrived. It didn't contain a word about the Conseil de Recherches or similar matters. Also the ominous enumeration of congresses was omitted. So I believe that you will have no difficulty with your efforts, for which we are very grateful.

Whether this letter will reach you in Bologna. It was a bit delayed.

Next week I will look more closely into the Michels [288] case and write to you to Amsterdam.

Hopefully you will soon recover. Schönflies would gladly put a few pounds of bacon at your disposal!

With cordial greetings

Your
A. Sommerfeld.

[Signed autograph – in Brouwer]

---

[286] *In treuer Freundschaft.*   [287] DMV.   [288] Michels, Amsterdam physicist.

28-04-12b

To R. von Mises — 12.IV.1928[b]   Rapallo

Dear Mises [Lieber Mises]

I first have talked to Pincherle, and subsequently corresponded with him. The result is as follows. The gentlemen in Bologna will send a new circular, in which neither the Union [289] nor the *real* congress will be mentioned, but on the contrary it announces a closing session of all congress attendants *and discussion about time, place and modality of the next congress* on the last congress day. So our hosts will organize the congress independently of the Union, and will clearly show this independence, and maintain it towards everybody.

However, they cannot make the facts go away, that the initiative for this congress was taken by the Union, and that the Union will have a meeting simultaneously with the Bologna congress. Just as little can they take the responsibility upon themselves that the Union will not try in Bologna to gain influence on the congress and on this closing session. Under these circumstances it seems to me that adherents and opponents of the Union can equally well take part in the congress, the latter with the intention that they will fight the Union if it should interfere with the congress, and if possible to destroy it. Moreover the congress participants that oppose the Union can continue their struggle against the Union during the months before the congress without being disloyal to the congress.

I spoke also with Levi-Cività in Rome and with Cipota in Palermo, and I have the impression that in Italy hardly anyone takes the Union seriously.

I hope to come to Berlin in the beginning of May, to discuss the matter once more with Schmidt, Bieberbach and you (and if possible also with Planck) on the basis of my correspondence with Pincherle.

Please inform Hahn and Ehrenhaft too about the above situation. Cordial greetings from your

Brouwer

[Signed autograph, postcard – in Mises]

---

[289] Mathematical Union.

1928-07-03

From H. Bohr — 3.VII.1928   **p.t. Fynshav Als**
                                                              Dänemark

Dear Mr. Brouwer, [Lieber Herr Brouwer]

Many thanks for your letter, which I got just forwarded from Göttingen. As you have perhaps learnt, I too have had an exchange of letters with Prof. Pincherle, and I even have written on May 26 a long letter (in the name of my good friend prof. Hardy and myself) to Pincherle, and received a detailed answer of Pincherle — unfortunately accidentally much delayed. Hardy and I expressed as strongly as possible our opinion that it would be absolutely necessary that 'the congress will be in every respect on a completely international footing and that the German participants have no different position from the others.' As you will probably know, Hardy and I have waged the same fight against the Conseil Internationale [290] as you in Holland, and we wrote also in our letter to Pincherle, how sad we were that such a Conseil was established, which carried unjustifiably the name 'international'. Also we have fought with all means against joining the Union (in Denmark I would certainly have succeeded to obstruct this joining, if Nörlund hadn't formed a committee in favor of joining, independently from our academy and Math. Society).

Actually, the point of view of Hardy and me was in principle the natural one, namely that we didn't want to have anything to do with a congress that like the congress in Bologna was so tied in its early history to the Union. [291] But when we thought (just as I see from your letter you thought) that we should try to help to make all mathematicians of the world come together in Bologna, it was important for us that we heard from all sides that the leading Italian mathematicians, Pincherle, Levi-Civita and so on, were internationally minded in the true sense, but foremost that we heard that in several circles in Germany people were prepared out of deep interest for the internationality of science, to ignore the foolish and sad previous history, *when only the congress itself would be fully international and would meet completely independent of the Union.* Pincherle's answer, through his letter to Picard (I speak now only about the actual contents and not about the form), of which I have received a copy from Pincherle, and most of all because of the new circular which explicitly gives completely equal rights to all real

---

[290] Conseil Int. d. Recherches.   [291] Mathematical Union.

participants (voting rights etc.), the congress has in my view been certainly and factually put on an international footing. From your exchange of letters with Pincherle I see with deep regret that you think that Pincherle did not achieve everything he promised you. Quite apart from this more personal question I find that since it has even been successfully arranged that the congress determines the place of the next congress, the Union is now — even with respect to questions that are not directly connected with this congress — so completely cut out, and 'we' internationally minded (i.e. people like you, Hardy etc.) have in fact won so completely, that I from my point of view would think it would neither naturally nor for the future look good, if the congress now would be sabotaged from the side of the Union opponents.

It would all have been much easier if we, who are of completely the same mind for these questions, would have contacted each other sooner, but because we were so outraged about the establishment of the Conseil and didn't want have anything to do with it, we have somewhat pushed away all questions connected to it. But I would think it just too sad when in the end the instigators of this corporation that is science unworthy, would attain that we, the opponents of the Conseil, having reached the point to score a complete victory, cannot come to agreement about relatively small questions and formalities, and that thereby a division would come between us like-minded.

With the most cordial greetings

Sincerely yours [292]
Harald Bohr

[Signed autograph – in Brouwer]

28-07-17

To A. Heyting — 17.VII.1928   Laren

Dear Mr. Heyting [Waarde Heer Heyting]

Your manuscript [293] has interested me very much, and I am sorry that you have to rush me to send it back. In the future I would appreciate it, if you made a copy of your manuscripts before you send them to me, at least if

---

[292] *Ihr ergebener.*   [293] On the formalization of intuitionistic logic, the sequel to Heyting's prize winning essay, cf. *Mannoury to Brouwer 26.I.1927.*

you appreciate a more than superficial reading by me. Meanwhile I have already now formed such a high opinion of your work, that I ask you to write it in German for the Mathematische Annalen (and rather somewhat more extensive rather than abbreviated). Maybe you can make then an even sharper distinction between the *original* signs and those that are introduced by definitions (as *abbreviations* for other symbols). And perhaps the notion of 'Law' can be formalized (in view of § 13). But these are only inessential remarks.

As to your remark concerning the [paper in] Mathematische Annalen 93, [294] p. 245, at the occasion of my Berlin lectures several improvements of 'On the founding of intuitionistic mathematics' [295] have turned out to be necessary. Among others I assigned then to each property as 'species' the 'identity with an arbitrary thing that possesses the relevant property'. I started then from the species of order zero, by which I mean either a given element of a spread or the identity with an arbitrary element of a given spread. A better way of treatment may however be, to introduce next to the things themselves, the 'species of identical things', and to consider the latter in the first place, similar to the manner in which in topological set theory not the points themselves are studied, but the point cores.

The Berlin lectures will soon appear in print. [296] If the publication is delayed then I will send you reprints of 'On the founding of intuitionistic mathematics' with the main improvements.

With friendly greetings

Your
L.E.J. Brouwer

[Signed autograph –in Heyting]

1928-09-27

From H. Härlen — 27.IX.1928   **Eislingen/Fils** [297]

Dear Professor, [Sehr geehrter Herr Professor]

Below I allow me to give you a *Report about Bologna*. I must state in advance that I can only render my subjective impressions, and that I don't

---

[294] [Brouwer 1925].   [295] *Zur Begründung der intuitionistischen Mathematik.*
[296] They appeared posthumously in 1991.   [297] The last page(s) of the letter are missing.

claim in the least to be complete, only that I report on more or less accidental observations of mine. Moreover, I will not treat the mathematically interesting things.

On arrival in Bologna (Sunday, September 2) I first was struck by a German poster, pointing to the information stand for congress participants. There seemed to be more German than French and English posters in the railway station. In the information stand a German speaking lady. In these external appearances the German was quite satisfactorily taken into account.

Monday, September 3: Opening session, very splendidly done by city and state. Speech of the *podesta*: [298] Welcome in the name of the town, the fascist town which is happy to show its foreign guests the achievements of fascism. Praise of fascism. Then welcoming speech by the *rector* to the guests that had responded to the *invitation of the university*. Then opening speech by *Pincherle*. Short report about the previous history of the congress which was the reason for the university to take it in its own hands. Clear effort to offend nobody. All the same he mentioned 'discordant voices, coming from diametrically opposing sides', and also that the exclusion of some nations in Strassbourg and Toronto were explained, if not *justified* by the 'morning after the war'. And later, that this state of mind nowadays wasn't *justified* anymore.

After Pincherle *Birkhoff* spoke in French and in English, and he thanked the Italian mathematicians for their work to create a *truly international* congress. Quite a few remarked that these thanks were not repeated in German. Afterwards speech by the *minister of education* about the significance of mathematics.

Afternoon: 1st session. Choice of the chairman. Proposal of the meeting: Pincherle, adopted with great applause. Then Pincherle makes proposals for vice-chairmen, adopted by acclamation: for Belgium: de la Vallée Poussin, France: Hadamard, Germany: Hilbert (very strong applause, very striking), Switzerland: Fehr (as representative of the education committee), England: Young (*as board member of the Union*), United States: Birkhoff (representative of the government), Scandinavia: Bohr, Spain & South America: Rey Pastor, Poland: Sierpiński, Russia: Lusin. An error by me in this list, especially in the order, is possible.

After this the first talk by *Hilbert*, who is greeted with a storm of applause. Frequent repetitions; his ability to concentrate clearly much influ-

---

[298] An old city governing position, going back to the Middle Ages, comparable to mayor; revived by Mussolini in 1926.

enced by physical suffering. Contents essentially known from recent publications. Great applause. — *Hadamard* is also greeted with great applause, and his talk is also very good in presentation— much more effective than the one of Hilbert. With Hadamard the applause afterwards was much stronger than beforehand. With Hilbert the applause was almost only for the person, with Hadamard also for the talk.

The longer talks that were given:

by Germans 3, French 3, English 1, Americans 2, Russians 1, Italians 6. Lusin and Birkhoff spoke in French, the other speakers in their mother tongues. Of the more than 400 section talks the most part were French, then came Italian, and at distance German and English. Among the participants Germany was strongly represented, also Poland, Hungary, Switzerland, Scandinavia and younger Frenchmen. Of the older ones many, among whom Borel and Painlevé, seemed to be absent because of external circumstances. Noticeably weak was the participation of England. Also the United States were weakly represented.

The participants received insignia on ribbons in the Italian colors. It would have been more tactful when they would have chosen the colors of Bologna. Not only for us Germans is it an ordeal to have to wear the colors of Italy, but also for a few other countries, e.g. Yugoslavs, Swiss and maybe the French. — In the concert given on the occasion of the congress the Italian national anthem and the fascist hymn were played, with Italian manifestations. Such manifestations occurred also at the breakfast organized by the city. At breakfast every menu was decorated with a small Italian flag. It was clearly expected that we would wear these flags, as was done at least by the Italians.

During this breakfast I entered a discussion with Mr. *Stoilow* (Romania) about the position of the Germans at the congress and their attitude to the Union. Mr. Stoilow told that Picard as chairman of the Conseils des Recherches could not take part — in his own words — in the congress because the invitation two years ago to Germany to join had remained unanswered. He moreover mentioned that the French were afraid that we would establish a German Union. I rejected this curious fear and represented the point of view: precondition for international cooperation is that the past should be thoroughly stowed away. Violations of one or the other side during or after the war are to be explained by war psychosis and should be considered as dealt with. The mentalities of peoples are too different, so every nation should show the greatest restraint and consideration. A Romanian whom I didn't know and who entered the conversation, recalled the manifest of the 84 [299] German scholars of 1914, which apparently also today gives offense. Because I don't know the manifest I didn't take a position on it, but I pointed out the situation of Germany then. I added that if in this manifest there are places that can only be explained by the situation of Germany at that time, but that are not justified today, then undoubtedly those scholars would not subscribe to that manifest today. All the time I stressed very much that in Germany the wish for a rapprochement *in all circles* is dominant, also in 'nationalistic' circles, provided this rapprochement does not include a humiliation.— Just as little as the manifest can a corresponding manifestation of the opposite side (Painlevé's introduction speech for the Conseil) constitute a basis for international cooperation. Hence before there is any question of Germany's joining, the Conseil has to base itself on a new foundation, or better yet, a whole new organization should be established.— Essentially Mr. Stoilow had to recognize my point of view, when he also observed that he as Romanian wasn't so sensitive in these matters and that he was amazed about our sensitivity. I have the impression that a rapprochement with the French is possible, even if there are maybe great difficulties to be overcome.

9. The breakfast was Saturday afternoon. In the afternoon an invitation from the Union to its members was distributed for a meeting on Sunday, which should take place during one of the general talks. In the evening I heard from Mr. *von Kerékjártó*, that Prague was considered for the next congress site. The invitation also came from the German university in Prague, because it expected a strengthening of German culture in Bohemia. For Hungary participation in a Prague congress is impossible, because of the situation of the Hungarian minority in Cechoslovakia. Mr. von Kerékjártó pleaded for Switzerland. Even though the idea of the Germans from Prague appealed to me, I have to admit that with the present situation in the world only a congress in a truly neutral country like Switzerland is possible.— Until now I haven't used your file for reasons I already informed you about, except that I told to some gentlemen the matters related to the final session. I left it further in the hands of Mr. von Kerékjártó.

The final session in Florence on Monday, September 10, started with a welcome by a Florentine magistrate and a talk by *Birkhoff*. Then the choice of the next meeting place followed. *Pincherle* announced that an invitation from Switzerland had come. So he proposed Switzerland. The proposal

---

[299] Most likely 'of the 93'. This was manifest signed by 93 prominent German scholars, who reports about misconduct of the German military. C.f. [Van Dalen 1999] p. 337.

was adopted with great applause. The *representative of Switzerland* (name unknown to me) presented his invitation first in German, then in French, and then thanked in the German language the Italians and especially Pincherle for the magnificent course of the congress.

After the final session Mr. Stoilow told me in short about the Union meeting. Pincherle had resigned as chairman of the board, but remained a board member. A request to appoint a committee to clear up the relations with Germany, is superseded because the board has been charged with that problem. A suggestion of Holland as next congress location was mentioned. But because of uncertainty about your position this idea was abandoned, and also no further proposals have been made.— Because of the method used in Florence, the Union is for the time being without any influence on the organization of future congresses.

About the mood during the congress it must be said that overall it was good. The relations between the subjects of different nations were friendly. Where there were dangerous moments, one really managed very well to take away all conflict matter.

Finally, let me say a few words about us Germans. For us the trip to Bologna was very taxing because of the German-Italian relations. The Italians celebrate the date of their declaration of war [300] as a national holiday; they know that for us that war declaration has a special meaning. But what is much worse, is the situation in South Tirol. I know the situation from own experiences, and I have to say that they are much worse than one can imagine from even the most detailed press reports. Such a horrible brutality against a minority has no precedent in the entire civilized world. In view of this fact it is actually impossible that a German accepts the hospitality of the Italian government. That this was the case in Bologna, was because we had no influence on the choice of the congress location and because of the role of mediator of the Italians, a rejection would be misunderstood.— Whether the trip to Lake Ledro has gone through, I don't know. Most congress participants joined the trip to Ravenna. Incidentally, Lake Ledro lies in territory that is undoubtedly Italian [301]

[Typescript – in Brouwer]

----

[300] May 23, 1915 Italy declared war, in the hope to gain pieces of territory such as South Tirol, i.e. the region around Bolzano.   [301] Document breaks off here.

Editorial supplement

H. Härlen to Ms. I. Gawehn [302] — *27.IX.1928*

Dear Miss Dr. Gawehn! [303]

Would you please tell Professor Brouwer:

To my report I must still add that a committee of representatives of the whole world has deliberated about the site of the next congress. So the meeting in Florence received an already established proposal. How the choice of the representatives for this committee was made I don't know. I only know that *Landau* belonged to it (he is said to have proposed Jerusalem) and probably also *Hahn*.

With friendly greeting

Yours sincerely [304]
H. Härlen

[Signed typescript, postcard – in Brouwer]

———————

28-10-25a

From D. Hilbert — 25.X.1928[a]   **Göttingen** [305]

Dear Colleague, [Sehr geehrter Herr Kollege]

Because it is not possible for me, given the incompatibility of our views on fundamental questions, to cooperate with you, I have asked the members of the editorial board of the Mathematische Annalen for the authorization, and received that authorization from Messrs. Blumenthal and Carathéodory, to inform you that henceforth we will forego your cooperation in the editing of the Annalen, and that consequently we will delete your name from the cover page.

---

[302] Brouwer's assistant.   [303] 'Sehr geehrtes Fräulein Dr. Gawehn'. In handwriting preceding the typescript.   [304] *Ihr sehr ergebener.*   [305] The letter was not opened by Brouwer, see [Van Dalen 2005] section 15.3. The text is taken from *Hilbert to Einstein 25.X.1928.*

At the same time I thank you in the name of the editorial board of the Annalen for your past activities in the interest of our journal.

Sincerely yours [306]

D. Hilbert

[Typescript copy – in Einstein]

1928-11-02a

To O. Blumenthal — 2.XI.1928[a]   Laren

Dear Colleague, [Werter Kollege]

On October 27 I received simultaneously a '*Kennisgeving*' [307] concerning two registered letters from Göttingen and a telegram from Erhard Schmidt, [308] which made me postpone the collection of the letters for the time being, but to wait with that until the visit of Carathéodory that was announced in the telegram.

During this visit, which took place on October 30, both letters were present, unopened, and from the statements of Carathéodory I gathered:

about one of the letters (which had no sender's address on it).

1. That the communication in this letter should have, according to the rules, either several signatures or yours.

2. That in the letter the name Carathéodory is mentioned not in accordance with the facts (but that Carathéodory will not disavow the letter, should I have learned the contents).

3. That the sender of the letter would within a few weeks probably seriously regret sending it.

Thereupon I have decided not to open or read the letter.

about the second letter.

1. That your name as sender on the envelope was incorrect and that the letter was written by Carathéodory.

2. That Carathéodory regretted the contents of the letter.

---

[306] *und ergebenst.*   [307] Notification (from the postal office).   [308] *Schmidt to Brouwer 27.X.1928.*

Thereupon I gave the letter back to Carathéodory unopened.

Furthermore Carathéodory informed me that the board of chief editors of the Mathematische Annalen planned to remove me from the board of editors of the Annalen (and that it felt legally entitled to do so). This because Hilbert wished that removal, and because his state of health demanded indulgence. Carathéodory asked me, out of compassion with Hilbert, who was in such a state that one could not hold him accountable for his misdemeanor, that I would accept this infuriating insult with equanimity and without resistance.

With respect to this plea of Carathéodory I have made a reservation to decide after calm deliberation. Today I have decided. You find enclosed the copy of a letter to Carathéodory. [309]

Your
(signed) Brouwer.

[Carbon copy – in Brouwer]

28-11-02b

To C. Carathéodory — 2.XI.1928[b]   Laren [310]

Dear Colleague, [Werter Kollege]

After careful consideration and extensive consultations I have to take the point of view, that the plea you directed to me, namely to treat Hilbert as of unsound mind, could only be complied with, if it would have reached me in writing, and in fact jointly from Hilbert's wife and his family doctor.

Your
(signed) Brouwer.

To Prof. C. Carathéodory.

[Typescript copy – in Brouwer]

----

[309] *Brouwer to Carathéodory 2.XI.1928.*   [310] A copy was enclosed in *Brouwer to Blumenthal 2.XI.1928*

1928-11-05a

To Eds. Mathematische Annalen — 5.XI.1928[a]   Laren

To the publisher and the editors of the Mathematische Annalen.
[An Verleger und Redakteure der Mathematischen Annalen]

From information communicated to me by one of the chief editors of the Mathematische Annalen on the occasion of a visit on 30-10-1928 I gather the following:

1. That during the last years, as a consequence of differences of opinion between me and Hilbert, which had nothing to do with the editing of the Mathematische Annalen (my turning down of the offer of a chair in Göttingen, conflict between formalism and intuitionism, difference in opinion concerning the moral position of the Bologna congress), Hilbert had developed a continuously increasing anger against me.

2. That lately Hilbert had repeatedly announced his intention to remove me from the board of editors of the Mathematische Annalen, and this with the argument that he could no longer 'cooperate' [311] with me.

3. That this argument was only a pretext, because in the editorial board of the Mathematische Annalen there has never been a cooperation between Hilbert and me (just as there has been no cooperation between me and various other editors). I have not even exchanged any letters with Hilbert since many years and that I have only superficially talked to him (the last time in July 1926). [312]

4. That the real grounds lie in the wish, dictated by Hilbert's anger, to harm and damage me in some way.

5. That the equal rights among the editors (repeatedly stressed by the editorial board within and outside the board [19] allow a fulfillment of Hilbert's

---

[19] From the editorial obituary of Felix Klein, written by Carathéodory 'He (Klein) has taken care that the various schools of mathematics were represented in the editorial board and that the editors operated with equal rights alongside of himself—He has (...) never heeded his own person, always had kept in view the goal to be achieved.' (From a letter from Blumenthal to me, 13-9-1927). 'I believe that you overestimate the meaning of the distinction between editors in large and small print. It seems to me that we all have equal rights. In particular we can speak for the *Annalenredaktion* if and only if we have made sure

---

[311] *zusammenarbeiten.*   [312] Brouwer lectured on July 22 in Göttingen on *Überall und scheinbar überall definierte Funktionen* (Functions that are defined everywhere and functions that are defined apparently everywhere). At that occasion there was a reconciliation between Brouwer and Hilbert, see [Alexandrov 1969] and [Van Dalen 2005], p. 571.

will only in so far that from the total board a majority should vote for my expulsion. That such a majority is scarcely to be thought of, since I belong to the most active members of the editorial board of the Mathematische Annalen, since no editor ever had the slightest objection against the manner in which I fulfill my editorial activities, and since my departure from the board, both for the future contents and for the future status of the Annalen, would mean a definite loss.

6. That, however, the often proclaimed equal rights, from the point of view of the chief editors, was only a mask, now to be thrown off. That as a matter of fact the chief editors wanted (and considered themselves legally competent) to take it upon themselves to remove me from the editorial board.

7. That Carathéodory and Blumenthal explain their cooperation in this undertaking by the fact that they estimate the advantages of it for Hilbert's state of health higher than my rights and honor and professional prospects, [313] and than the moral prestige and scientific status of the Mathematische Annalen, that are to be sacrificed.

I now appeal to your sense of chivalry and most of all to your respect for Felix Klein's memory, and I beg you to act in such a way, that either the chief editors abandon this undertaking, or that the remaining editors split off and carry on the tradition of Klein in the management of the journal by themselves.

L.E.J. Brouwer

[typescript copy – in Brouwer]

---

of the approval of the editors interested in the matter under consideration. — Although I too take the distinction between the two kinds of editors to be more typographical than factual (I make an exception for myself as managing editor), I understand your wish for a better typographical make up very well. You know that I personally warmly support it. However, we can for the time being, as long as Hilbert's health is in such a shaky state as it is now, change nothing in the editorial board. I thus cordially beg you, to leave your wish for later. At the right moment I will certainly and gladly bring it out.

[313] *Wirkungsmöglichkeiten.*

1928-11-06b

O. Blumenthal to Editors Math. Annalen — 6.XI.1928^b Aachen

To the editors of the Mathematische Annalen. [An die Redakteure der
Mathematischen Annalen]

Dear Colleague,

I accidentally learned that in the affair, you are familiar with, B r o u w e r
has written a letter to the joint editors and the publisher. I beg you not to
answer this letter, before you have received from me an extensive exposition
of some new events, which appear to me essential for judging the situation.
You will receive this exposition within a few days.

Best greetings,

Your
O. Blumenthal

[Copy of signed typescript – in MA collection]

1928-11-16a

O. Blumenthal to Eds. Math. Annalen — 16.XI.1928^a Aachen

To the publisher and the editors of the Mathematische Annalen
[An Verleger und Redakteure der Mathematische Annalen] [314]

As manager of the editorial board of the Annalen I feel obliged to make a
rejoinder to Brouwer's circular to the publisher and editors of the Mathematische Annalen. [315] I base my explanations partly on letters from Hilbert,
Carathéodory and Brouwer, and partly on a long and detailed conversation
that I had with Hilbert in Bologna.

---

[314] According to *Blumenthal to Courant 12.XI.1928* no copy was intended for Brouwer.
Eventually Blumenthal sent a copy of the final version to Brouwer (see *Blumenthal to Bohr
& Courant 4.XII.1928*). In the Brouwer archive there are typescript copies of the circular.
It is plausible that one of the editors – most likely Bieberbach, but possibly Carathéodory –
sent a copy to Brouwer.   [315] *Brouwer to Publisher and editors Math. Annalen, 5.XI.1928.*

I want to remark in advance that the formulation of Brouwer's letter is misleading: one might get from it the impression that the editor that visited Brouwer on October 30 (Carathéodory) has drafted statements 1–7. This is of course not the case with any of them, they are rather the viewpoints that Brouwer has formed for himself.

In the following I give a brief representation of the events and I will go into Brouwer's letter in the appropriate places.

1. Hilbert's letter and his reasons.

The letter that Hilbert sent to Brouwer on October 25, is as follows.

Dear Colleague!

> Because it is not possible for me, given the incompatibility of our views on fundamental questions, to cooperate with you, I have asked the members of the editorial board of the Mathematische Annalen for the authorization, and received that authorization from Messrs. Blumenthal and Carathéodory, to inform you that henceforth we will forego your cooperation in the editing of the Annalen, and that consequently we will delete your name from the cover page.
>
> At the same time I thank you in the name of the editorial board of the Annalen for your past activities in the interest of our journal.
>
> Sincerely yours
> D. Hilbert.'

This letter has not been opened by Brouwer, as I must remark already here, and motivate later. He was, however, informed by Cara [316] about its contents, more specifically about the reasons of Hilbert's actions mentioned in the first sentence. Brouwer's points 2 and 3 refer to this. About this I want to say the following:

On point 2 and 3. Brouwer interprets the idea of cooperation in an extrinsic sense (point 3). This is a complete misjudgment of the true interpretation. It is rather so, that Hilbert has acquired the firm conviction that Brouwer's actions are damaging for the Annalen, and that he therefore cannot take the responsibility to act as chief editor in a board to which Brouwer belongs. So it is in no way a pretext.

---

[316] Nickname for Carathéodory.

On point 1 and 4. The grounds given by Brouwer for Hilbert's acts don't apply. The reason given in point 4 is spiteful and hence needs not be answered. Also the scientific difference of opinion concerning the foundations of mathematics plays no role. More specifically, it is not correct, as Brouwer seems to suggest in point 5, that the mathematical direction he represents will in the future be heard less in the Annalen. Also Brouwer's circular letter before the Bologna congress, by the wording of which Hilbert felt insulted, has only together with other, maybe more important factors, brought about the decision. The motives are lying much deeper. I give them in my formulation, but I am certain that they completely represent Hilbert's meaning.

Felix Klein has been until his resignation from the editorial board [317] a kind of highest authority among us, who was called upon in difficult cases or who acted on his own initiative to support important decisions (e.g. the transfer of the Annalen to the Springer Verlag), or to resolve differences within the editorial board. It is good and necessary that in a numerous board such as ours there is such a higher authority, who is not concerned with the details but keeps an eye on the general context and feels responsible for it. After the death of Klein [318] Hilbert has thought himself obliged to fulfill this function, and he already has acted in this sense, and I at least have personally always recognized him as such. Hilbert has seen in Brouwer an obstinate, unpredictable and dominant character. He was afraid that once he would resign from the editorial board, Brouwer would bend it to his will and he has considered this such a serious danger for the Annalen, that he wanted to counteract him when he still could do so. Probably under the influence of his recent illness he felt obliged, in the interest of the Annalen, to effectuate Brouwer's exit from the board, and to implement this measure right away and with all his energy.

Cara and I who have been friends with Brouwer for many years, had to recognize the objective correctness of Hilbert's objections to Brouwer's editorial activities. Although Brouwer was a very conscientious and active editor, he was really difficult in his contacts with the management and meted out difficulties to authors that were hard to bear. For example, manuscripts that had been sent to him for refereeing were stored for months, because he, on principle, first had copies made of all articles refereed by him. (I just recently had an example of that.) There is no doubt whatsoever that Klein's resignation from the editorial board goes back to Brouwer's rude behavior (although in an affair where Brouwer formally was right [319]). The further

---

[317] 1924.   [318] 1925.   [319] The Mohrmann affair.

developments (see below) have shown that Hilbert was even more right than we thought at that time.

Because we could not ignore the factual justification of Hilbert's point of view, and because we saw ourselves confronted with his irrevocable determination, we consented to Brouwer's removal from the board. Only we wished — unjustified, as I see now — a milder form, in that Brouwer should be persuaded to resign his editorship himself. But Hilbert could not be persuaded, and finally we have, although reluctantly, decided to give him a free rein. Einstein has not consented, with the motivation that one should not take Brouwer's peculiarities seriously.

Point 5 and 6. I will not examine here in how far it is justified that the other editors were not informed in advance of Hilbert's plan. Formally the justification seems to be given by the distinction made on the Annalen cover between 'advisors' and 'editors'. [320]

II. The events after the letter was sent

On October 26 and 27 Cara and I were in Göttingen to discuss the situation. Then Cara went on to Berlin to discuss the matter. Although he objectively held the removal of Brouwer from the board for unavoidable, he decided in Berlin to make a last effort to come to an amicable agreement by softening the categorical form of the dismissal. So he came on the 30th [321] to Laren, after Brouwer had been telegraphically requested not to take any steps. Because Brouwer hadn't opened Hilbert's letter, Cara told him the contents (but not the formulation) and proposed to him to step down voluntarily from the board of editors of the Annalen, and leave the letter unopened. He thus wanted to prevent Brouwer to feel insulted by the form, and felt justified because it seemed to him that its rudeness was partly caused by Hilbert's ailing condition. He left Brouwer in the dark about the fact that in our opinion he should step down from the board, and asked him, out of compassion with Hilbert and his disease to resign by himself. Brouwer reserved a decision until after calm deliberation. He has left Hilbert's letter unopened and on November 2 wrote the following letter to Cara:

> Dear Colleague,
>
> After careful consideration and extensive consultations I have to take the point of view that the plea you directed to me, namely

---

[320] *'Mitwirkenden' und 'Herausgebern'*; the present day formulation would be 'associate editors and (chief) editors.   [321] October

to treat Hilbert as of unsound mind, could only be complied
with, if it would have reached me in writing, and in fact jointly
by Hilbert's wife and his family doctor.

Your Brouwer.'

For this horrible and repulsive letter, which Brouwer has communicated
also to me by means of a copy, I have only this one explanation, that Brouwer
(intentionally or involuntarily) has put together for himself the ugliest view,
from Cara's utterances and pleas. I have to admit — and Cara has written
the same to me —, that I have thoroughly misjudged Brouwer's character,
and that Hilbert understood him and judged him more accurately than we
did. I too am unable to cooperate further in the editorial board with the
writer of this letter, and I now also actively take Hilbert's side. I can't
understand that Brouwer after this letter can appeal in the final paragraph
of his circular to the chivalry of the editors and the memory of Felix Klein.

I ask you gentlemen either to speak out soon, or for your tacit approval
that from the next issue on Brouwer's name is omitted from the cover page
of the Annalen and that he receives no further Annalen-information.

Yours sincerely
O. Blumenthal

[Copy of signed typescript – in Einstein, typescript copy in Brouwer]

1928-12-22

D. Hilbert, F. Springer to Eds. Math. Annalen — after 22.XII.1928 Göttingen, Berlin [322]

[note on Brouwer's carbon copy in his handwriting:] received 27.12.28

Dear Sir, [Hochgeehrter Herr]

The editors until now of the Mathematische Annalen have together with
the publisher agreed that with the publication of the 100th volume the old
contract will be terminated and replaced by a new one at the publication of
volume 101.

---

[322] This letter is dated XII.1928; the same letter has been sent to Courant with the
date 22.XII.1928 [copy in Brouwer archive]; this suggests a date between 22.XII.1928, and
the date of delivery: 27.XII.1928.

At the same time a change will take place insofar that Carathéodory and Einstein have withdrawn themselves, and Hecke has joined.

The revision of the publisher's contract is combined with a fundamental change in the manner of management. It has been shown desirable, that for the acceptance or rejection of articles only the real editors take the full responsibility, and that they will be satisfied with soliciting referee reports from colleagues outside, without burdening them with a final responsibility. Accordingly only the names of the responsible editors will be shown on the title page, starting from volume 101.

The publisher and the editors use this occasion, to express our warm thanks to all those who have until now regularly taken part in the publication of the Annalen as associate editor, [323] for the rendered exceptionally valuable work, and to combine this with an appeal, that the cooperation in the form of referee reports also in the future will not be refused. Independent of this the publisher wishes to show his gratitude for the shown help, by making available to all of the gentlemen concerned a free copy of the Annalen, as before.

For the Editorial board of the Mathematische Annalen
    (signed) D. Hilbert

For the Publisher of the Mathematische Annalen
    (signed) F. Springer

[Signed typescript – in Einstein; signed carbon copy of typescript – in Brouwer]

28-12-23b

R. Courant, H. Bohr to C. Carathéodory — 23.XII.1928[b]
                                                            Göttingen [324]

Dear Carathéodory, [Lieber Carathéodory]

Many thanks for your letter of December 19, and most of all for the announcement of your visit.— The Annalen matter is now formally wrapped up: the new contract has been signed and the circular of Springer and Hilbert has been sent. Bohr and I are like you very happy about the conclusion of

---

[323] *Mitwirkende.* [324] There is a draft and a (presumably) final version with a letter of Bohr appended. The corrections in the text are clearly Bohr's.

this affair that has worried us so much during the last month and that made great demands on our thoughts and time for work.

Our satisfaction about the solution of the crisis would be even greater if not a couple of phrasings of your letter worried us because they suggest a possibility for new misunderstandings. It is about the question what the real motives of Hilbert were.

When I first heard of Hilbert's intention, the immediate reaction was a shock. [325] Because at first Hilbert also did not [326] explain his motives to me, I have only gradually understood these clearly. But now, where it is again possible to speak calmly and in detail with Hilbert, all doubts have vanished that Hilbert's motives were absolutely objective, based on his sense of responsibility for the Annalen, and moreover on his understanding that Brouwer's personality could be dangerous for the Annalen when Hilbert wouldn't be able anymore to act as a counterbalance. Hilbert has stressed again and again to us that he has no personal feelings of hate, anger or offense against Brouwer, and that he rather deemed a factual separation necessary and that he wanted to carry that through with all his strength. The more radical solution to abolish the whole advisory board was immediately and eagerly adopted by Hilbert, not only because he thought it objectively useful, but also he was very happy with it because thereby the personal edge against Brouwer was taken from the whole action.

So it is nothing less than a construction after the fact, if one now, at the winding up of the matter, stresses these objective motives, even though the first step taken by Hilbert under such singular circumstances could create a different impression.

To point emphatically to this state of affairs, seems — in the very first place because of Hilbert — to be our duty. In the whole affair we have acted in his name, and we cannot admit that a version about his intentions becomes public that does him no justice. When already you accept such a view, what should we expect from those who are farther removed? Our responsibility to Hilbert on this point is all the greater, because until now Hilbert hasn't been informed about all details of the development of the conflict; more specifically he is totally unaware of your visit to Laren and the outrageous representation of that by Brouwer. [327] So he doesn't know that the reproach of subjectivity and personal wish for revenge has been

---

[325] In draft: 'a mild shock'   [326] The word corresponding to 'not' is missing, but comparison with the draft learns that this is a copying error, caused by a slight rephrasing. [327] In draft this part runs slightly different: 'in particular nothing about your visit to Laren and the distorted representation as reported by Brouwer, and he does not suspect, that the reproach ...'

raised against him; he cannot defend himself against that and we must take that task upon us as long as we haven't informed Hilbert about all details, which we would so much like to avoid in the interest of all concerned.

It remains to take the future relations between German mathematicians among each other into consideration. When some of the colleagues do not learn to understand, what is really at the bottom of Hilbert's mind, then the bad feelings won't go away and can erupt here and there. When such a latent tension — which won't come from the circle of Hilbert — in the future is to be avoided, then we must use this moment now to remove any unjustified ugly appearance from the matter, and enter into a basis of mutual understanding and trust. It would be very gratifying and reassuring if you could help with it, that all concerned, especially also the Berlin colleagues, take this attitude.

Many cordial greetings and also Christmas wishes from house to house

Your
Courant

———————————

Dear Carathéodory, [328]

I add two words to Courant's letter. First to say how much I and my wife are looking forward to seeing you and your wife in January in Göttingen. But secondly also, because I want to tell you of my own accord personally how much you have, in my opinion, misunderstood Hilbert, when you think that he wanted Brouwer removed from the board, just because he felt personally insulted. I had never doubted that you, like and me in the discussion with you, were quite clear about it that Hilbert (correctly or not) thought *that Brouwer's stay* in the board *would constitute a danger for the future.* When you are not completely convinced yourself then the only right thing to do is really that you ask Hilbert quite openly about his reasons, because Hilbert — without him knowing it, so he can't defend himself — is first considered of unsound mind and then as not-objective, [329] this is a situation which I, as representative of Hilbert, in the long run cannot bear standing by idly.

With best greeting, sincerely yours [330]
H. Bohr

[Copy of carbon copy – in MA collection]

———————————

[328] Lieber Carathéodory.   [329] *'unzurechnungsfähig', 'unsachlich'.*   [330] *Mit den besten Grüssen, Ihr ergebener.*

[*Editorial supplement:* Carathéodory re Hilbert's motives. From *Carathéodory to Courant* — *19.XII.1928.*]

I am tremendously happy about the final settlement of the Annalen Affair and also about the fact that Hilbert has acknowledged that I have done the best possible for him. I have admired from the beginning the strength with which he attacked Brouwer. He has, however, indicated as the sole grounds for his decision at the time, that Brouwer had insulted him; I would find it unworthy if one would construe after the fact, that he was motivated by impersonal grounds.

1929-01-23

To Editors Mathematische Annalen [331] — 23.I.1929   Laren

To Messrs. Bieberbach, Bohr, Carathéodory, Courant, von Dyck,
Einstein, Hoelder, von Karman, Sommerfeld.   [An die Herren
Bieberbach, Bohr, Carathéodory, Courant, von Dyck, Einstein,
Hoelder, von Karman, Sommerfeld.]

Because I persist for the time being in the interest of the decorum of the mathematical community in the point of view that I expressed in my circular of December 23, 1928, namely to await the result of Carathódory's efforts, and only correct the erroneous impressions contained in Blumenthal's circular, if the possibility of a rectification by the other side cannot be counted on anymore, I restrict myself right now to take position on the Hilbert-Springer circular of December 1928, which I only received after I had sent my circular of December 23, 1928.

1. The Mathematische Annalen constitute a spiritual heritage, a common spiritual property of the whole editorial board, which has got together to serve the collective progress of mathematics without regard for personal scientific activity. The so-called chief editorial board was established by free choice of the joint editors [20] and occupied a merely representative position

---

[20]This character of appointment doesn't change by the fact that usually a formal choice by a majority vote is replaced by informal discussions within the total board.

[331]Blumenthal and Hilbert excluded.

with respect to the public. [21] The formal right with respect to contracts with the publisher constitutes therefore for the chief editors not an inherent possession, but something that has been entrusted to them. And if Messrs. Hilbert and Blumenthal purloin these entrusted goods from their principals, then they commit a misappropriation, also when this accidentally cannot be challenged legally. [332]

2. The role of Blumenthal as revealed in the Hilbert-Springer circular can be described as a breach of trust and faith on the following grounds:

Firstly, Blumental has in his quality of managing editor repeatedly and in the most unambiguous way acknowledged the structure of our circle as explained above. An even clearer example than the one mentioned in my circular of November 5, 1928, is the following statement in a letter of October 12, 1924: 'The editorial board of the Annalen was from the outset a democratically organized institution where all editors have equal rights. We would like to uphold this principle or rather to revive it.'

Secondly, in the summer of 1925, when in my opinion the amount of irregularities committed by Blumenthal as manager had become excessive because of a very serious infringement, and I demanded a full session of the whole board to discuss this and to prevent repeats, I only relinquished this request on the explicit announcement of Blumenthal's plan to stay on as manager at most until volume 100. [22] Volume 100 has just now, on December 28, 1928, been wound up.

According to the above the editors of the Annalen have to recognize as the contents proper of the Hilbert-Springer circular, that Hilbert and Blumenthal as editors and Springer as publisher have thus advocated their dismissal. The remaining editors therefore have the task to further administer the inheritance of Felix Klein together with a new publisher and continue

---

[21]If this interpretation of the structure of our circle hadn't since 1914 been repeatedly emphasized to me by several editors, especially by our leader Felix Klein (who also took this most conscientiously into account during the handling of several incidents in which he and I were involved), then the responsibility experienced by me as editor and also the activity I took upon me would never have reached the magnitude which in fact existed and actually is known among my co-editors only to Blumenthal.   [22]When I reminded him orally of this in August 1927, I received from Blumenthal the evasive answer that it was very difficult, as long as Hilbert was alive, to change anything in the board. Blumenthal himself has given a striking refutation of this pretext.

[332]Observe   the   similarity   to   Brouwer's   comment   on   consistency   proofs, [Brouwer 1923b] p. 3.

the Klein tradition of running a mathematical journal.

(signed) L.E.J. Brouwer

[Carbon copy of typescript – in Brouwer]

1929-04-30

To Editors Mathematische Annalen — 30.IV.1929   **Laren**

To the publisher and the editors of the Mathematische Annalen.   [An Verleger und Redakteure der Mathematischen Annalen]

1.

To my amazement and disappointment, notwithstanding my demand, so far no rectification from the other side, of the false expositions contained in the Blumenthal circular of November 16, 1928 has appeared. My amazement and disappointment concern most of all the circumstance that Carathéodory did not feel it his duty of honor to gainsay Blumenthal's statements concerning his visit to me on October 30, 1928, and to confirm the statements in my circular of November, 5, 1928.

Therefore I take the floor myself.

The points 1–7 formulated in my above mentioned circular are not, as the Blumenthal circular falsely pretends, 'viewpoints that Brouwer has formed for himself', but *viewpoints that during the mentioned visit came up between me and Carathéodory in mutual agreement, i.e. that each time was enunciated by one of us and accepted by the other.*

To substantiate this I provide details about the visit of Carathéodory, pointing out that I defend myself against Blumenthal's slander, and how I was driven to the general statements in my circular of November 5, 1928 concerning the earlier mentioned visit, by the necessity to defend myself against Hilbert's attack that was announced in the course of visit.

As was already stated in the annex of my circular of December 23, 1928, I received on October 27, 1928 simultaneously a 'Notice' of two registered letters from Göttingen and the following telegram from Berlin that made me to collect the letters at a later time: 'Professor Brouwer. Laren N.H.— Please do not undertake anything until you have talked to Carathéodory, who must inform you about a matter unknown to you with the greatest consequences. The matter is completely different from what you must believe from the letters received. Carathéodory comes to Amsterdam on Monday. Erhard Schmidt.'

During his visit on October 30, 1928, Carathéodory informed me first, while the two letters that just had been collected were lying unopened before us, that the 'matter unknown to you with the greatest consequences' consisted of the following: recently the taking of a wrong medicine had produced in Hilbert such a state that he on the one hand 'could not be taken seriously anymore at all' (words of Carathéodory),[23] and that on the other hand the slightest resistance to his will could be fatal to him.

In this situation the idea had come up to remove me from the editorial board of the Annalen and he wanted to carry out this idea with all means.— It was evident that the realization of Hilbert's plans would constitute a grievous injustice. In order not to endanger Hilbert's life, he (Carathéodory) begged me not to undertake anything against this for the time being. Hopefully Hilbert would soon use the right medicine again, and as a consequence of the improvement of his situation, come to better views, before anything definitive had happened.

One of the closed letters present was from Hilbert. The statement in it, that Hilbert fired me as editor, 'authorized by Blumenthal and Carathéodory' were unjustified; because when he (Carathéodory) after his return from America had been requested in writing by Hilbert for this authorization, he answered: he would in principle not put any obstacle in Hilbert's way, but he would come to Göttingen to discuss the matter. When he arrived in Göttingen, he heard from Blumenthal that Hilbert had already dispatched his letter of dismissal, under reference to the mentioned authorization. In the subsequent discussion of half an hour with Hilbert the matter was not touched on, as little then as today.[24] — With reference to the second letter (which carried on the envelope Blumenthal's name as sender), this was written by him (Carathéodory), and in this he asked me to resign voluntarily

---

[23]One could think for a moment that communicating such utterances is somewhat incorrect, because naturally one assumes a certain degree of confidentiality with reference to these. But the assumption of confidentiality and the ensuing solidarity can certainly not, insofar as they have not become null and void because of the further course of the conversation as sketched below, be brought into agreement with Carathéodory's later silence upon Blumenthal's false impressions. Moreover, also justified scruples must in the case at hand, where it concerns the clarification of a scandalous calumny and robbing someone's position, yield — in analogy to the case of hearing witnesses in a criminal process.
[24]Einstein, too, was asked by Hilbert for authorization, but he refused.

from the board of editors out of consideration with Hilbert's state of health. But now he regretted the formulation of this letter.

Thereupon I have returned this latter letter closed to Carathéodory, and I have told him that I considered my possible removal from the editorial board not only a grievous injustice, but also a serious damage to my scope of action, and as an insult of my honor in the public opinion; and that, if this unheard of event really would come to happen, my honor and my scope of action could only be restored by a most extensive appeal to the public, and hence that an atrocity committed against me would result in a public scandal.— Carathéodory answered that he had been prepared for such a standpoint on my part, that in his opinion the Annalen would be ruined through the realization of the plans hatched against me, and that he himself already had taken the decision to resign from the board, a decision which actually — again out of consideration with Hilbert's state of health — could for the time being not be carried out.

The further course of the discussion then brought the seven points mentioned in my circular of November 5, 1928.

With respect to the desired consideration for Hilbert's state of health of me by Carathéodory, I expressed my opinion that in case there was a direct risk of Hilbert's life, it would be a crime to be an accessory to see him ending his life with a crime; but on the other hand unreasonable tolerance could increase his petulance and lust for power in way that could put the happiness of his life in danger. I promised however that I would discuss this last psychological question with appropriate acquaintances. In case my point of view would not change after closer consideration, then yielding to Carathéodory's plea to undertake nothing for the time being against the realization of Hilbert's plans, would be equivalent for me to the probability that these plans would be cancelled without active interference by me.— The discussion closed with Carathéodory repeated pointing at Hilbert's terrible situation, and the words that he (Carathéodory) under these circumstances 'appealed to my mercy'.

During this discussion of two hours in the morning of October 30, Carathéodory's attitude was all the time that of a confidant, friend and ally, who advised me on the possibilities and means to prevent a calamity. The discussion seemed to be concluded in full agreement, notwithstanding the tentative differences in our opinion on details of the affair. Accordingly, Carathéodory stayed still several hours together with me and a few guests, who were invited because of him, who all had the impression of an untrammeled atmosphere. Only at the farewell, when I was alone again with Carathéodory, I expressed the thought that occurred to me only at that moment, that since Hilbert had been able to face Einstein's objections to his plan, he also could bear without any danger a repudiation of the unjustified authorization in his letter to me. Only when I didn't get a logical answer from Carathéodory to this remark, but only (maybe to be attributed to the agitation of the farewell) received as answer cries like 'What should one do' and 'I don't want to kill people', I started to feel surprise, uncertainty, and irritation with respect to Carathéodory's attitude, which, in a complete change of mood, on my side found their expression in remarks like 'I don't understand you anymore', 'I consider this visit as a final parting' and 'I am sorry for you'.

The impressions that Carathéodory's visit left with me were basically confirmed 14 days later at the occasion of a discussion with Erhard Schmidt in Berlin, but completed in the following manner. I heard in the course of that discussion: [25]

1. That Carathéodory had visited me at the instigation of Schmidt.

2. That the aim of this visit, in Schmidt's opinion, mainly had been this: to offer me in advance some satisfaction for the planned injustice to me, and in fact in the form of a open admission of the circumstance why I had to forego the protection of my co-editors against this injustice (Hilbert's state of health).

3. That according to remarks of Carathéodory to Schmidt, Hilbert's wrath against me was caused, even more than the three points mentioned in the first point of my circular of November 5, 1928, by my obstruction of the invitation of French mathematicians to contribute to the Riemann volume of the Mathematische Annalen.

Concerning the matter of satisfaction, the thought then came up between Schmidt and me that for a public insult a private satisfaction of course is insufficient, and that Carathéodory at least had the duty to make this private satisfaction a public one from the moment that this could be done without damage for the situation of Hilbert's health.

2.

From the arguments in Blumenthal's circular of November 16, 1928, under the caption 'Hilbert's letter and his reasons', I have gathered *that for the treacherous attack on me, apart from Hilbert's wrath, there had been a second reason: a strong desire of Blumenthal to remove me from the board.*

---

[25] Although I am aware of the confidential atmosphere of the talk with Schmidt, I must with respect to the communicability of its contents consider the argument valid that I gave at the end of footnote 1) above.

Because the purported 'grounds' that lie in my activity as editor, which Hilbert — suddenly proclaimed as the supreme authority by Blumenthal, ignoring all claimed equal rights — should have had for his action against me, could only have been suggested to him by Blumenthal himself.

1. Because the complaint brought against me, when traced back to Hilbert, would degenerate into an anecdote. Indeed, for years already he counts so little as editor [333] that it even has proved dangerous for the orderly handling of the business to submit manuscripts to him. Consequently Hilbert himself doesn't dare to mention this 'ground' in his dismissal letter, the content of which has become known through Blumenthal's circular of October 25, 1928.

2. Because Blumenthal is the only one who, except me, can judge my total activity as an editor.

If therefore Blumenthal, before as well as after the start of the campaign against me, is responsible for the complaints raised in his circular, then I claim furthermore that those are to be considered as mere pretexts, behind which is Blumenthal's above mentioned desire. In connection with the nullity of Blumenthal's accusations, to be explained below, the fact comes to the fore that Blumenthal might by my removal be liberated from the following inconveniences:

1. The obligation to fulfill his promise mentioned in my circular of January 23, 1929, to resign from the management after the winding up of volume 100. [334]

2. My frequent admonishments concerning the arbitrariness in the management and the fact that this is damaging for the Annalen.

I now proceed to the discussion of Blumenthal's accusations. I am blamed for the following:

1. That I have been rude in my behavior as an editor.

2. That I should have caused Klein's resignation.

3. That manuscripts sometimes remained for months in storage with me.

4. That I made on principle a copy of each manuscript that was submitted to me.

---

[333] Cf. *Blumenthal to Courant 9.II.1929.*   [334] In defense of Blumenthal it should be pointed out that he tried to withdraw in 1925 from his editorial position (*Blumenthal to Hilbert 15.XI.1925*). His attempt was vigorously suppressed by Hilbert (*Hilbert to Blumenthal 18.XI.1925*). See [Van Dalen 2005] p. 626

Ad 1. One can very well speak about a reality corresponding to the word 'rude', [335] if the meaning is determined as follows: the will to integrity (duty to people), extended by the will to clarity (fate of the mathematician).— It came with me to an expression of this will, whenever the honor and the prestige of the Annalen were at stake. (Incidentally, among these were cases where Blumenthal himself had called me in.) In those cases neither the vanity of the authors, nor the tendency of Blumenthal to please everybody, could be taken into consideration.— When I occasionally made my will prevail against that of the manager, then the latter must have found no support from his colleagues in the board, or he had reasons not to elicit such support.

Ad 2. The event to which Blumenthal refers in his statement on Klein's resignation, hardly can be any other than the following: I had a discussion with Klein about an article that I had already dealt with, whose author [336] had appealed to Klein as chief editor in the matter of changes demanded by me, and in an oral discussion he had made his views sound reasonable. When I talked it over with Klein, he understood that the author was wrong (not formally as Blumenthal would have it, but in matter of content), and that he therefore could not honor his given promise. During the further course of this talk, Klein expressed his view that the manner in which the chief editors were mentioned on the cover apparently gave the wrong impression to the public and he personally could, insofar he himself was involved, hardly bear the responsibility for this impression.— Some time later he resigned as chief editor.— Such a behavior speaks as much in favor of Klein, as it speaks against Hilbert, who with a much smaller share in the editorial activity than Klein's at the time of his resignation, used the possibility to deploy the inner weakness of his position for its outer confirmation.

Ad 3. Because I spent on average about one thousand hours per year on my editorial activities, it is almost obvious that submitted manuscripts usually remained for months in my possession. Only the word 'stored' is misleading, because never were articles temporarily forgotten by me or even lost without a trace (as has happened with Hilbert), but they constituted each time the object of the most intensive editorial activity, by which their content usually was substantially influenced. As I have kept manuscripts longer than the normal deadline for printing only in the extremely rare

---

[335] The German word 'schroff' can mean all kinds of things like abrupt, blunt, brusque, curt, gruff, harsh, inaccessible, and is translated here by 'rude'. [336] Brouwer refers to Mohrmann. See *Blumenthal to Brouwer 23.VII.1924, Brouwer to Klein 29.XI.1923*. Mohrmann had gone over Brouwer's head to Klein in the matter of a paper of his. See [Van Dalen 2005] p. 631.

cases where very large defects came to light, the articles were taken care of by me much better than if they had been 'stored' with Blumenthal during the same time.— Blumenthal held, by the way, until recently the opinion that my method was normal and conscientious, otherwise he would not have asked me for refereeing, even in the case of articles where I could not all be counted as expert, considering their subject.

Ad 4. Although Blumenthal can give an 'example' of my 'basic method' of making a copy of every submitted manuscript, and although I consider such an act as an elementary right of a refereeing editor, since many years it has come to that only in cases where an article seemed quite acceptable, but only after revision or after considerable extension. Then I considered this measure a duty with respect to the historiography of mathematics, indeed because the possibility should be taken into account of an incorrect reference to the submission date.

I challenge Blumenthal to produce the Annalen archive, especially with the complete correspondence between him and me. I claim that precisely these documents will refute his accusations in the most complete manner.

(signed) L.E.J. Brouwer.

[Carbon copy – in Brouwer]

Editorial comment

The following letter is part of a long series of exchanges concerning the priority of the theory of dimension. Its history is complicated and drawn out. The major players are Brouwer, Menger and Urysohn. The last one acknowledged Brouwer's claims, but since he died in 1924, his view played no role in the discussions. There is no doubt that Menger, already during his stay in Amsterdam, developed the conviction that his role in dimension-theory was not given its rightful place by Brouwer and the Russian topologists. The reader should consult [Van Dalen 2005], ch.12, and section 15.5 for the historical background. The present letter and its sequel are concerned with the discussion in which Menger's book *Dimension Theory* (1928) was the first volley. Brouwer reacted in the paper *On the historiography of dimension theory* (1928). Hahn and Brouwer at one point decided that the conflict should be closed with a reply from Menger in the Proceedings of the Dutch academy (where Brouwer's paper as published). In spite of the efforts of Hahn, all attempts at a reconciliation failed; resulting in the end in plain hostility and irreconcilable differences.

)29-07-11

To H. Hahn — 11.VII.1929   Laren

Dear colleague Hahn, [Lieber Kollege Hahn]

Many thanks for sending me the manuscript of Menger and for your accompanying letter. I think that the manuscript really turned out well; both the general structure and the treatment of most details seem fit to me to bring the matter to a conclusion in that way, so neither from me nor from Moscow objections will be necessary. I hope that I can restrict myself to expressing in a short postscript (which I will show you beforehand) to the Menger note, the hope that the discussion that took place may be a useful contribution to clarification of the historical development of dimension theory, and to observe that the attentive reader can see from the reading of *both* notes that there are hardly any essential points of difference between Menger and me left. Naturally I will furthermore see to it that in the review of Menger's book [337] in the Jahresbericht of the D.M.V., the meanwhile obtained clarification of the situation and agreement between Menger and me will timely be taken into account.

I would like to discuss with you in person a few details in Menger's manuscripts that seem amenable to improvement as soon as possible, and I will arrange my travel plans (leading southwards anyway) accordingly. So please tell me until what day you will still be in Vienna, and which address you will have after your leave from there. It would be best to meet you in Vienna, where I have a chance to take a look at the relevant documents that I don't know yet (just as I on my part have, by the way, to show you some more documents).

Concerning the Menger documents in my safekeeping, this safekeeping is explicitly mentioned by me in the Amsterdam Proceedings, [338] so that I think it is more appropriate towards the public that in the future I myself

---

[337] [Menger 1928b].   [338] [Brouwer 1928d].

also will function as trustee for these items. But maybe we can find a fitting modus to meet your wishes in this. We can discuss this point too in person.

With warm greetings, hoping to meet you again [339] soon

Your
L.E.J. Brouwer

[Typescript draft with handwritten corrections – in Brouwer [340]]

1929-08-09a

To H. Hahn — 9.VIII.1929[a]   **Brussels**
                                                         Brüssel

Dear Colleague Hahn [Lieber Kollege Hahn]

As a consequence of an disruption in carrying through my travel plan, your letter from Belagio has reached me after considerable delay. The interruption was caused by a great calamity: four days ago my briefcase [341] which also contained my scientific diary was stolen from me on the front platform of a Brussels' tram, by a pickpocket, and both the police and the detectives consider the case as hopeless. Since in this diary my collected scientific thoughts and ideas of the last three years, which have largely disappeared from my memory, and of which only a few have already found a registration elsewhere, had been recorded, this event means for my scientific personality a serious personal mutilation [342]), roughly the same as what 'decapitation' (elimination of the central process) means for a pine tree. To my amazement, I remain so far, fairly calm under this blow of fate; I believe, however, from certain phenomena, that I have nonetheless suffered a nervous collapse, the consequences of which will perhaps only later become visible, together with a disorganization of my scientific thoughts.

In my present condition, my power of judgement is, as you will understand, at the moment somewhat uncertain; and it is with this reservation, that I believe to have to consider the counterproposals of Menger that are

---

[339] *Mit herzlichem Gruss auf hoffentlich baldiges Wiedersehen.* [340] Carbon copy of the letter itself also in Brouwer. [341] Brouwer uses *Brieftasche* (wallet); it is more likely that he was carrying a small type of briefcase that was very common at the time, than a wallet. [342] *Verstümmelung*

contained in your letter, as unacceptable (in particular in as far as according to these the slip of the pen of my Crelle paper has not been freed from the implicit doubt contained in Menger's book[26]).

With respect to these counterproposals I also have, for the time being, to take back my liberty concerning the postscript planned by me. As soon as I have regained somewhat my balance, I will write to you in extenso on this matter; as a follow up, we will be able, as I hope, to have a definitive fruitful discussion in the Tessin; the problem of the mutually satisfactory version has indeed its objective solution.

Anyway, even in the most unfavorable case that we should not discover the solution, and that therefore the postscript had to be given up, I would not consider the situation as desperate. The main thing is that Menger rehabilitates himself, by representing his disputed views to the public in a chivalrous way in person, and in the same journal where he was attacked, and to explain these, even when in the conflict with me, these should retain their one-sided character, in an acceptable way.

With warm greetings I remain
always your [343]

(signed) L.E.J. Brouwer

[Carbon copy of typescript – in Brouwer]

Re enclosure

[Of the enclosure two carbon copies have been preserved in the Brouwer archive. The first one only contains the texts of two letters from Blumenthal (*Blumenthal to Brouwer 3.II.1912, Blumenthal to Brouwer 12.II.1912*); at the top of the first one finds in Brouwer's handwriting '*Copies, enclosures to the letter of Brouwer to Hahn of 9.VIII.1929*'; the second one contains the same text and is preceded by the following lines:]

Enclosure to the letter of Brouwer to Hahn of 9.VIII.1929, containing a copy of documents (known to Menger since the year 1925), from

---

[26]To facilitate clearing up of this point of difference, I send at the same time to Menger the document that is enclosed here in copy.

[343] *Mit herzlichen Grüssen verbleibe ich – stets Ihr.*

which it appears that I have extensively refereed a paper of Lennes, which was intended as an extension of the paper published in 1911 in the American Journal of Mathematics 33 by the same author: Curves in non-metrical analysis situs with an application in the calculus of variations (these documents offer a rebuttal of the insinuation, contained both in Menger's book 'Dimensionstheorie', as in the note that was submitted to the Amsterdam Proceedings [344] on July 1, 1929, that I could not have known in 1913, at the time of writing my paper on dimension theory in Crelle's journal, the above mentioned paper of Lennes of 1911). [345]

1929-10-07

From A. Heyting — 7.X.1929   Enschede

Dear Professor, [Hooggeleerde Heer]

I am most grateful for the sending of the documents about the coup in the editorial board of the Mathematische Annalen. [346] I summarize in what follows my opinion about some important points.

Anybody who has in recent years taken a look at an issue of the Mathematische Annalen, could recognise in it the important results of your activity as editor. If he moreover knows from experience, that you always took an interest in helping to make each article appear in the best possible form, then he must share your indignation about the attempt to remove you from the board of editors, and admit that the term 'grievous injustice' is a correct qualification.

The conditions that you put to Mr. Carathéodory in your letter of November 2, [347] are logical and correct; only about the question whether the form of this letter was fortunate, a difference of opinion is possible.

I share your view about the effectuated change of the editorial board, as expressed in your circular of January 23. For the many who kept primarily in touch with contemporary mathematical research through this journal, the fact that it has lost now a great deal of its representative character,

---

[344] *KNAW, Proceedings.*   [345] The upshot of Menger's claim was that Brouwer was not aware of the modern definition of connectedness   [346] Brouwer had put together a file of documents relevant to the Mathematische Annalen conflict. The Brouwer archive contains presumably most of the material he collected.   [347] *Brouwer to Carathéodory 2.XI.1928.*

constitutes a heavy blow. I want to support your attempt to fill the void thus created by establishing a new journal to my best ability, even though I am of the opinion that it is in general undesirable to increase the number of mathematical journals, and that the Annalen with their important historical tradition will not be replaced easily.

I appreciate that you don't want to expose my manuscripts that have been deposited with you, to an indeterminate delay. I will consider it an honor if they can be published in the *Bericht* of the Berlin Academy. I hope that I can soon send you the third article [348] which has to be revised because of your changes in '*Zur Begruendung der intuitionistischen Mathematik*'.

Many thanks for the improved copy of the above mentioned article. My own copy shows so many traces of frequent use, that I cannot send it back to you. I have copied all changes and return the copy that has been amended by you.

[Carbon copy – in Heyting]

29-10-26

From T. de Donder — 26.X.1929   **Brussels**

5 Rue de l'Aurore, Bruxelles
Université Libre de Bruxelles, [349]
Faculté des Sciences, 50,
Avenue des Nations

Dear colleague, [Très honoré Collègue]

I have had the honor to present in 1927 and 1928 several notes, written by Messrs. Barzin, A. Errera, Glivenko, Paul Lévy, etc., to the Royal Academy of Belgium (Science Division). These notes refer to your new logistic system. By presenting these notes, I nourished the hope to stimulate discussions that would throw more light on your ideas.

A recent article by Messrs. Barzin and A. Errera '*Sur le principe du tiers exclu*' (Bruxelles; Archives de la Société Belge de Philosophie, 1929) [350] gives me the impression that your ideas have been erroneously interpreted. You certainly would render a great service to Science by letting me know

---

[348] [Heyting 1930a]   [349] [letterhead]   [350] On the principle of the excluded third, [Barzin, M 1929].

what you think about the articles mentioned, *more in particular those of Messrs. Barzin and A. Errera.* I would please me very much to present your note to the Royal Academy of Belgium (Science Section); if you prefer, you can write *in Dutch.*

Sincerely yours [351]

T. De Donder

[Signed autograph – in Brouwer]

––––––––––––––

[351] *Vieullez agréer, Monsieur et très honoré Collègue, l'expression de mes sentiments les meilleurs.*

Chapter 5

1930 – 1939

To H. Hahn — 10.I.1930   **Laren**

Dear colleague Hahn, [Lieber Kollege Hahn]

The two points of the planned note of Menger, for the deletion of which
I make an appeal to the discretion of the editors of the Monatshefte, are the
following:

First footnote [9]) on p. 6 together with the words in the editorial comment on p. 1: 'the points of difference that concern the Monatshefte itself'.
This complex suggests to the reader the opinion that the Monatshefte were
involved in isolated points of difference existing between Menger and me,
and that I therefore have partly yielded in the points of difference between
Menger and me by means of my correction that appeared in the Amsterdam
Proceedings. As you know, both of these contradict the facts. If nonetheless
the editors should leave this complex (which deals only with a matter that
concerns only the editors and not Menger) as it is, and use the excuse on
p. 8 that Brouwer's correction in the Proceedings only became known after
the layout of Menger's note was made, then this excuse fails for two reasons: *first,* the objections related to typesetting are untenable, because in
Menger's letter to the Amsterdam Academy he makes a proposal that goes
much further in this respect, namely to destroy the whole note which was
meant for the Monatshefte and which already had been typeset; *second,* the
correction in the proceedings was already orally promised to you on July 22,
1929, that is, immediately after you had showed me that the dating on the

D. van Dalen, *The Selected Correspondence of L.E.J. Brouwer,*
Sources and Studies in the History of Mathematics and Physical Sciences,
DOI 10.1007/978-0-85729-537-8_5, © Springer-Verlag London Limited 2011

reprints of Menger's note in vol. 34 of the Monatshefte did not correspond to the date of the issue itself (and we at the same time agreed that this matter was irrelevant for the matter of the conflict between me and Menger).

Second the paragraph on p. 5, l. 7–17, which insinuates without a shred of evidence that I have high-handedly modified articles of others in parts that referred to me. [1] Already in my previous letter I have declared that this accusation of Menger is absolutely slanderous and totally unfounded (just like the rumor he spread last spring that I perpetrated something similar in the case of an article of Vietoris that had appeared in the Annalen. — Menger is suffering from delusions in these things. Here in Laren he has behaved repeatedly exactly like this). Moreover the paragraph insinuates without a shred of evidence that there is no certainty about the exact agreement between Urysohn's proof of the main theorems of dimension theory as they appeared in the Fundamenta, and the manuscript submitted in March 1925 to the editors of the Fundamenta. Also this implicitly suggested accusation of Menger is absolutely slanderous and totally unfounded: the relevant manuscript in Urysohn's own handwriting is *still available*, and has been left by Alexandroff in the spring of 1925 in Menger's hands for an arbitrary period. In my opinion it is illegitimate for the editorial board to print the mentioned paragraph of Menger *after* the statements above, without demanding from him that he at the same time makes public the proofs of his statements. Because after my above statements, the editors should at least clearly see the danger that in their columns one works according to the basic principle 'calomniez toujours, il en restera quelque chose'. [2] As an aside it must be remarked that it is moreover ridiculous in itself when Menger says that the opinion that Urysohn is a 'follower' [3] can be supported by the single reference in Urysohn's Memoir to my Crelle article (Fundamenta Mathematicae 7, p. 37, footnote 3)). Indeed, in this reference is a mere *mention*, which refers much less clearly and explicitly than for example Menger himself in his first articles and more specifically in his *Bericht* on dimension theory. This ridiculous remark appears thus indeed in the light of a *pretext* to weave the above mentioned slanderous insinuations into the text.

After what I have said you will understand that a possible publication of both points mentioned above will make it impossible to publish my continuum talk [4] in your journal, and not only because of the suspicion these points cast on me, but also because an eventual printing of these points unavoidably constitutes an announcement of an unfriendly, if not downright

---

[1] Menger's accusation of editorial abuses.   [2] Always slander, something will stick.
[3] *Fortsetzer* in the text, i.e. a person who continues the work of someone.   [4] Brouwer's second Vienna lecture.

aggressive attitude of the Monatshefte towards me.

With cordial greeting [5]

your

(signed) L.E.J. Brouwer

P.S. When you date this note of Menger by July 1929, [6] then to be logical you must date the submission of my continuum talk March 14, 1928, because the changes in the latter with respect to the shorthand notes of my speech are proportionally less than the differences of the first manuscript of Menger, compared to what he submitted to the Proceedings in July.

(signed) LEJ B.

[Carbon copy of typescript – in Brouwer]

30-01-11a

To H. Hahn — 11.I.1930[a]   Laren

Dear colleague Hahn, [Lieber Kollege Hahn]

With reference to the publication of Urysohn's memoir in Fundamenta Mathematicae 7 and 8 I can inform you of the following:

The first half, which appeared in Fundamenta Mathematicae 7, was typeset and corrected during Urysohn's life. Urysohn himself has introduced several changes during this time, among which also, *by a letter to Kuratowski,* the only reference in the memoir to my Crelle article, namely a rewriting of Fundamenta Mathematicae 7, p. 37, footnote [3]) into the present form. Further improvements had been written down by Urysohn, but not yet sent in when he died. When Alexandroff and I, a couple of weeks after Urysohn's death, approached the editors of the Fundamenta to finish the correction of this part of the memoir, taking Urysohn's notes into account, it turned out that the article had been in the mean time already almost completely printed. Only on two proof sheets (p. 33–48 and p. 129–137) we have been able to correct the text as it was (for example, Urysohn's planned reference to my report in Crelle 153 and Menger's note in the Monatshefte 33 were

---

[5] *Mit herzlichem Gruss – Ihr.*   [6] the date of the manuscript that was submitted to Brouwer.

inserted in the above mentioned footnote); for the remaining sheets only the most pressing improvements were introduced in an extensive list of errata at the end of the volume.

However, the second half, that appeared in Fundamenta Mathematicae 8, was typeset only after Urysohn's death and corrected by Alexandroff and me, it has in a certain sense a posthumous character. The original hand-written manuscript of Urysohn of this half which was submitted in March 1923 to Fundamenta Mathematicae is still in meticulous safekeeping with Alexandroff, and he and I bear full public responsibility for the authorship of Urysohn's text in Fundamenta Mathematicae 8.

Cordial greetings from

Your
(signed) L.E.J. Brouwer

[Carbon copy of typescript – in Brouwer]

1930-06-07

To Monatshefte f. Mathematik und Physik — 7.VI.1930   Laren

To the editors of the Monatshefte fuer Mathematik und Physik   Wien [An die Redaktion der Monatshefte fuer Mathematik und Physik, Wien]

Hereby I confirm once more explicitly that it is impossible for me to consent to publication of the lecture I gave in Vienna 'The structure of the continuum' in the Monatshefte fuer Mathematik und Physik. I am however prepared to publish this talk (possibly using the already typeset text of it) as a special publication of the Committee for guest lectures of foreign scholars of the exact sciences or also in a suitable *other* (mathematical or philo-sophical or general scientific) journal published by the Academic Publishing Company.

Sincerely yours

[carbon copy of typescript – in Brouwer]

)30-06-10a

To H. Hopf — 10.VI.1930[a]   Laren [7]

Dear colleague, [Sehr verehrter Herr Kollege]

The undersigned is planning to publish soon an international mathematical journal, which will appear with the Noordhoff company in Groningen, where also the *Revue Semestrielle des Publications Mathématiques* appears. He would like to ask you most kindly to declare your willingness *in principle* to join the editorial board of this journal. [8]

In case he has the honor to receive from you the favorable answer he hopes for, the undersigned will soon send you a list of all scholars who likewise have announced their provisional joining, and also the draft of the bylaws of the editorial board together with a contract with the publishing company.

Sincerely yours [9]

L.E.J. Brouwer

Member of the Academy of Sciences at Amsterdam.
Professor of the University of Amsterdam.

[Signed typescript – in Hopf]

––––––––––

30-08-03

To H. Freudenthal — 3.VIII.1930   Laren

Dear Mr. Freudenthal, [Sehr geehrter Herr Freudenthal]

Hereby I would like to ask you whether you would be interested to accept next winter a position of assistant with me (which includes a Habilitation). [10] Essentially your task will be to support the publication of the

––

[7] Copies of this letter were sent to the prospective editors of *Compositio Mathematica* in English, French, and German   [8] For the history of this journal see [Van Dalen 2005] section 16.6 and [Van Dalen and Remmert 2006].   [9] *Mit ausgezeichneter Hochachtung – Ihr ganz ergebener.*Mit   [10] There was and is no such thing as a *Habilitation* in Holland. From other correspondence it appears that Brouwer considered the admission as 'privaat docent' as an equivalent of the German *Habilitation.*

new journal 'Compositio Mathematica'. Moreover you will have to give a
one hour lecture on a subject of your own choice, and occasionally help students with the preparation of seminar talks or in some other way in case of
a difficulty. Your salary will amount to 3000 to 3500 Dutch guilders.

I already discussed last winter the possibility to attract you for this
position with prof. Bieberbach. I would be most pleased to receive soon a
favorable reply.

With best greetings

Your [11]
L.E.J. Brouwer

[Signed typescript – in Freudenthal]

1930-09-20

To A. Heyting — 20.IX.1930   **Laren**

Dear Mr. Heyting, [Waarde Heer Heyting]

Thank you very much for your letter of the 16th of this month; I infer
from it, that you must have been satisfied with your talk in Koenigsberg, [12]
and I gladly share in that satisfaction.

I shall read your note in the Bulletin of the Belgian Academy, [13] and I
hope to return to the question you asked in your letter. Meanwhile I enclose
a copy of a letter I just received from *Regierungsrat* Dr. Kerkhof, editor
of *'Forschungen und Fortschritte'* and also the enclosures to that letter. I
would appreciate it very much if you would write the article requested; [14]
if you so wish, I am quite willing to have a look at it before it appears.

With friendly greetings

Your
L.E.J. Brouwer

[Signed autograph – in Heyting]

---

[11] *Bestens gruessend – Ihr ergebenster.*   [12] *6. Deutschen Physiker- und Mathe-
matikertagung,* (4-7.IX); *Tagung für exakte Erkenntnislehre* (5-7.IX). See [Heyting 1931a].
[13] [Heyting 1930b].   [14] [Heyting 1931b].

30-10-09

To Th. De Donder — 9.X.1930   Laren

My dear colleague, [Mon cher collègue]

While preparing a note on intuitionism for the Bulletin de l'Académie
Royale de Belgique, I was pleasantly surprised to see appear an article by
my student Mr. Heyting, which elucidated in a masterly way the points
that I had wanted to clarify myself. I believe that after the note of Mr.
Heyting [15] not much remains to say about these matters, and that the
reader of the publications of your Academy will know well enough what to
think of the ideas of Messrs. Barzin and Errera, which, aside from the great
interest they present are nonetheless untenable in their essential purport. I
will investigate whether anything can be added to the note of Heyting which
can make the general ideas of intuitionist logic more profound, and in the
affirmative case I will not hesitate to compose a note and I will be happy to
send it to you.

Sincerely yours [16]

[Carbon copy – in Brouwer]

31-04-24

From R. Carnap — 24.IV.1931   Vienna

Stauffergasse 4, Wien

Dear professor, [Sehr verehrter Herr Professor]

Enclosed I send my publication list, as you asked, and also my curriculum vitae. Yesterday I sent you (in two packages as printed material) my
publications number 1,2,4,5,7,8,9,11,12. The subjects of my lectures and
exercises here are mentioned in my cv. For your further orientation I might
remark that the number of registered attendants of my lectures in the last
five semesters were as follows: 32, 44, 55, 154 (introduction), 113.

---

[15] [Heyting 1930b].   [16] *Agréez, mon cher collègue, l'expression de mes sentiments
cordiaux.*

I also enclose a review of the book by Kaufmann. [17]
Sincerely yours [18]

R. Carnap

[Signed typescript – in Brouwer]

Enclosure 1

Prof.Dr. Rudolf Carnap
Wien XIII/5
Stauffergasse 4

C u r r i c u l u m   V i t a e.
Born May 18, 1891 in Rinsdorf near Barmen. Final examination in
the humanistic gymnasium [19] in Jena in 1910. Study of physics,
mathematics, philosophy and psychology at the universities of Jena
and Freiburg im Breisgau, interrupted 1914–18 by military service.
State examination (high school teacher examination) Jena, autumn
1919, in mathematics, physics and introduction to philosophy. Ph.D.
in Jena, February 1921 in philosophy. Habilitation as Privatdozent
November 1926 at the University of Vienna. September 1930 title of
extraordinary professor.

During my studies my major subject was physics. My teachers were
Max Wien in Jena, Himstedt in Freiburg i.B. and most of all Baedeker
in Jena, under whom I started with experimental work which was terminated at the outbreak of war. After the war I didn't do anymore experimental work, but mainly occupied myself with theoretical physics
until the exam for my teacher's diploma (physics and mathematics for
senior high school). For my dissertation I took as subject (stimulated
by Bauch) philosophy (see number 1 in the literature list), and all
other publications also are in this field. My articles concern mostly
logical and epistemological problems in the foundations of mathematics and physics. The essential stimuli for this are due to my personal
teacher Frege and the writings of Russell. Later my field of interest

---

[17] [Kaufmann 1930].   [18] *Mit ergebenstem Gruss.*   [19] Standard term for a high school
with Latin and Greek.

expanded; my articles were on the system of formal logic and general epistemology (my main article '*Der logische Aufbau der Welt*' belongs to this field) and comparative epistemology.

Together with Reichenbach I edit the journal *Erkenntnis* (continuation of *Annalen der Philosophie*), which is mostly dedicated to research about the philosophical foundations of science.

Lectures and exercises that I gave in the University of Vienna: Philosophical foundations of physics, Logic I and II; Problems of epistemology; Philosophical foundations of (mathematical and physical) geometrie; General axiomatics; Philosophical foundations of arithmetic; Development of theoretical philosophy since Descartes; Introduction to philosophy; Russell's epistemology; Discussions of selected problems in logic.

Prof.Dr. Rudolf Carnap
Wien, April 1931.

_____

Enclosure 2

Published writings:

1. Der Raum. Ein Beitrag zur Wissenschaftslehre. (Erg.-Heft 56 der Kantstudien, Berlin 1922). 87 pages (Diss.)

2. Ueber die Aufgabe der Physik und die Anwendung des Grundsatzes der Einfachstheit. Kantstud. XXVIII, p. 90–107, 1923.

3. Dreidimensionalität des Raumes und Kausalität. Ann. d. Philos., IV., p. 105–130, 1924.

4. Ueber die Abhängigkeit der Eigenschaften des Raumes von denen der Zeit. Kantstud. XXX, 331–345, 1925.

5. Physikalische Begriffsbildung. (vol. 39 of the collection 'Wissen und Wirke'; Karlsruhe 1926. 66 pages

6. Eigentliche und uneigentliche Begriffe. Symposion I, p. 355–374, 1927.

7. Der logische Aufbau der Welt. Berlin 1928 (now: F. Meiner, Leipzig). 290 pages

8. Scheinprobleme in der Philosophie. Das Fremdpsychische und der Realismusstreit. Berlin 1928 (now: F.Meiner, Leipzig). 46 pages

9. Abriss der Logistik, mit besonderer Berücksichtigung der Relationstheorie und ihrer Anwendungen. (Vol. 2 of the collection 'Schriften zur Wissensch. Weltauffassung', J. Springer, Wien 1929).

10. Die Mathematik als Zweig der Logik. Bl. f. Dt. Philos. IV, p. 298–310, 1930.

11. Die alte und die neue Logik. Erkenntnis, vol. I (=Ann. d. Philos., vol. IX), p. 12–26, 1930.

12. Bericht über Untersuchungen zur allg. Axiomatik, Erkenntnis I, p. 303–307, 1930.

13. Die logizistische Grundlegung der Mathematik. Erkenntnis II, Heft 2, 1931. (In print).

Unpublished manuscripts: [20]

14. Topology of the space time world. Axiomatically represented with the tools of the theory of symbolical relations. Part I.

15. Investigations into general axiomatics. Part I. The most important results have been communicated in (12).

16. The language of physics as universal scientific language.

17. Psychology in the language of physics.

18. Conquering metaphysics through logical analysis of language.

(16), (17), (18) will appear in: Erkenntnis II, 1931.

1931-05-20

From G. Feigl — 20.V.1931

Dear Professor, [Sehr verehrter, lieber Herr Professor]

On April 23 I have written to you in detail about the matter of the 'Revue-semestrielle — Jahrbuch' [21] and in the last two weeks I have anxiously waited for your answer. Just now I received a card from Freudenthal,

---

[20] All titles have been translated from the German.   [21] On behalf of the *Wiskundig Genootschap* Brouwer carried on preliminary negotiations with Feigl for a merger between the *Jahrbuch über die Fortschritte der Mathematik* and the *Revue semestrielle des publications mathématiques : rédigée sous les auspices de la Société Mathématique d'Amsterdam.*

saying that you did not receive my letter. So I send you two copies of my letter of April 23. Fortunately I still have three copies of the sample page mentioned in the letter available; I am sending them now to you, in the lost letter a larger number of copies were enclosed.

I had hoped, as we had planned during the conference of March 16, to come to Amsterdam during the Pentecost week [22] to talk about the details of the final agreement. Because of the loss of the letter there is now a delay, so I will possibly have to postpone this trip. Of course, I am most willing to come instead in one of the following weeks.

With cordial greetings from house to house

Gratefully yours [23]

Enclosed: two copies of the letter of April 24 and three copies of the sample page. [24]

[Carbon copy – in Freudenthal]

31-09-11

From M.J. Belinfante — 11.IX.1931   **Amsterdam**
                                                                  2de Jan Steenstraat 23

Dear Professor, [Hooggeleerde Heer]

In the enclosed manuscript, which I have to honor to bring to your notice, I tried to give an intuitionistic proof of the theorem of Picard. [25] I based myself on the proofs published by Landau. Although these demand changes and completions in several points, the main line of reasoning could be adopted without change. The theorems from the theory of complex functions that are used in this proof required a more thorough reworking. As a consequence only the last seven pages of the 34 pages of the manuscript are devoted to the theorem of Picard, and all the others are spent on the theory of complex functions.

After the introduction, which on page 2 contains by way of explanation of the contents a couple of examples of proof methods that are not allowed

---

[22] Whitsunday was May 24, 1931   [23] *Ihr dankbar ergebener.*   [24] Not extant.
[25] [Belinfante 1931].

from the intuitionistic point of view, the theorems I–V lead to the Cauchy result that $f(z)$ has p zeroes inside L, if $\frac{1}{2\pi i} \int_L \frac{f'(z)}{f(z)} dz = p$ (theorem VI p. 15). By using this one can prove the fundamental theorem of algebra (theorem VII p. 18) and its intuitionistic supplement (theorem VIII p. 18). The remaining theorems are used in the proof of the Picard theorem, among others the Weierstrass theorem (XI–XIV, p. 22–25).

I would like to present this research in my course, although it doesn't belong to the subject of infinite series. I would therefore be very pleased to hear from you if there is an objection to this.

Sincerely yours [26]

M.J. Belinfante

[Signed autograph – in Brouwer]

1931-09-20

From M.J. Belinfante — 20.IX.1931   **Amsterdam**
                                                    2de Jan Steenstraat

Dear professor, [Hooggeachte Heer]

In the enclosed addendum to the manuscript, 'On the elements of function theory and the theorems of Picard in intuitionistic mathematics', [27] converses of the theorem of Weierstrasz and of both theorems of Picard are given. The last two converses have a positive character, even though the proofs are largely negative. They agree more with the formulation of Landau than is the case with the theorems of the manuscript.

The numbering of the pages of the addendum continues that of the manuscript, so they can be added without any problem.

Sincerely yours
M. J. Belinfante

[Signed autograph – in Brouwer]

----

[26] *Met alle hoogachting, – Uw dienstwillige.*   [27] *Über die Elemente der Funktionentheorie und die Picardschen Sätze in der intuitionistischen Mathematik*, [Belinfante 1931].

From H. Freudenthal — 13.VIII.1932   Amsterdam

Dear professor, [Sehr verehrter Herr Professor]

Having just returned from my trip, I received a letter from Hopf. He writes to me that he has not received from you a reply to his request for a lecture about intuitionism on the Congress in Zürich. Because the congress is close at hand, Mr. Hopf asks me to settle the matter as quickly as possible. You told me some time ago that you would charge Heyting with it (or that you already had done so?). Is it possible that Heyting has not registered? Please tell me, so I can answer Hopf.

I already ordered a telephone, and I will get it in the next few days; then I will inform you immediately of the number.

With best greetings

Sincerely yours [28]
Hans Freudenthal

[Signed autograph – in Brouwer [29]]

To P.S. Alexandrov — 20.X.1932[a]   Amsterdam [30]

Dear Alexandroff, [Lieber Alexandroff]

Thank you for your beautiful booklet [31] and also thank you for your card from Ascona, the southern branch of Laren. Unfortunately I wasn't home when your card arrived, but was staying in Berlin; hence this belated answer. I am still observing with astonishment the process of dissolution of

---

[28] *Ihr sehr ergebener.*   [29] Handwritten copy in Freudenthal.   [30] Picture postcard of Berlin; posted in Amsterdam. Addressed: Herrn Prof.Dr. Paul Alexandrov aus Moskau, Ascona–Moscia. Casa Sole. Forwarded 22.X.1932.   [31] [Alexandroff 1932]

my life, which is developing with admirable universality and thoroughness;
I am curious whether there will be a new season.

Most cordially your Brouwer [32]

[Cor added in the margin:]
Cordial greetings from C. Jongejan.

[Signed autograph, postcard – in Alexandrov]

1932-12-01

From R. Courant — 1.XII.1932   **Göttingen**

Dear Mr. Brouwer, [Lieber Herr Brouwer]

The Göttingen *Gesellschaft der Wissenschaften* has received — probably in the end not without your initial cooperation — an amount of 5.000.–
Mark, the interest of which must be used every three years for a Urysohn
prize for geometric research, to be awarded in particular among young researchers. The *Gesellschaft der Wissenschaften* has accepted this foundation
and it has decided that the awarding is to be decided by a committee to
which you, Hopf, Weyl, possibly also Alexandrov and me — maybe also
Veblen — will belong.

Would you be so kind to inform me whether you agree with this proposal?
As soon as the *Gesellschaft der Wissenschaften* has arranged the formal part
of the foundation, I will inform you further. There is no hurry, because the
prize will be distributed for the first time if the interest has accumulated
enough, i.e. after about 3 years.

With friendly greetings

Sincerely yours

[Typescript copy – in Courant]

---

[32] *Herzlichst Ihr Brouwer.*

From J.G. van der Corput — 11.IV.1933   Rotterdam

Dear Colleague, [Geachte Collega]

Now that the evening has fallen, I don't expect a telegram anymore. The report written by me and Schaake has just been finished. [33] Our other activities left only little time for drafting these notes, and we are convinced that our formulations will not always sufficiently render your thoughts. The more is added or changed, the more it will be appreciated.

The next but last sentence in § 2 is justified by § 21, but it hasn't been mentioned there explicitly; moreover, the issue about inconsistency is not yet quite clear to us. The formulation in § 7 about the union of the species of Fermat numbers and the species of non-Fermat numbers is not satisfactory to us.

Coster, the Groningen physicist, has offered to have the report with your corrections typed by his typist, and then the students will make copies. The number of possible copies is practically unbounded, but I first want to discuss with you how far we will go. For, the speaker has the final decision about the manner in which the report will appear. The possible profits will flow into the coffers of our philosophical faculty association, [34] which will probably get your approval.

The typist has to type the report in the hours made available to her. She does have a number of hours in the coming months, but for example after the summer vacation we run the risk that she doesn't have enough hours available for this. If I get the report back soon, then she can start right away. But if it takes a few months, then we probably lose our typist.

I am afraid that you think the report too verbose. But this report is meant for mathematicians who want to get informed about the intuitionist direction in mathematics, but for whom the publications that have appeared until now present overly great difficulties.

I think that in those twelve lectures you have given an excellent introduction to intuitionism. Of course time didn't permit to prove the fundamental theorem mentioned in § 19 about fans [35] (much has been treated already). For the report this is a pity because it concerns here a theorem that is very important for intuitionism, and which will seem very strange to the reader.

---

[33] Notes of Brouwer's course 'Introduction to Intuitionism' in Groningen, March, April 1933.   [34] A student association   [35] finitary spreads (sets)

About the matter of the Wiskundig Genootschap [36] I will write as soon
as I have received the data that I expect one of these days.

With many greetings

tt

J.G. van der Corput

[Signed autograph – in Brouwer]

1933-05-14

From P. Ehrenfest — 14.V.1933

Dear Brouwer, [Waarde Brouwer]

In response to a request of a few non-Jewish German physicists of the
top level, I was in Berlin from Friday, May 5, to Monday, May 8, where you
just met me on my return trip. — I had the opportunity to speak there with
a great many of colleagues, also a few from small university towns, where
the moral atmosphere is completely unbearable, because there also *private*
life is affected in all possible ways

I feel the need to inform you about something concerning the MATHE-
MATICIANS, which I accidentally (but RELIABLY) heard. It is well possible
that some of it is accidentally unknown to you.

1. Because of a refusal by the authorities (I *believe* the university authorities) it is [impossible (ed.)] for the Russian mathematician KOL-
   MOGOROV, who was to come to Göttingen on a Rockefeller-Fellowship,
   to work there. —Now I know from several earlier cases how easily it
   can happen to young Russians that an EXIT permit which has been
   obtained with infinite trouble can be IRRETRIEVABLY lost because of
   small obstructions, and also how much this can be a destructive disappointment for the person concerned. So I have immediately and
   urgently asked my friends in Göttingen on the one hand and the Rockefeller people in Paris on the other hand to avoid any exit problems, by
   switching the fellowship for example to Paris.

---

[36] Dutch Mathematical Society

2. *In case I have understood correctly,* the real organizer and leader of the Göttingen institute, Courant [37] has been denied entrance to the institute. And, as I *believe,* also his most important assistant, Dr. Neugebauer [38].

3. Harald Bohr was first in Germany and after that in England, mainly with Hardy (and with Niels Bohr) exerting himself to organize support. I believe that Harald Bohr is now in back Copenhagen.

4. With all prudent restraint, the Rockefeller people are VERY grateful for any reliable information, I know that for sure. And I believe, they are, as far as that is compatible with their statute-riddled policies, at least helpful through their personal connections.

As far as the *physicists* are concerned, I get the following impression: all YOUNGER people (who are not full professors) have absolutely no possibility to work in Germany (neither in the academic profession, nor as high school teacher nor in industry). Also for literary work, practically all possibilities are excluded. Independently of the prestige they may have in their subject.— For the older professors of world fame the picture is still completely unclear (both for those that have been left alone until now, as for those 'provisionally on furlough' as for the 'voluntarily resigning' (!!!)) — It seems that one wants to keep very many of those in their position, *provided it is absolutely certain* that their moral fiber has been TOTALLY destroyed.

Even by only short invitations abroad one could morally help many younger and older people. ENORMOUSLY!!! We will soon start with that as far as physics is concerned. Obviously one must exert the utmost in tact and carefulness in correspondence and formulation of invitations.

[further pages missing].

[Typescript copy – in Ehrenfest]

33-11-23

To J.G. van der Corput — 23.XI.1933   **Laren**

Amice,

I would certainly rather like the idea to give another course in Groningen in early spring next year, were it not that that time is taken up for me with

---

[37] In text 'C'.   [38] In text 'Neug'.

lectures in Geneva and possibly yet another Swiss university. But I would be happy to reconsider the plan for a following year.

The plan mentioned in your letter concerning the financial side only became clear to me when I checked your earlier letters and then read in your letter of April 11 a sentence, which at the time had escaped me, that is where you ask me to donate the revenues of the sale of the lecture notes, in their further detailed form, to the coffers of the Groningen philosophical faculty association. Because I am on the one hand not familiar with the customs in such matters, and on the other hand my life the past few years is dominated by a financial affair [39] which most strictly forbids me vis-à-vis my creditors to make donations, I have talked the matter over with some colleagues to whom I felt that I could impute some business acumen, and the result is that I have reached the conviction that according to prevailing norms I have the right to receive at least two thirds of the revenues. So I propose to you that I take care of the production of the stenciled copies of the text as revised by me from begin to the end, and that the Groningen faculty association then offers the printed matter I send them for sale, and receive a provision of one third of the revenues after deduction of my advances.

Where I regret that my emergency situation forces me to this commercial attitude, you must on the other hand take into account that without this situation I probably would not lightly decide to give talks away from my place of work (and also not to travel two months during the summer as external examiner and expert [40]), because this all happens at the expense of my scientific researches, for which the conditions have been so drastically reduced since the theft also a few years ago of my scientific papers. [41]

Forgive me for being so detailed, which is essential for clarity, and believe me.

With collegial greetings [42]

your
(w.g.) L.E.J. Brouwer

[Signed carbon copy – in Brouwer]

---

[39] Brouwer refers to the Sodalitas Affair, see [Van Dalen 2005], section 16.3.
[40] Committees composed of university staff for a long time supervised the high-school final examinations. A member of such committees was called 'gecommitteerde and deskundige'
[41] C.f. Brouwer to Hahn 9.VIII.1929, [Van Dalen 2005] p. 656.   [42] collegialiter.

To Mr., Mrs. Erich Gutkind — 10.XII.1933 Berlin-Charlottenburg

Once more, how empty is Berlin without you! I am so eager to hear how
you have taken root in the new world [43] and how receptive you have found
the people there to your outlook. I have met your brother Erwin a few times
in Amsterdam. On his advice I am now here after I heard that the house in
Zehlendorf is already empty for many months and is about to be publicly
sold because of tax debts of the emigrated Steinbergs. It is questionable if
I am able to save anything.

Receive the embracement of your Bertus. [44]

[Signed autograph, postcard – in Brouwer]

To A. Heyting — 13.IV.1934   Laren

Dear Mr. Heyting, [Waarde Heer Heyting]

I haven't heard about your application any more; should you have occasion to apply for a position in Amsterdam, then you must let me know
immediately; then I can do something for you without waiting until I am
asked for information. I am convinced that also Prof. de Vries [45] would
make every effort for you.

On the occasion of the talks I recently gave in Geneva, your work has
been discussed quite a bit. I have promised several gentlemen that I would
ask you to send them your publications. They are Arnold Reymond, Professor of Philosophy in the University of Lausanne (Cerisiers 10, Pully –
Lausanne); F. Gonseth, Prof. of Mathematics in the Polytechnical School of
Zürich; F. Abauzit in Thonon, on the Lake of Geneva (France); R. Wavre,
Prof. of Mathematics in the university of Geneva; E. Claparède, Prof. of Psy-

---

[43] Gutkind emigrated to the US in 1933. [44] *Seit umarmt von eurem Bertus* [45] The
mathematician Hk. de Vries.

chology in the university of Geneva; G. Juvet, Prof. of Mathematics in the university of Lausanne; H. Reverdin, Prof. of Philosophy in the university of Geneva.

Hoping that you will render this service to the gentlemen named, I remain with friendly greetings,

Your

L.E.J. Brouwer

[Signed autograph – in Heyting]

1934-06-19

From E. Tornier — 19.VI.1934   **Göttingen** [46]
                                        Mathematisches Institut der Universität
                                        Bunsenstrasze 3/5

Dear Professor, [Hochgeehrter Herr Professor]

As you know, the flooding of Germany by alien races, in particular also in the teaching staff of the mathematical institute here, has led to insufferable situations.

I allow myself the request whether you, whom many German mathematicians consider as one of the greatest researchers with a typical Germanic attitude, would be willing to help lay a new foundation for the old fame of Göttingen's mathematics.

I firmly believe that you will have here, both because of the scientific resources and because of number and quality of the listeners, a satisfactory sphere of action.

I ask you whether you would be so kind as to inform us whether you would in principle be inclined to enter into negotiations about a call to Göttingen. I can add to this, that as I know, my joy of seeing you maybe forever in Göttingen is shared by the responsible official in the ministry,

---

[46] This is the first letter in a prolonged correspondence which from Brouwer's side was motivated by the necessity to get the cooperation of various officials and offices to transfer the German Marks, obtained through the sale of his Berlin house into guilders. We may safely assume that Brouwer did not seriously consider the offer, nothing came of it. See [Van Dalen 2005], section 16.7 and *Brouwer to Mayor of Amsterdam — 8.X.1946*.

namely ministerial director Professor Dr. Vahlen.

Sincerely yours [47]
E. Tornier
temporary managing director
of the Mathematical Institute
Göttingen

[Signed typescript – in Brouwer [48]]

34-08-03

From H. Hasse — 3.VIII.1934   **Tübingen** [49]

Dear Colleague, [Sehr verehrter Herr Kollege]

Many thanks for your telegram, which reached me just now at Knopp's.
I am glad that there is a prospect of meeting you next week. My address
is on the other side. It should be too large a detour for you to visit me
in Partenkirchen. I am gladly willing to come to Göttingen on a day determined by you, or to come to Göttingen on a day determined by me,
or meet you at some suitable place. Please let me know in time about
your travel plans, so that I can make arrangements. Tornier is in Berlin
until next Tuesday, he probably will have written you so. Meanwhile I express the hope that all negotiations will go to your satisfaction and that
the result will be the fulfillment of our wishes, and I remain with friendly
greetings

Yours truly [50]
Hasse

ab 4.8 Partenkirchen (Oberbayern), Hotel Gibson

[Signed autograph, postcard – in Brouwer]

---

[47] *Ihr sehr ergebener.*   [48] Original partially burned in the fire at Brouwer's cottage,
cf. [Van Dalen 2005] p. 750. Various copies in archive.   [49] Place - postmark.   [50] *Ihr
sehr ergebener.*

1934-08-13

To H. Hasse — 13.VIII.1934   Laren

Dear colleague, [Sehr verehrter Herr Kollege]

Unfortunately I have to inform you that I have for some time to stay in bed as a consequence of a dog bite, and this will take another 10 or 14 days. As soon as I have an idea of the point in time that I am fit to travel, I will make an appointment with you on the basis of your letter of the third of this month.

With friendly greetings

Yours truly [51]

[Typed copy – in Brouwer]

1934-10-12

From H. Hasse — 12.X.1934   Göttingen
Mathematisches Institut der Universität
Bunsenstraße 3/5

Dear Colleague, [Sehr verehrter Herr Kollege]

At this time we have to prepare the announcements for the lectures of the winter semester. When you were here, you mentioned that you would definitely come here in winter, [52] and then wanted to lecture on intuitionistic mathematics. I would be grateful if you could provide me as soon as possible with more detailed descriptions about the title of the course that is to be announced, and the number of hours per week, so that I can fill out the announcement for you. As far as the time is concerned, the best thing is to tell me on the basis of the enclosed copy of the lecture time table, how you wish it. Also, in case you want to give further lectures

---

[51] *Ihr ganz ergebener.*   [52] I.e. *Winter Semester.*

or problem sessions, or seminars, I'd like to ask for more precise indications.

When may we expect you here? The beginning of the lectures is settled on Thursday, November 1. If I can help you with anything else, I am happy to be any time at your service.

With friendly greetings

Your
Hasse

[Signed typescript – in Brouwer]

34-11-07

From F. Gonseth — 7.XI.1934   **Zürich**

Dear Sir, [Bien cher Monsieur]

First of all, I want to apologize for the long delay in checking the shorthand notes of the fifth and sixth lectures of your course in Geneva. [53] The main cause is the state of my sight, which forced me to take rather inconvenient precautions.

Concerning my work, I must admit that I am not very satisfied with it. I just have checked the 26 pages of the fifth lecture, and from the point of view of the language the corrected text is not yet completely satisfactory. However, I cannot decide to edit it more radically, because I would not be sure of respecting your intentions.

The shorthand itself has been taken in a very imperfect way. Certain of your sentences are completely disfigured and I cannot take it upon me to put them back in their unimpaired state.

Don't you think that for the sixth lecture the following method would give better results? You shall begin to re-establish the parts corrupted by the stenographer, and to shorten the text in order to change spoken language into written language. With this text, *this time authentic*, it would be rather easy for us to correct the linguistic irregularities that would have

---

[53] Brouwer's Geneva lectures on intuitionism, March 5–14 1934. Six lectures held.

escaped you. I think that the result would be in all respects more satisfactory.

I preserve the best and most lively memories of the sessions in Geneva and allow myself to assure you of my total sympathy. [54]

F. Gonseth

[Signed typescript – in Brouwer]

1935-01-08

From L. Bieberbach — 8.I.1935   Berlin [55]

copy.
Bieberbach to Brouwer   Berlin 8.1.35.

Dear Brouwer, [Lieber Brouwer]

Please remove my name from the title page of issue 3, from the title page of volume 1 and of all the following issues, and erase my name from the list of editors. [56]

Unfortunately your communication of January 6, 1935 does not take into account the objections made by me, and neither the conditions that I have attached to my further staying on the editorial board of Compositio. My national feelings forbid me to belong to an editorial committee in which there are so many representatives of the international Jewry and in particular also emigrants.

Because it appears impossible to you, to introduce a change in that, nothing further remains for me than to withdraw myself. I understand that you were faced with problems when you wanted to fulfill my conditions. But I hope that you will appreciate my attitude and you will not blame me for frankly taking position, for the sake of our old friendship.

---

[54] *...je me permets de vous assurer de ma plus entière sympathie.* [55] Copies to Doetsch. Brouwer's reaction, see *Brouwer to Bieberbach 15.III.1935.* [56] The letter is part of the correspondence concerning the founding of the new journal, Compositio Mathematica. For Bieberbach's resignation from the board of Compositio, see [Van Dalen and Remmert 2006] and [Van Dalen 2005], section 16.6.

I am happy to recognize the good intentions of the proclamation planned by you. But unfortunately it ignores my objections. Indeed, it does not concern just my personal protection, but most of all the treatment that the international Jewry metes out to my fatherland and which of course determines my attitude. It is also not about possible public manifestations of individual editors, but about the silent tenacious struggle which is conducted under the surface and anonymously against my fatherland and my countrymen under the leadership of the international Jewry.

With cordial greetings [57]

[Typescript copy – in Doetsch]

35-02-05

To H. Hasse — 5.II.1935   **Laren**

Dear colleague, [Verehrter Herr Kollege]

To enlighten you about the reason for my continued silence and absence, I send you in copy my two last letters to Mr. Prof. Bachér. [58]

Hopefully, the expected rescue action from Berlin will take place in time, so as to prevent the definitive destruction of my mental activity. A scientific corpse would be of no use in Göttingen.

Sincerely yours [59]

(sgd.) L.E.J. Brouwer

[Typescript copy – in Brouwer]

---

[57] *Mit herzlichen Grüssen.*   [58] Bachér was attached to the *Preussisches Ministerium für Wissenschaft, Kunst und Volksbildung.* The correspondence with Bacher (and Vahlen and Kerkhof) was part of Brouwer's attempt to transfer the money from the sale of his Berlin house to Holland. At the time the transfer of currency was made almost impossible. [59] *Mit hochachtungsvollem Gruss – Ihr ergebenster.*

1935-03-20

To G. Doetsch — 20.III.1935   Laren [60]

Dear friend Doetsch, [Lieber Freund Doetsch]

You know that Bieberbach because of his extreme position has resigned from the management committee and the board of editors of Compositio Mathematica. It would please me very much, if you were under these circumstances willing to take over the position of Bieberbach as representative of Germany.

With cordial greetings [61]
Your
sgd. Brouwer

[Signed typescript – in Doetsch]

1936-03-20

To P.S. Alexandrov, H. Hopf — 20.III.1936   Laren

In the first place I express with thanks my delight over the exquisite book [62] with which you have associated my name.

In the second place, after all German nationals [63] have resigned from the board of *Compositio Mathematica* a vacancy has arisen, more specifically in the editorial committee. I am of the opinion that this position (thus next to Julia, Whittaker, De Donder and me) would be best filled by one of you both, I hope from the bottom of my heart that one of you is willing to satisfy my strong aspiration in this matter.

Warmest greetings for you [64]

Your
L.E.J. Brouwer

[Signed copy – in Hopf]

---

[60] Addressed: Prof. Dr. Gustav Doetsch, Freiburg i. Breisgau.   [61] *Mit herzlichem Gruss*   [62] [Alexandroff and Hopf 1935] — L.E.J. BROUWER GEWIDMET (Dedicated to L.E.J. Brouwer).   [63] *Reichsdeutschen.*   [64] *Es grüsst Euch herzlichst.*

To A. Heyting — 30.III.1936 Laren ^{⟨65⟩}

Dear Mr. Heyting, [Waarde Heer Heyting.]

   May I, in connection with your enclosed manuscript, ⟨66⟩ bring to your attention, that what you say under point 4 is not correct. An *arbitrary* continuous function can be a *point core* of a *topological space* (as I have described in my 'Intuitionistic introduction of the notion of dimension' ⟨67⟩) of continuous functions. Functions determined by a *law* in that topological space are the 'sharp' point cores; just like the numbers $\frac{1}{2}, \pi, e$ etc. are 'sharp' – i.e. determined by a law – point cores of a number continuum. On the unit interval one can for example define the very simple topological space of continuous functions $y = \sum \pm \frac{x^n}{n!}$, where the choice of the sign remains free for each n.

   It is possible that the above is not clearly emphasized in my writings (in the *first* introduction of the intuitionistic function concept I have restricted myself to functions determined by a law); in any case, in my lectures and talks I have emphasized already for some time that an *arbitrary* continuous function emerges just as much by 'Free becoming' ⟨68⟩ as an *arbitrary* point of the continuum.

   You would do me a pleasure if you would modify the enclosed manuscript somewhat, taking the above into account.

   Just to be sure, I also enclose the manuscript of Freudenthal, ⟨69⟩ to which it will be published as a sequel.

   With friendly greetings

Your
L.E.J. Brouwer

[Signed typescript – in Heyting]

––––––––––

⟨65⟩ This letter is cited in [Brouwer 1942b].   ⟨66⟩ [Heyting 1936].   ⟨67⟩ *Intuitionistische Einführung des Dimensionsbegriffes.*   ⟨68⟩ *freies Werden*   ⟨69⟩ [Freudenthal 1936].

1936-04-10

To A. Heyting — 10.IV.1936   **Laren**

Dear Mr. Heyting, [Waarde Heer Heyting]

One can collect in the following way the *totality* of unitary bounded continuous functions of the unit interval into a *spread*, which assigns to certain κ-intervals of the unit interval uniquely determined λ-intervals, and do this in a *normal* manner, i.e. such that when a κ-interval α is part of a κ-interval β, the λ-interval assigned to α is part of what is assigned to β; when α borders on β, the λ-interval assigned to α will cover the one assigned to β wholly or in part; and when α and β are of the same length the λ-intervals assigned to both also have the same length.

This assignment is done 'in free generation' [70] as follows: Suppose that after the $3n$-*th* choice *all* κ_{p_n}-intervals of the unit continuum have been assigned λ_{q_n}-intervals in a normal manner. The $(3n + 1)$-*st* choice then produces an arbitrary natural number $p_{n+1} > p_n$, the $(3n + 2)$-*nd* choice an arbitrary natural number $q_{n+1} > q_n$, and the $(3n + 3)$-*rd* choice an arbitrary determination (from the finite number of possibilities for it) of an assignment, happening in a normal way and normally fitting to already existing assignments, of $\lambda_{q_{n+1}}$-intervals to *all* $\kappa_{p_{n+1}}$-intervals of the unit continuum. [71]

With friendly greetings

Your
L.E.J. Brouwer

[Signed typescript – in Heyting]

————————

1939-10-04a

To H. Freudenthal — 4.X.1939[a]   **Blaricum** [72]

Dear Freudenthal, [Waarde Freudenthal]

Your letter, which I received last Saturday, has to my delight had a relieving effect on our relations which had turned into a predicament as a

---

[70] *in freiem Werden*   [71] See also [Brouwer 1942a].   [72] This letter is one of the series that dealt with the "triangulation problem", which had been solved by Brouwer for the case of differentiable manifolds. Freudenthal subsequently gave another solution. The matter caused at the time a great deal of friction between the two correspondents. See [Van Dalen 2005], p. 724 ff.

consequence of the letter you sent me one month ago. Where furthermore your disposition, that manifests itself in your letter, seems to open the path to recovery of mutual agreement of our intentions; in the interest of prevention of new disruptions of this agreement I consider it desirable to be more precise about a few points.

Accepting your letter of September 1 was already forbidden by the national honor. Indeed, such would create in the Netherlands a situation, where a President of the Wiskundig Genootschap, [73] who establishes a mathematical result with a proof, which is expounded in detail in a lecture at a General Meeting of this Society, but which is not published immediately thereafter, is forced to protect himself against theft of priority by his listeners by either recording his talk during the meeting by means of a dictaphone, or by depositing the manuscript used in the talk under seal with a notary or another institution authorized in such matters, under penalty of possibly later having to bear the high costs of a chemical analysis to establish the age of his manuscript, in order to defend his priority. [74]

I have indeed, already on July 4—when you [75] indicated to me that you wanted to send me your reflections concerning the triangulation problem, Of which you had made notes following my talk, — immediately underlined that a written communication of these considerations *to third parties* would be admissible only after publication of my talk, or together with a reference to such publication or with a description of the contents of that talk. Pursuant to this you immediately declared that the planned dispatch was merely intended as a private communication exclusively regarding the two of us.

It is self-evident that the reception on last September 1 of your notification, which is diametrically in contradiction with your attitude of July 4, namely that you now wished to regard a manuscript by your hand concerning the triangulation problem as having been submitted, had to be perfectly dumbfounding to me, and had to obstruct every mutual understanding and cooperation between us.

For indeed, more than any other listener to my talk, especially my highest ranking (moreover pre-eminently expert and interested) official assistant should, rather than taking the position that he had forgotten essential parts

---

[73] Dutch Mathematical Society   [74] Note the implicit reference to the Menger affair.
[75] In the original text Brouwer uses here and in the sequel the slightly archaic and formal pronoun (Ge, or Gij, capitalized) rather than the common 'U', which was used in the first paragraph. 'Gij' could be translated as 'thou'. There was a fine distinction between the two. At the time 'Ge' was reserved for formal purposes, whereas 'U' was used on a somewhat friendly, personal basis. For a modern Dutchman these distinctions have been almost totally lost; even official organizations address strangers in writing as 'jij' (the equivalent of the german 'du').

of my talk, feel obliged at the occasion of possible attempts of theft of priority by third parties, to make the above mentioned chemical analysis for the determination of the age of the original manuscript superfluous by his personal expert testimony.

In order to dispel the incident entirely, I would like to request from you to confirm once again *explicitly* (as it seems the letter received from you last Saturday has already done implicitly)

first, that the point of view according to which written communication to third parties about investigations into the triangulation problem made after my talk of April 24, 1937, which was attended by you, can only happen posterior to publication of that talk, or together with a reference to such a publication or with a description of the contents of the talk, is agreed to by you;

second, that until now nothing has happened that would conflict with this point of view.

I would appreciate it to receive, together with this confirmation, word that your ideas on the triangulation problem have been handed over to Prof. Rosenthal in a closed envelope addressed to me and that Prof. Rosenthal [76] has been authorized by you to receive for you my proof as presented in my talk of April 1937, in exchange for handing over this envelope; he will receive this proof from me (or possibly from Miss Jongejan, as, being prevented by illness, I will not be coming to Amsterdam for several days).

With friendly greetings

Your
L.E.J. Brouwer

[Signed typescript – in Freudenthal]

1939-10-09

From H. Freudenthal — 9.X.1939   **Amsterdam**

Dear Professor, [Hoog Geëerde Professor]

With reference to your letter of October 4, I inform you that I subscribe to its contents starting from the last paragraph of page 2 until the end, and

---

[76] Rosenthal had left Germany. On his way to the USA, he had stopped for some time in Holland.

that I have handed over the manuscript to Prof. Rosenthal in the required modus. However, I very much regret to see in the mutual relations trust replaced by legal formulas.

With the remaining content of your letter I can not fully agree.

In the first place I am not satisfied with the representation of the facts as given by you; among other things I am bothered by a gap: the letter of July 4, which you don't mention, definitely displaced in my feelings the matter from the sphere of teacher and student (or at least of assistantship) to a highly official sphere.

Secondly I am disturbed by a passage which could be interpreted as an implicit accusation of me.

In the third place I don't agree with the entire tenor of your exposition: I don't see in which respect *I* should have brought you in a more difficult position in the defense of your priority (which has never been contested by me and is even emphasized in my manuscript) than anybody would have done who proves the theorem concerned independently from you and publishes it.

Finally I want to note that I cannot consider your point of view taken in the paragraph starting on the bottom of p. 2 as a *general* rule (which fact should not diminish my agreement in the *special* case). Indeed, the consequence of your point of view would be that one could block entire fields of research, by supplying an oral statement about one's researches and then failing to publish them, and so make it impossible for others to publish their researches. There are enough examples to show how nefarious such a possibility could be for the development of science, even if one excludes all cases where the oral communication turned out to be premature or incorrect.

With wishes for your recovery and greetings [77]

[Carbon copy – in Freudenthal]

---

[77] *Met wenschen van goede beterschap en groeten ben ik.*

Chapter 6

1940 – 1949

40-05-08

H. Freudenthal et al., circular — 8.V.1940

Confidential

Dear Sir, [Sehr geehrter Herr]

On February 27, 1941, L.E.J. Brouwer reaches the age of 60. The wish exists to present to him on this day a Festschrift in the pages of Compositio.

The fundamental articles of Brouwer have exerted such an influence on the creative activity of many mathematicians — topologists and foundational researchers — that these will be happy to consider themselves as Brouwer's students. To such mathematicians we now turn, asking them to contribute an article that bears witness of this influence.

Please let us know, *as soon as possible*, through one of the undersigned, whether we may expect such a contribution from you. Please also indicate the title of your article, the approximate number of pages and the presumable date of completion!

We would gratefully welcome your cooperation.

P. Alexandroff,	Moskow (USSR), Staropimenowski 8, 5.
H. Freudenthal,	Newtonstraat 75, Amsterdam (Nederland).
A. Heyting,	Oudblaricummerweg 5, Laren NH. (Nederland).
H. Hopf	Zürich-Zollikon (Schweiz). Alte Landstrasse 37.

[Typescript – in Hopf]

D. van Dalen, *The Selected Correspondence of L.E.J. Brouwer*,
Sources and Studies in the History of Mathematics and Physical Sciences,
DOI 10.1007/978-0-85729-537-8_6, © Springer-Verlag London Limited 2011

1940-08-10

From H. Freudenthal — 10.VIII.1940

Dear Professor, [Hooggeëerde Professor]

I received your letter of August 9, which surprised me very much.

You informed me on June 26 that, for the time being, no more issues of Compositio Mathematica should appear. [1] Meanwhile I had received a communication to the same effect (dated June 15) from Noordhoff, who also didn't want to bring out issues of Compositio Mathematica for the time being. However, Noordhoff wanted to go on receiving manuscripts for Compositio Mathematica, awaiting further developments, so that the printers could be supplied with typesetting jobs. I did not want to comply with this request of Noordhoff, without your explicit permission. Unfortunately you did not answer my repeated questions in this matter at all.

Because Noordhoff agreed with you to halt for the time being the publication of Compositio Mathematica, and because you have not taken any decision at all with respect to the further wishes of Noordhoff, I don't understand that you speak in your letter of August 9 about a resistance of Noordhoff to your decision of June 26, and that you think it necessary to invoke the authority of Mr. Wijdenes, to convert Noordhoff to a decision that Noordhoff already had taken on June 15.

After a substantial correspondence, a conference between you and Mr. Wijdenes and several not particularly enlightening telephone conversations, I still don't know more than two months ago. I still don't know whether I should comply with Noordhoff's request for new manuscripts or not. In the hope that after an instruction from you in this matter, the case can be considered as settled, I remain, with many greetings

Sincerely yours [2]

[Carbon copy – in Freudenthal]

---

[1] After a brief but brave resistance to a superior attacker, May 10–15, the Dutch army surrendered. Holland had become occupied territory. The continuation of Compositio was suspended for the duration of the war. [2] Uw dienstwillige.

40-11-30

From H. Freudenthal — 30.XI.1940

Dear Professor, [Zeer Geëerde Professor]

Several students asked me to facilitate the continuation of their studies in a manner to be further arranged, or still to conduct their exams. [3] I have explained to these students that I am, as it is, willing to cooperate in any manner,

But that I consider it impossible to invoke, against the obvious intention of the measures taken, formal circumstances, such as that I have not been discharged from my function as private docent, or that the exams do not have an official character. [4] I would be pleased if in the interest of the students at least a satisfactory transitional arrangement could be made, but I think this can only be done by a higher authority.

I don't have to assure you of in so many words, that I be happy to continue carrying out all activities that can be considered as being of a private nature.

With many greetings

Yours sincerely [5]

[Carbon copy – in Freudenthal]

41-01-14

To H. Freudenthal — 14.I.1941   Blaricum

Dear Freudenthal, [Waarde Freudenthal]

Hereby I inform you that Dr. E.M. Bruins will substitute for the analysis courses. [6]

---

[3] Freudenthal had been dismissed on November 23 as part of the general dismissal of all Jews from public offices (including institutions of education). He no longer had access to the university, and continuing educational activities would have put him at considerable risk.   [4] The 'exams' mentioned here are the so-called 'tentamens', i.e. examinations on the material of a specific course. When a student had passed all the 'tentamens', he was qualified to take the real examination.   [5] *Met vele groeten – Uw dienstwillige.*
[6] Bruins was a physicist, his appointment eventually led to universal embarrassment. See [Van Dalen 2005], p. 786, 794.

With friendly greetings

Your
L.E.J. Brouwer

[Signed typescript – in Freudenthal]

1941-04-19

From G.F.C. Griss — 19.IV.1941   **Gouda**

Crabethstraat 69 [7]

Dear Professor Brouwer, [Hooggeachte Professor Brouwer]

As agreed, I send you an exposition of my views on the concept of negation in intuitionistic mathematics. [8]

To start with, I would like to stipulate the following brief formulation:

Showing that something is not true, i.e. showing the incorrectness of a sup-
position is not an intuitively clear act. For it is impossible to have an
intuitively clear concept of an assumption that later turns out to be even
wrong. One must maintain the demand that only building things up from
the foundations makes sense in intuitionistic mathematics.

Although this point of view seems clear and indisputable to me, I will
try to justify it in more detail, and then I will show the consequences in
some cases and finally I show that also practical considerations can lead to
the same view.

1. Although my ideas about the foundations of mathematics are not completely identical to yours, the differences are unimportant for what
   follows, so, for example, I can agree completely with your considerations in the *Tijdschrift voor Wijsbegeerte*, 2nd volume, 1908. [9] Let
   me just remark that the concept of negation does not explicitly occur
   in the formulation of the foundations of mathematics, but only in the
   examination of the validity of the logical principles. You say there:

---

[7] Address on envelope.   [8] A first exposition of Griss' ideas on negationless mathematics. For publications see [Griss 1944, Griss 1946, Griss 1950, Griss 1951].   [9] Journal for
Philosophy, [Brouwer 1908b].

'The principle of contradiction is just as little in dispute: the execution of the fitting of a system a in a particular way into a system b, and finding that this fitting turns out to be impossible are mutually exclusive'

What does impossibility of a 'fitting in' mean here?

In the first place this can mean that one assumes the possibility of fitting, and that this assumption leads to a contradiction. This manner far exceeds the construction of mathematical systems on the basis of the ur-intuition, and as I remarked in the beginning, one cannot clearly obtain a conception of it. If one still accepts it, then one takes in principle a similar step, as when one accepts the principle of the excluded third. An element of arbitrariness enters in our idea about what is and what is not admissible in mathematics, if one does not stick strictly to the requirement that one only builds up mathematical systems from the foundations which are given in the ur-intuition.

Another meaning which can be given to 'finding that this fitting of a system a into a system b turns out to be impossible' might be this: that the system a demonstrably differs (in that case this concept has to be defined) from every system that can be fitted into b. One asks for example whether e is an algebraic number and one finds that e is positively transcendent so e demonstrably differs from each algebraic number. If need be, one can even answer the question whether e is algebraic by: e is not algebraic, but then we have assigned a new meaning to the word 'not'.

2. The consequence of my view is of course that in intuitionistic mathematics all negative propositions have to be replaced as much as possible by positive ones.

As first example I take the beginning of the Set Theory as given by you in Mathematische Annalen 93. [10] In the definition of set, negation is used several times, but this definition can easily be freed from negations. In the concepts equal or identical, species and subspecies, negation does not enter. The concept of different has to be defined in a positive way: Two elements of sets are called (demonstrably or positively) different, if a number n is known such that at the nth choice a different sign occurs in these elements. Two sets (species of the first order, species of the nth order) are called different if at least one of these

[10] [Brouwer 1925].

sets (species) contains an element that differs from each element of the other set (species). In the sequel 'differ' means 'differ demonstrably of positively'; whenever confusion is possible, I explicitly distinguish differing negatively and positively.

M is called a proper subspecies of N, if an element of N is known that differs from each element of the subspecies M. Two species M and N are called mutually disjoint if each element of M differs from each element of N.

After the union of two or more species has been defined, one can give the definition of half-identical species, for instance as follows:

> Two species A and $\mathfrak{S}(B,C)$ [11] are half-identical if $\mathfrak{S}$ is a subspecies of A, an element of A that differs from each element of B is element of C and an element of A that differs from each element of C is element of B.

Similar definitions can be given for congruent species, and for unions $\mathfrak{S}_1$ and $\mathfrak{S}_2$ of respectively m and n species. The examples you give for congruent and half-identical species can be left unchanged.

The properties: 'A proper subspecies of a finite species E is not equinumerous [12] with E' and 'a finite species is not infinite' can easily be formulated positively, while the property: 'each reducible infinite species U contains proper subspecies equinumerous with U' remains valid.

Now I treat no. 144 of the Problems of the Wiskundig Genootschap, volume 16: When of two triangles have one side, the angle opposite that side and the sum of the two other sides equal, then it is impossible that those two triangles are not congruent.

We first prove:

> If two triangles [have] two sides and the angle opposite one of those sides equal, while the sum of the angles opposite the other sides [13] differs positively from 180°, then the triangles are congruent.

Given: For $\triangle ABC$ and $\triangle A'B'C'$ is $AB = A'B', BC = B'C', \angle C = \angle C'$ and $\angle A + \angle A' \# 180°$. [14] Show: $\triangle ABC \cong \triangle A'B'C'$

---

[11] Following Brouwer, Griss denotes the union of B and C by $\mathfrak{S}(B,C)$.
[12] 'gelijkmachtig' in text.   [13] i.e. the other side in each of the two triangles.   [14] # denotes the apartness for real numbers.

Proof: According to the sine rule $\sin A = \sin A'$, so $\sin A - \sin A' = 0$
or: $2 \sin \frac{1}{2}(A - A') \cos \frac{1}{2}(A + A') = 0$
The last factor differs positively from 0, so $\sin \frac{1}{2}(A - A') = 0$, hence
$\angle A = \angle A'$. So $\triangle ABC \cong \triangle A'B'C'$

Instead of the desired property we have now the following:

> If two triangles have a side, the angle opposite that side and
> the sum of both other sides in common, while it is known
> about one of the adjacent angles in both those triangles that
> they are equal or (positively) differing, then the triangles are
> congruent.
>
> We only have to prove the case that the adjacent angles
> differ.

Given: For $\triangle ABC$ and $\triangle A'B'C'$: $AC = A'C', \angle B = \angle B', AB+BC =$
$A'B' + B'C'$, while $\angle A \# \angle C'$. To prove: $\triangle ABC \cong \triangle A'B'C'$.
Proof: Extend AB with $BD = BC$ and $A'B'$ with $B'D' = B'C'$,
then $\triangle ACD \cong \triangle A'C'D'$, because $AD = A'D', AC = A'C', \angle D =$
$\angle D'$, while $\angle ACD + \angle A'C'D' = (\angle C + \frac{1}{2}\angle B) + (\angle C' + \frac{1}{2}\angle B') =$
$\angle C + \angle B + \angle C' \# \angle C + \angle B + \angle A$, so $\angle ACD + \angle A'C'D' \# 180°$. From
$\triangle ACD \cong \triangle A'C'D'$ it follows immediately that $\triangle ABC \cong \triangle A'B'C'$.

If one uses negations, the last property entails what was asked (not
the converse of course): If the triangles weren't congruent then it would
be true of two angles that they would be both equal and negatively
different, which is impossible.

This fits in with the remark that in Cartesian geometry also a few
changes have to be made; for example the definition of parallel lines
must become:

> Two lines are called parallel, if each point of one line differs
> (locally) [15] from each point of the other line.

3.   Finally I make two more practical remarks which are also significant
for those who don't agree with the more fundamental exposition.

No real number a is known about which it has been proved that it
cannot possibly be equal to 0 $(a \neq 0)$, while at the same time it has
not been proven that the number differs positively from 0 $(a \# 0)$. If we
compare the two properties: 'from $ab = 0$ and $(a \neq 0)$ follows $b = 0$'
and 'from $ab = 0$ and $(a \# 0)$ follows $b = 0$', then no real numbers a

---

[15] '*plaatselijk verschillen*' and '*örtlich verschieden*' were introduced by Brouwer for his
strong inequality – no known as 'apart'.

are known, for which the first property may, and the second property may not be applied. The negative concept 'distinct' for real numbers is also of no practical use in intuitionistic mathematics. If theories about that notion are considered, then it is only to preserve again as much as possible of classical mathematics. More specifically this aim occurs in prize problem 13 of the Wiskundig Genootschap for this year:

> It is asked to give to an as large as possible part of classical analysis an intuitionistically correct 'weak' interpretation containing only 'stable' propositions (where a proposition in intuitionistic logic is to be called 'stable', if it equivalent to its double negation).'

In my opinion the most essential part of intuitionistic mathematics is in this way relegated to the background.

In many cases a result is formulated negatively, although the proof makes it just as well possible to use a positive formulation. An example is the property mentioned in no. 2, that every reducibly infinite species U contains subspecies of the same power as U if the concept of proper subspecies is formulated negatively. Also some other results of no. 2 can serve as examples.

With friendly greetings,
sincerely [16]

G.F.C. Griss

[Signed autograph – in Brouwer]

———————

1942-05-26

From H. Freudenthal — 26.V.1941   **Amsterdam**

Dear Professor, [Zeer Geëerde Professor]

Mr. J. de Groot did not receive page proofs either of his Proceedings note, which was supposed to be presented by you to the April meeting of

---
[16] *Met vriendelijke groeten en de meeste hoogachting.*

the Academy. [17] After a telephone inquiry with the Academy I found that my note has not been submitted.

Especially in the interest of Mr. de Groot I would like to ask you respectfully to inform me about the fate envisioned by you of the two notes intended for the April meeting (those of J. de Groot and me) and the two intended for the May meeting (by J. de Groot and by J. de Groot and F. Loonstra). I would like to ask you also not to postpone this communication, because it is possible that one of the parties concerned could or would want to arrange for other destinations of his work before coming Saturday. [18]

With many greetings

Sincerely
Hans Freudenthal

[Signed handwritten draft/copy – in Freudenthal]

42-05-29

To H. Freudenthal — 29.V.1942   **Roosendaal**

Dear Freudenthal, [Waarde Freudenthal]

Touring the country for the final examinations of the gymnasium, I received your forwarded letter of the 26th of this month. You and the other two gentlemen involved [19] seem to forget that the members of the Academy [20] are a board of journal editors for the Proceedings, and that editorial boards are not printing automatons.

With friendly greetings

L.E.J. Brouwer

[Signed autograph – in Freudenthal]

---

[17] [De Groot 1941].   [18] The present letter was written on the preceding Monday. [19] J. de Groot and F. Loonstra.   [20] KNAW.

1942-05-30

To H. Freudenthal — 30.V.1942   Blaricum

Dear Freudenthal, [Waarde Freudenthal]

Hereby I return the manuscripts of you and the other two impatient Gentlemen [21] to 'arrange for other destinations'.
With friendly greetings

L.E.J. Brouwer

[Signed autograph – in Freudenthal]

1942-05-31

From H. Freudenthal — 31.V.1942   Amsterdam

Dear Professor, [Zeer Geëerde Professor]

I received your letters of May 24 and May 29.

1. I have taken care that the books of the Mathematics Institute have immediately been returned. Fortunately I have so far experienced no other problems in using non-public libraries.

2. I do not remember a case as mentioned in point 2 of your letter of May 24. [22] Nonetheless, I have declared to Mr. Bruins to be prepared to supply him with all desired information about the whereabouts or conjectured whereabouts of items of the library.

3. Although I see little use in organizing a formal transfer of the archive of Compositio Mathematica, which I haven't touched since $1\frac{1}{2}$ years, I am quite willing to cooperate if you insist on this formality.

Consequently, I request you to supply me with the necessary permission from the supreme authorities to be allowed to appear for this purpose in the Mathematical Institute, or to have the cabinet in the assistant's room,

---

[21] J. de Groot, F. Loonstra. [22] Brouwer had asked Freudenthal to try to get students to return books of the reading room, that were borrowed (allegedly) with Freudenthal's permission.

where the archive is, brought to a place where that transfer can take place without such a permission. [23] Unfortunately, it is not clear to me, what the subordinate clause 'which in its entirety should be located in the Mathematics Institute' refers to. That part of the archive of Compositio Mathematica over which I had any authority, has been without interruption in the Mathematics Institute since 1935, to wit in the cabinet of the 'Assistant's Room' — unless very temporarily some articles were taken out to study them quietly at home or to take care of the current affairs of Compositio Mathematica from my vacation accomodation. According to what Mr. Bruins has assured me, that part of the archive is still in the same place, as he was firmly convinced, untouched in the same state in which I left it a year and a half ago. Among the five mathematicians whose names have appeared on the cover page of Compositio Mathematica, there will perhaps be some who have a greater number of recollections of Compositio Mathematica.

4. I have preferred not to transmit your statement of May 29 to the other interested parties. They would perhaps have little appreciation for insults, that one can only bear out of respect for a great mathematician.

5. Just now [24] I receive your express letter of yesterday, postmarked today, May 31, containing the manuscripts of the four notes to be presented.

In April I sent you two notes to be presented to the Academy, [25] one of Mr. de Groot and one of me. On May 21 I sent two more notes — of de Groot, and of Loonstra and de Groot, all accompanied by recommendations. Also I informed you that until now I have not received page proofs of my note. Instead of an answer I received on May 26 a letter, the tone of which made me suspect that my note had not been presented and that it also would not be presented — a conjecture the first half of which was confirmed by a telephone conversation with the administration of the Academy and the rest by your letter of today. Unfortunately my effort failed to get a decision from you about the fate of the notes in advance of the meeting of the Academy. I am afraid that authors concerned — in view of the facts and also in view of your entire manner towards me — will hardly consider your

---

[23] Crossed out part, replaced by 'Although ... permission.': 'I don't understand what purpose the formal transfer of the archive of Compositio Mathematica, which I didn't touch at all since one and half years, would serve. However, if you wish such a formality, I entreat you to ask the Reich Commissioner [i.e. the highest authority in the Netherlands, Seyss-Inquart] for permission, so I am allowed to appear in the Mathematics Institute for this purpose.'   [24] Crossed out: — on Sunday Afternoon —   [25] KNAW.

motivation for not presenting and for returning the manuscripts as anything but a pretext. So I ask you most urgently to suggest me how you might motivate your refusal to the parties concerned and others.

With many greetings

[Handwritten draft/copy – in Freudenthal] [26]

_____

1944-05-20

From E.J. Dijksterhuis — 20.V.1944   **Oisterwijk**

Dear professor Brouwer, [Hooggeachte professor Brouwer]

The idea to include an obligatory examination in history of mathematics in the requirements for the doctoral examination [27] with a major in mathematics has my full approval. The course material which should suffice for this category of listeners, and which is necessary for those who take philosophy as a major or minor subject, could be taught in the general two hour lecture, and the third hour would then be assigned to this latter group.

It shocked me very much to hear on April 29, that you have been the victim of a fire for the second time now. I hope that the damage and the shock for your wife and you have remained within moderate bounds.

With polite greetings

Sincerely yours [28]
E.J. Dijksterhuis

[Signed typescript – in Brouwer]

_____

_____

[26] A typescript copy of the letter was enclosed in *Freudenthal to Van der Corput*, *24.VIII.1945.* [27] Comparable to M.Sc. degree, not part of a Ph.D. examination. [28] *Met beleefde groeten en hoogachting – Uw dienstwillige.*

45-07-17a

To Committee of Restoration — 17.VII.1945[a]   Amsterdam [29]

The declaration of loyalty 1943

When a company of civilized travelers is overpowered by superstitious cannibals, their behavior, in particular their spoken or sign communication with their captors will be directed at their liberation. For a method they will have to rely on cunning, cheating, and dissimulation for communications as well as proposals, and promises. Honesty, chivalry and demonstrative proclamations will not only have to be rejected because of their being contrary to the goal, but also lack rational content: the essential commitment of the meaning of a word, gesture, or sign required for an honest rapport is in fact only possible on the basis of tacit cooperation of the interacting parties as 'good understander' and this cooperation can only derive its moral orientation from the (in this case lacking) common orientation of will. [1]

Such a situation existed in the Netherlands during the occupation. The *manner* in which the enemy attacked us, and in which he subsequently had trampled good faith and human rights, had on the one hand exterminated any common orientation of will or respect, on the other hand exclusively oriented on the following goals: 1) to serve the occupying forces as little as possible, 2) to obstruct the occupying forces as much as possible, 3) to safeguard our national heritage as well as possible against destructive intervention of the occupying force. And in this framework the language and sign communication of the Dutch population with *this* enemy was for communications, proposals, and promises was thrown back, on the basis of above mentioned arguments, on cunning, deception, and dissimulation, on the other hand, chivalry and demonstrative proclamation had become practically unacceptable and had lost rational content.

In view of this exposition it is thus not correct that, as was said at the time, the signing or not signing of the declaration of loyalty by the Dutch

---

[1]compare my lecture *Willen, weten, spreken* (Will, Know, Speech)(in *De uitdrukkingwijze der wetenschap* (Way of expression of Science), Groningen: Noordhoff 1933, in particular under section I.4.

[29] Addressed: College van Herstel voor de Gemeentelijke Universiteit te Amsterdam. This is a note, presented to the committee at the interview of 17.VII.1945. The note contains a defense of Brouwer's position with respect to the 'declaration of loyalty' in the Senate meeting of the UVA of 26.III.1943.

students involved ethical or idealistic goods of the Dutch national community. On the contrary, there was the possibility that a general signing would have the consequence that 1) a smaller part of the Dutch potential would serve the enemy; 2) the students working in the resistance movement would obtain more favorable conditions for their activity; 3) it would be less detrimental to the health and intellectual education of the Dutch students. This had at the time to result in the conclusion that the signing of the declaration of loyalty would serve the interest of the fatherland as well as that of the students. And I felt that I could not suppress this conclusion, when it thrust itself on me, because the tradition — in particular held in high esteem by the Dutch national community — that the ventilation of a sincere opinion is not only an inalienable right, but in cases that touch on the general interest moreover an undeniable duty, represents to me one of the most treasured goods, on account of which I felt, even against personal interests, lastingly connected with the Netherlands.

L.E.J. Brouwer

[Signed typescript – in GAA; signed typewritten copy in Brouwer.]

1945-12-01

From Hk. de Vries – 1.XII.1945   Benjamina [30]

Amice,

Thank you for your letter of October 10, 1945, which I received recently. I had already heard from Henk that you dropped in at Mannoury's place, just when he showed a letter from me to Gerrit; so you know *also* about how the Wife and I have struggled through all these years, not directly in contact with the misery of the war, but all the time, and still, coping with financial problems that poison our lives. First I had a lot of hassle with my pension, but at least that has been settled now, even though the State of the Netherlands now still owes me a large sum and bluntly refuses to pay, but then I had negotiated a nice annuity from the Hollandsche Sociëteit [31] (van Haaften), to support myself in the last years of my life. Good Lord! half of

---

[30] Settlement in former Palestina, now Northern Israel.   [31] An insurance company.

the payments have been stolen by that stinking vermin, and the other half which van Haaften managed to save are at my disposal on Heerengracht 475. [32] But getting them, no! Can *you* give me a nice definition of an annuity that hasn't paid a cent since January 1940? And so I have money on the Heerengracht, and we live here in very straightened circumstances. That's nowadays the quiet and peaceful old age.

About an examination of a certain C. Kramer, I don't recall anything; I can neither deny, nor confirm it. I don't know whether I knew a C. Kramer personally during the last part of my existence in Amsterdam; if he has taken the exam at all, it is stupid of him not to have demanded a proof, because he knew I would be leaving.

I had heard already about Belinfante [33] and Koppers, [34] and just yesterday I received a letter from Van Pommeren, [35] in which he in great length described his experiences, so I also know what *he* had to enjoy. One thing is even more depressing and nasty than the other. That Weitzenböck [36] has sneaked out, [37] I have also heard; a pity that they didn't get to him in time, because *he* really deserved it. I always hated his guts. And I heard also that a couple of you guys have been suspended, including our good Stomps, who seems to take it very seriously, because he always did everything what he could do to assist the Jews, for example the whole Heimans family, and he even hid a Jew in his house. That man truly didn't have to be suspended!

And so life goes on again, I am curious whether it will stay such a mess as it is now, or get better, or even worse.

With cordial greetings, also from the Wife.

Yours truly
Hk. de Vries

[Signed autograph – in Brouwer]

---

[32] The address of the insurance company. [33] The Portuguese-Jewish private docent, PhD student of Brouwer, who died in Auschwitz, 14 oktober 1944. [34] Misspelled as 'Coppens'. The former janitor of the mathematics institute. [35] The former beadle. [36] Specialist in invariant theory, had strong nazi sympathies. See [Van Dalen 2005], p. 774. [37] Weitzenböck was in fact almost immediately arrested.

1945-12-03

To Committee of Restoration — 3.XII.1945   Blaricum [38]

Dear Sirs, [Mijne Heeren]

Now that the investigation into my behavior during the occupation apparently still has not been concluded, I think it is opportune to mention to you — apart from the sanctioning and protecting of an underground operation in the Mathematics Department during almost the entire time of the occupation, which has already been brought to your attention — a few more facts, which perhaps can also make it plausible that my acts during the occupation which are subjected to your criticism, or which will be subjected to your criticism, in the end have taken place with the aim and under the reasonable assumption, that thereby ultimately the Dutch interest would be served.

I have immediately suspended at the beginning of the occupation both printing and publishing, also for publications of fellow Dutchmen, of the international mathematical journal Compositio Mathematica, which was founded by me, and in 1940 still directed by me, [2] because for me it was out of the question to see my editorial management subjected to control by the occupying force.

When in the summer of 1942 [39] the establishment of a 'Cooperative of scientific organizations in the Netherlands' was being prepared, which in my view was nothing but an instrument to submit and register the free Dutch scientific activities contrary to the Dutch national character — which, without having any real usefulness, would only facilitate possible attempts for nazification of Dutch scientific activities — I have, after having experienced that the mentioned establishment could no longer be stopped, exerted myself, to give the statutes a character that was as harmless as possible. When later the mentioned Cooperative

---

[2] at that time with the cooperation of 46 scholars from all countries of scientific importance, with the exception of Germany, which (as well as the enclosed copy of the press release dated January 16, 1937) [Brouwer publicly supported Van Anrooy's refusal to play the 'Horst Wessel-lied', *De Tribune*, 16.I.1937 — added as enclosure, see *16.I.1937* in the collected correspondence.] may be an indication that even before the war the Nazi spirit had raised normal feelings of revulsion in me.

[38] Adressed: College van Herstel voor de Gemeentelijke Universiteit te Amsterdam.
[39] Correct year '1940' — see *Brouwer to Committee of Restoration 30.XII.1945.*

started, in my view needlessly, to draw public attention, by publishing a survey of the research done in this country through the years 1938–1942, under the title 'Scientific research [40] in the Netherlands', I have first tried, with respect to this publication, to dissuade the Wiskundig Genootschap from cooperation, and when I didn't succeed, I have at least personally observed complete abstinence.

I have of course never accepted invitations for participation to scientific congresses in enemy countries during the occupation. In one case, when the invitation reached me through the representative of the government of the country concerned, I had to express my views very explicitly.

In the beginning of 1944 most of the telephone connections in het Gooi [41] were cut off. Shortly after that I met a fellow villager with whom I was at friendly terms, a businessman, who told me that he had visited the office of the Beauftragten für das Post- und Fernmeldewesen [42] to plead the reconnection of his telephone. He then showed me a form for the request to re-establish the telephone connection, containing a kind of loyalty declaration, which he had to sign to get his telephone back. He had a few more copies of this form with him, and he offered me one for signing, and he said he was willing to take care that it was returned together with his form. Whereupon I answered that such signatures in my opinion where permissible only if done collectively and in general terms, and never in the interest of one's own personal advantage. Whereupon after further expostulations my fellow-villager also abandoned his plans to sign.

During the occupation I have held the view that persons and groups that were not in a direct official relation with the authorities of the occupation authorities and who had no means of power with respect to these authorities (which was for example in fact the case with miners and medical men) should neither direct requests nor admonitions to the occupation authorities.

For such concessions could only derive their meaning and content from an existing basis of mutual understanding between both parties, and hence would implicitly recognize the existence of such a basis, which then must act as encouragement to the ever present ambition of the occupation force

---

[40] i.e. in the exact sciences.   [41] The region containing among other places Laren and Blaricum.   [42] The (German) title for the supervisor of mail, telephone and telegraph affairs.

to penetrate, and which could supply reasons or pretexts for new measures of nazification.

This view has been the cause that during the whole period of the occupation I have not had personal contact with the occupation authorities, other then by force, such as under duress, house searches and interrogations by the police.

The only purpose that would have given me the liberty to contact the occupation authorities on my own accord, would have been, as far as I can see, deception of the occupation authorities in the service of the Dutch interest. But indications of possibilities in this direction have not come my way.

Sincerely yours [43]
L.E.J. Brouwer

[Signed typescript – in GAA; Signed typescript copy in Brouwer.]

1945-12-06a

To J. G. van der Corput — 6.XII.1945[a]   Blaricum

Dear van der Corput, [44]   [Waarde van der Corput]

To my regret my health situation, which has been unstable for some time (maybe caused by the injustice committed against me), has come again into a critical phase, and now I am unable to come to Amsterdam because of an asthmatic bronchitis. Maybe you can write to me a few words about the present stage of the plans to establish the 'Mathematical Center'.

With friendly greetings

t.t.
L.E.J. Brouwer

[Signed autograph – in Corput]

----

[43] hoogachtend.   [44] Van der Corput had been appointed in Amsterdam; he was also made chairman of a committee for the reorganization of the Dutch mathematics departments, including the founding of a national centre for mathematical research.

)45-12-12

To Committee of Restoration — after 12.XII.1945 [45]

Memorandum

The complaints that were raised against me by anonymous sources with the Restoration Committee of the University of Amsterdam and that have been discussed with me by this Committee concern:

1. The document concerning the student declaration of loyalty, put on the bulletin board of the Mathematical Institute on April 8, 1943 by the joint mathematics teachers, having the Dutch nationality. In the discussion with the Committee I have emphasized that the point of view expressed in this document is tied to the day it was dated, and that it was on view there only during the time that the possibility of a *general* signing by the students could be reckoned with, and that it was moreover removed once temporarily during this time when the expectation seemed justified that the Senate would lay down and make public its point of view in this matter. It is possible that at the discussion with the Committee, when the chronology of the events was understandably not completely at my disposition, I may have mentioned from memory, as probable date of re-posting after its removal, the beginning of May (the minutes of the meeting were never shown to me); in any case I have sent later a written communication to the Committee, in which April 19 is mentioned as the date of re-posting. Of this statement the Committee has taken no notice, as appears from the letter of the Minister dated December 11, 1945.

For the text of the notice see enclosure 1. [46] For the motivation of the content see enclosure 2. Compare also enclosure 3. [47]

It is remarkable that there have been no complaints in this matter against several other professors of the same university who have voiced the same or even less strict points of view in writing or orally (albeit in a manner which was formally different from the one of the mathematicians).

2. Obstruction of the resistance; as such were put forward by the Committee, aside from the above mentioned position with respect to the declaration of loyalty

---

[45] In view of the dates given below, this note was submitted after December 12. The Memorandum may equally well have been addressed to the Minister of Education. Original enclosures not included.   [46] Text of the notice posted by Brouwer, Heyting and Bruins on the bulletin board of the Mathematical Institute.   [47] See *Brouwer to Committee of Restoration, 17.VII.1945.*

a) My submission of an amendment to the concept letter of the Senate to the General Secretary of Education, Teaching and Protection of Culture dated March 26, 1943, to the effect of omitting a passage in which the threat of a strike was made. About this I have argued before the Committee that the statements of the Rector Magnificus at the time indicated as the goal of that letter, not only to the letter, but also in fact, to obtain the changes deemed necessary in the Ordinances and Decisions of March 10 and 11, 1943. I was convinced that the chances of reaching this set aim (aside from the circumstance that the threat of a strike probably could not be carried out) were subjected to a diminishing by this passage, that could not be justified.

b) The continuation of the work in the Mathematics Institute after May 1943. For the motivation of this see enclosure 4. [48] Compare also enclosure 5.

3. My contributions to the Nederlandsche Volksdienst. [49] In this matter I have put forward to the Committee that this had happened exclusively in the interest of Mayor Klaarenbeek's staying on as long as possible, after having received a circular from his hand exhorting to cooperation, and after it had to be deduced from a speech of the then Governor [50] of the Province of North Holland that in his territory the retaining of mayors would to a large degree be related to how well the Winterhulp and the Volksdienst [51] were functioning in their municipalities. Already in the light of the protection that the very actively patriotic Blaricum police enjoyed from mayor Klaarenbeek, it was my opinion that the objections to the hardly useful and in any case (at least initially) more ridiculous than harmful Volksdienst should be overlooked.

It is remarkable that about this no complaints have been raised against a professor and a lector of the same university who live very close to me and who just like me, and just as long as I have, have contributed to the Volksdienst. Also mayor Klaarenbeek himself has contributed just as long as I did, and also he has been interviewed by a Purification Committee. No measure against him has been taken, on the contrary Her Majesty the Queen has called him to a high honorary office per January 1, 1946.

---

[48] See *Brouwer to Committee of Restoration, 20.VIII.1945 and 3.XII.1945.*   [49] The National Socialist substitute for the various Dutch social organizations.   [50] *Commissaris,* under Dutch law this used to be the *Commissaris van de Koningin* (of the Queen). [51] both national socialist institutions replacing the traditional institutions.

For further details see enclosures 6 and 7. [52] As enclosure 8 is added
my letter to the restoration committee dated December 3, 1945.

L.E..J Brouwer

[Signed typescript copy – in Brouwer]

›46-01-07

To Mayor of Amsterdam — 7.I.1946   Amsterdam [53]

To Mayor and Aldermen of Amsterdam [Aan Burgemeester en
Wethouders van Amsterdam]

At the occasion of the interview that I had on December 27, 1945 with the
alderman for Education, it has become apparent to me that your College has
received from the Restoration Committee two proposals for appointments
of lecturers in mathematics, with which the members of the regular mathematical teaching staff of the university, i.e. Dr. Heyting and Dr. Bruins and
the undersigned, were unfamiliar, and the preparation of which has taken
place without cooperation from our side.
It is concerning these appointment recommendations that I feel obliged
to ask your attention for the following expositions.

I

Some months ago a Committee of mathematics professors [54] was established by the Minister of Education, Arts and Sciences, which according
to statements by its members has been charged to encourage the filling of
the many at present vacant chairs of mathematics in the country with such
scholars, that the flourishing of mathematical sciences in the country as a
whole is served as well as possible. In the discussions with members of the
committee I have frankly made it known, that for me the cooperation in my
Amsterdam working environment with conservator Dr. Freudenthal has over
the years become so difficult and that it had such a paralyzing influence on
my working energy, that any authority that could put this cooperation to an
end without personal disadvantage for Mr. Freudenthal, should do so in the
general interest. So in my view the aforesaid Committee has the duty to do

---

[52] See *Brouwer to Committee of Restoration — 30. VIII.1945.*   [53] Adressed: mayor
and aldermen of Amsterdam.   [54] The so-called Van der Corput Committee, cf.
[Van Dalen 2005] p. 801.

everything in its power to have the mathematical vacancies in the different Dutch Universities filled in such a way that, either to Mr. Freudenthal, if he would be acceptable for the Netherlands, or to me a suitable position outside of Amsterdam is assigned.

In this connection a member of the committee voiced his fear to me that Freudenthal might not want to leave Amsterdam and would reject any nomination elsewhere. Naturally, with such a standpoint of Mr. Freudenthal, it would become substantially more difficult for the Committee to find a solution, consistent with the general interest. And Mr. Freudenthal would be fatefully encouraged in such an attitude, if just now the municipality of Amsterdam would offer him a lecturer's position.

Under these circumstances I would like to suggest that you suspend the planned appointment of Mr. Freudenthal as lecturer in Amsterdam (where, by the way, his teaching assignment should be 'analysis, group theory and topology') at least until full clarity will have been obtained about both his willingness to accept a position outside of Amsterdam and the possibility to find such a position for him among the existing vacancies.

Indeed an incompetent party has put forward an argument in favor of an immediate appointment of Mr. Freudenthal as lecturer, which is on the basis of a passage in the Acts of the Amsterdam Municipal Council of the year 1937 the City of Amsterdam bound by a promise; but this argument is based on a completely wrong interpretation of that passage, as a closer inspection of the files involving this earlier matter will undoubtedly confirm to you.

Concerning the origin of the circumstances that make further cooperation between Mr. Freudenthal and me difficult if not impossible, your Committee may consult my exposition [55] dated August 28/30, 1945, a copy of which is hereby enclosed.

II

As far as Dr. Bruins is concerned, his teaching assignment should in my opinion indicate the subjects that he teaches nowadays, i.e. applied and propaedeutic mathematics.

Of applied mathematics (a field of research that, as we hope, will later become of great importance in the Central Institute for Mathematical Research, to be established in Amsterdam), the parts that are taught by Dr. Bruins concern such mathematical theories as are important for research in physics, like higher numerical methods of calculation and the mathematical foundations of quantum theory. For this kind of teaching Dr. Bruins is

---

[55] *Brouwer to Committee of Restoration, 28. VIII. 1945.*

particularly well-suited, because on the one hand he is in the first place, by nature as well as by education and interest, a mathematician, and on the other hand he has held a position in the physics laboratory for several years.

Dr. Bruins has been charged at the time with teaching propaedeutic mathematics, i.e. the mathematics for students in chemistry, mineralogy and psychology, for whom this subject is an auxiliary science. He was the successor of Prof. Pannekoek for the mathematical part of his teaching assignment. (The title of lecturer, that was consequently granted him, would have been better given the predicate of propaedeutic mathematics, rather than analysis.) As experience has shown, this kind of teaching too is in excellent hands with Dr. Bruins, because of his clear and simple way of presenting things and his easy personal accessibility for the students.

III

As far as mechanics is concerned, the Faculty of Mathematics and Physics has taken the point of view after Prof. van der Waals retired, that because of the present state of science, the teaching of mathematical physics necessarily has to be divided between two teachers (a necessity which has become even much more urgent with the prospect of the establishment of the Central Institute for Mathematics Research) and that as soon as there are two teachers available for mathematical physics, one of them must be charged with mechanics. In anticipation of this, my temporary teaching assignment for mechanics was at the time continued for the time being, with the understanding that aforementioned task would, as soon as the vacancy for first assistant in the Mathematics Institute was filled, be taken over temporarily by the first assistant until the definitive arrangement was made.

This first assistantship of the Mathematics Institute was offered by me to Dr. F. Loonstra in The Hague in 1943, and also accepted by him, but with the prospect of the liberation of the fatherland, which seemed all the time imminent he has repeatedly in his communications about the progress of his preparatory studies for the teaching assignment intended for him asked me to postpone the actual submission of the proposal for his appointment for just a little longer. Meanwhile, if I hadn't been out of circulation for quite some time after the liberation of the fatherland, the proposal to appoint Dr. Loonstra would have reached your Council already in August.

However, if in relation to changed circumstances Dr. Loonstra would now no longer be available, then one might consider transfering my temporary teaching assignment in mechanics, after a few months to settle things, to Dr. Bruins, awaiting definitive fulfillment. However then one should consider the danger that Dr. Bruins would be overburdened and that he consequently on the one hand could not fulfill his proper teaching assignment, for which he is well-nigh irreplaceable, and on the other hand would not have enough time to continue his scientific researches.

Also at an earlier occasion, when the transfer of the teaching of mechanics to Dr. Bruins was intended, this plan was abandoned in order to prevent overburdening this exceptionally dedicated teacher.

The Scientific Director [56] of the Mathematical Institute
(w.g.) L.E.J. Brouwer

[Signed carbon copy – in Brouwer [57]]

1946-01-10b

To J. G. van der Corput — 10.I.1946[b]   **Blaricum**

Amice,

I hope that the moving of your books and journals to the Institute has been carried out successfully, and that you have been able to find them a temporary place in the building. Your information regarding that matter has meanwhile inspired me to write a letter to the alderman for Education, [58] a copy of which is enclosed.

Now that I have been reinstated in my function, don't you think it rightful and in the interest of further developments, that as yet a place is assigned to me in the coordination committee instituted by the Minister of Education [59] that you preside over?

With friendly greetings

t.t.
L.E.J. Brouwer

[Signed autograph – in Corput]

---

[56] 'Hoogleraar-directeur', the usual title for the professor who was the head of an academic institution.   [57] Also in Corput. The letter (or note) *Brouwer to College van Herstel 28/30.VIII.1945* was added as an enclosure.   [58] *Brouwer to Alderman for Education 9.I.1946.*   [59] The so-called 'Committee van der Corput', see [Van Dalen 2005] p. 801.

946-01-23

To J. Clay — 23.I.1946

Blaricum

Dear Colleague, [Waarde Collega]

With reference to your letter of December 3, I must emphasize that in my draft for a letter from the Faculty to the Restoration Committee of November 7, 1945, the words *'that the Minister of Education wishes to encourage, that at the University of Amsterdam an institute is established for mathematical scientific research'* are an accurate rendering of an oral communication by yourself on this matter.

It was this statement which was the point of departure for our discussion on October 12 last year, which resulted in our agreement to postpone any possible proposals for the enlargement of the mathematical teaching staff of the Faculty until after the return of normal relations and forms of management, and to support jointly the candidacy of Van der Corput, in exchange for your withdrawing your initial candidacy of Van der Waerden. Only after agreement between us had been reached on these points, I have acquiesced at the end of the discussion in the preparation of the nomination-Van der Corput, which was planned for later, in the meantime advice would be sollicited, so that at that later point in time there would be no unnecessary loss of time; this after you had informed me that in your opinion the mentioned custom of the Faculty had such obligatory traditional rights, that even in cases where the choice of the Faculty was known in advance no departure from this custom was allowed, and after you had promised me that the letters soliciting advice would be sent out at a time determined by me and with a text approved by me. [60]

Naturally, the aforementioned discussion of October 12 inspired in me the confidence that with respect to the matter at hand further negotiations would take place on a basis of reasonableness and consistency and in an openness maintained by all parties concerned. And equally reassuring in this respect were the discussions between me and Van der Corput in that same month of October, during which I from the beginning and categorically took the position that my initiative and cooperation with the candidacy of Van der Corput for the chair of Weitzenböck was based on the expectation that Van der Corput agreed with my view that offering an position in Amsterdam to Van der Waerden, could not be considered until after a thorough

---

[60] Note in the margin in Van der Corput's hand: 'incorrect'.

investigation of the question whether he was acceptable from a national point of view, [61] and after the Amsterdam institute for mathematical research [62] (which is now still only in a rudimentary planning stage) would have been established and shown to be viable. After Van der Corput had in this context taken my offer into consideration, I reckoned that he would in the interest of his candidacy, where his own activity was concerned, stick to the above, and that he would not depart from that without first consulting me, and that he certainly would refuse to make his candidacy subservient to a purpose that was contrary to that to which it owed, among other things, its origin.

My initial confidence, established in this manner, was gradually disturbed, without turning into distrust, in the period until New Year by the following series of events:

First the dispatch of the letters soliciting advice was carried out without consulting me and with a text not approved by me. [63] Subsequently you convened nevertheless a faculty meeting to nominate Van der Corput, and you sent this nomination to the Committee of Restoration with a motivation which clashed with the spirit of our agreement and with a motivation that risked misunderstanding because of its incompleteness. Next you and Van der Corput informed me of your joint plan to obtain already now guarantees for the future, that after the start of the planned institute for mathematical research, Van der Waerden would be assigned a function there. In the light of the earlier discussions this communication understandably surprised me, but because it didn't yet sound dangerous by itself, I finally answered after some exchanges of thoughts with Van der Corput, with a statement that even though it was unacceptable for me to cooperate with Van der Waerden in a university where we would share localities and facilities and responsibilities for teaching, exams and granting Ph.D. degrees, I would not object to collegial ties in the much looser relationship of a research institute.

Also, I felt not yet alarmed after the faculty meeting of December 1 last year, the convening of which I got to know on November 30 from colleague Aten, in which according to your mentioned letter of December 3, it was decided *'that the Faculty will propose Van der Waerden, if it is certain that the Mathematical Institute* [64] *will be established in Amsterdam,* and in

---

[61] Van de Waerden's record as a professor in Leipzig was definitely frowned upon. There was opposition to his appointment from political sides, and the government. [62] i.e. the future Mathematical Centre. [63] Marginal note in Van der Corput's handwriting: 'insinuation! nonsense! of course other reason!' [64] i.e. the future Mathematisch Centre, not the already existing Mathematical Institute of the UvA.

connection with this you gave a promise to colleague Aten on December 2, repeated at a considerably later time, that as a consequence of that decision no letters would be sent by the faculty that would not in draft be subjected in advance to my approval. [65] For one can of course only speak of the 'certainty' of the establishing of the 'Mathematical Institute' in Amsterdam after the necessary funds have been granted, and before this, it cannot at all be assumed with certainty that our representative bodies will and can bear responsibility, because of the destitute state of the public coffers and the fact that for the time being there is no certainty about the usefulness of the planned institute. Until now not even a clear description of the aim and the modus operandi of this institute has been given. Only one part of the plan has been formulated clearly at this moment, namely the establishment in Amsterdam of a laboratory for applied mathematics, but the realization of this has not progressed further than asking prof. Vening Meinesz to use his influence in America to obtain the necessary monetary means.

Even though in this matter my confidence had been gradually upset during the last few months, and was replaced by a state of uneasiness, I had until the last moment not the faintest inkling of the preparations that had been kept hidden from me until the sudden attempt to appoint Van der Waerden during the Council session of the 16th of this month, an event that I also experience as a sudden assault on me personally, which cannot fail to fill me with distrust towards the other persons involved in this matter.

In spite of my request to you in my letter of November 30 last year, the matter under discussion has now indeed set foot on roads that are just as conflicting with the agreements between the two of us as fateful for the mathematics department of the Amsterdam university. I appeal to your cooperation to turn it as yet aside from these roads.

With collegial greetings

L.E.J. Brouwer

[Signed typewritten copy – in Corput]

---

[65] Marginal note of Van der Corput: 'statement of facts'

1946-05-01b

From M. Minnaert — 1.V.1946[b]   **Utrecht**
                                                    Sterrewacht Sonnenborgh
                                                       der Rijksuniversiteit
                                                            Zonnenburg 2

Dear Brouwer, [Waarde Brouwer]

I have succeeded in making a telephone connection with Dr. Freudenthal.
I have put the matters to him just as we had discussed together. Understandably, he could not answer immediately: he informed me that recently
he had adopted the point of view that a lecturer's position in Amsterdam
should first be established before he could consider other appointments.
However, it was conceivable that he would change of opinion.

I have urged him to send me his answer well in time, before May 8.
He would do so in writing, because it is difficult to get a connection by
telephone. As soon as I know more, I will phone you.

Many greetings from

M. Minnaert

[Signed autograph – in Brouwer]

1946-08-03b

To A. Dresden — 3.VIII.1946[b] [66]

Many thanks for your letter of August 3, which unfortunately reached
me only a few days ago. Yes, the problem of ensuring peace occupies here
too a lot of the available brain power. The very first requirement seems to
me that the division of the earth into different regions with separate centers
of military powers should be abolished. Awaiting that, in my opinion, at
least the United States should immediately unite itself in military respect
with the other American nations, the British Empire, Scandinavia, Switzerland, Austria, Italy, Spain, Portugal, France, Belgium and the Netherlands

---

[66] Undated; obviously August or September 1946; reply to *Dresden to Brouwer,
3.VIII.1946.* Document incomplete.

to a single state. I have sent the postcard you sent me, together with a few important names among my relations, to the Committee for foreign correspondence.

Meanwhile I would very much like to emigrate now (unlike before) from the Netherlands, because I'm afraid that we will have to wait for a few more years for the establishment of a unified state. If at the moment the possibility of employment in America would materialize for me, I would seize that possibility with both hands. Because the confused situation during the first months after the liberation has brought here (also in scientific and university circles) people into office ... [67]

[Handwritten draft – in Brouwer]

46-10-08a

To Mayor of Amsterdam — 8.X.1946[a]   Blaricum [68]

Dear Sirs, [Edelachtbare Heeren]

Allow me to call the attention of your Council to the following matter:

On the municipal budget for 1946, that was approved this summer by the city council, there appears an item of f 25,000, for which the explanation in the concept budget submitted by your Council to the City Council reads as follows:

> It is proposed to allot for the year 1946 a subsidy of at most
> f 25,000 to the Mathematical Institute Foundation, for a mathe-
> matical institute that shall take the place of the European Center
> for Mathematics in Göttingen [69]

Because there is in Amsterdam no other mathematical institute than the one of the University of Amsterdam, which has, in so far as the municipal

---

[67] The draft breaks off here.   [68] Addressed: Burgemeester en Wethouders van Amsterdam, (mayor and aldermen of Amsterdam). A shorthand copy of this letter is in the van der Corput archive, probably dictated by telephone by a member of the City hall staff. One may conjecture that Van der Corput (one of the founders of the Mathematisch Centrum) attached a more than routine interest in the matter   [69] I.e. mayor and aldermen proposed to subsidize an institute for mathematics, comparable to the pre-1933 institute in Göttingen; the formulation suggests a new institute, but it remains silent on its relation to the existing mathematics institute of the university.

finances permitted, been designated since 1920 to be organized on the same footing as the mathematical institute of the university of Göttingen, which functioned until the last world war not only as a European but also as a global center of mathematics, the above explanation must have created with the city council the impression that the city finances now finally permit a beginning of the fulfillment of the promises received by me in 1920, and which since then I have been prompting — repeatedly, but in vain.

For, when I was offered in 1920 a chair in Göttingen, which had been held from 1886 to 1913 by Felix Klein, under whose leadership Göttingen had acquired its function as a world center for mathematics, and when acceptance of this call would have meant an important improvement for me, not only in affluence but also in the opportunity to do scientific research and in international influence of the results thereof, I have nonetheless complied with the pressure from the board of the city council of Amsterdam to stay here, after it was promised to me by the then mayor and chairman-curator that in the first place the mathematical teaching staff of the University of Amsterdam would be immediately be given such an extension, along lines to be indicated by me, that Amsterdam could become the center for the practicing of mathematics in the Netherlands, and in the second place that as soon as the Amsterdam municipal finances would allow this, a mathematical institute, to be placed under my direction, would be established at the University of Amsterdam, which would be as similar as possible, both in size and in organization, to the Göttingen institute. [70]

Of these two promises the first one was at the time immediately fulfilled; between 1920 and 1934 indeed several times a start was made to realize the second one, but under influences that never became clear to me, these introductory measures each time ended in nothing.

In that period there was a marked influx of foreign mathematicians to Amsterdam, which was I took care of during several years. However, in the absence of an institute and appropriate facilities for directing a group of studying foreigners, the required personal, mental and financial sacrifices became in the long run too much for me (especially after an indispensable source of my income, which was as such discussed in the negotiations of 1920, was strongly diminished as a consequence of municipal expropriation [71]), this hospitality had to come to an end. In this manner the board of the city council not only victimized me personally, but it also nipped the interna-

---

[70] The promises of the mayor in 1920 were in fact related to the offer of a chair in Berlin. The Göttingen offer played no role in the available correspondence. Cf. *Mayor of Amsterdam to Brouwer 12.II.1920* and [Van Dalen 1999] section 8.4.   [71] of the pharmacy at the Overtoom, see [Van Dalen 2005] p. 559.

tional mathematical center, that was emerging all by itself in Amsterdam, in the bud.

Perhaps because the honoring of the promises made to me in Amsterdam failed to be realized and the vanishing of my Amsterdam school for foreigners had attracted international attention, a chair in Göttingen was again offered to me in 1934, [72] which was from the beginning unacceptable to me notwithstanding the high pay and excellent facilities, because of the then established form of government in Germany, but I only formally rejected it after having once more explained in detail to the Chairman of the Board at the time how much I had been disappointed by the municipal council, and after having heard from his mouth that he deeply regretted this whole state of affairs, and that from the side of the municipal governing board everything would be done that could be done to set the matter right.

And indeed, soon after my second rejection of Göttingen a mathematical institute has been established in the Amsterdam university, and put under my supervision. But the location, the organization and the facilities of this institute have remained so far below my minimum requirements, that it fell short of its purpose, had little use for scientific activity, and it gave me personally worries and vexation, whereas it wasn't in the least advantageous, neither for my teaching, nor for my research. And after my repeated complaints the municipal governing board now and then held out perspectives of partial remedies; preparatory measures which were then taken again came so far to nothing.

On the basis of the historical exposition above and in combination with the passage from the concept budget for 1946, quoted at the beginning of this letter, the allocation of this budgetary post can hardly be interpreted in any other way than that the amount mentioned should — either with or without an intermediary foundation — be spent for the mathematical institute of the University of Amsterdam, and for purposes to be determined in consultation with the director of this institute.

Anyway, on the basis of the above exposition it would be in my opinion unacceptable in whatever way, to take away my leadership of the Amsterdam mathematical enterprise, after I had brought it, under difficult circumstances, sacrificing a great many many personal interests, to its present level. Unfortunately there are indications that plans in that direction exist in certain circles, and also that preparatory actions in that direction already have taken place under the protection of the smoke curtain of the liberation-

---

[72] Cf. [Van Dalen 2005] section 16.7.

confusion. If this scheme should succeed (quod consules avertant $^{(73)}$), then I believe that this would write a page in the history of science that will not fail to attract the astonished attention of future generations.

Sincerely yours

The director of the Mathematical Institute
(w.g.) L.E.J. Brouwer

[Carbon copy, typewritten signature – in Brouwer $^{(74)}$]

1947-08-21

To D. van Dantzig — 21.VIII.1947   Blaricum

Dear Van Dantzig, [Waarde Van Dantzig]

In answer to your letter of the 19th of this month I would like to draw your attention to the fact that intuitionism does not recognize axioms, and hence that it never uses them, and that more in particular it never uses the comprehension axiom and at most takes interest in what respect assertions from classical mathematics based on the use of the comprehension axiom can be assigned any intuitionistic meaning, and whether this meaning can be recognized as true or false. Only in very special cases such assertions turn out to be meaningful and true.

Intuitionistic mathematics possesses perfect precision, but on the contrary, intuitionistic language is vague and fallible; in different environments, at different times, different intuitionistic languages might become the preferred one to use; of all terms in all those languages one always will be able to say that they 'are in need of further precision'.

Calling a mathematical property a species $^{(75)}$ can just as little be based on an axiom, as calling an indivisible natural number a 'prime number'.

As to the two sentences in your article, quoted in your letter, mentioning the comprehension axiom, I would like to give you to consider to delete the

---

$^{(73)}$ What the consuls may avert, a variation of 'videant consules ne quid detrimenti respublica capiat' (May the consuls avert it that the state suffers harm), the standard formula in republican Rome for declaring a state of emergency. $^{(74)}$ Also in Ministerie van Onderwijs. $^{(75)}$ Brouwer uses here the German word, *Spezies*.

word 'unrestricted' in the first one (p. 4), and that you leave out the second one (p. 11, note 17) altogether. [76]

As to the reprints of my publications that you asked for, of many of those even my own private copies have been lost by the fire, so I can fulfill your request only to a very limited extent. But I still have the three mentioned in my letter of the 17th of this month, and I will send them to you tomorrow. I hope to send a few others soon.

With friendly greetings

t.t.

L.E.J. Brouwer

[Signed typescript – in Dantzig]

47-09-03

From D. van Dantzig — 3.IX.1947   Amsterdam

Dear Brouwer, [Waarde Brouwer]

Many thanks for your explications and your reprints.

Although I do not completely agree with you, I have crossed out the words you indicated, because they are not essential for my argument.

Should you feel like continuing the discussion a bit more, then I am quite willing to indicate in somewhat more detail what the basis for my deviating opinion is. For now I restrict myself to the matter of making the notion of species more precise.

Your analogy between the definitions of 'species' and 'prime number' does not apply, because I (and I think most mathematicians) do have a clear idea about what is meant by a 'natural number' and when that is called 'indivisible', but *not* about what you mean by a *'property'* and when you call that 'mathematical'. This latter predicate doesn't occur, if I recall correctly, in your definition. Is it a 'property' of a set (in your sense), leaving aside that it is 'mathematical' or not, that a certain person has constructed it on a certain day? Or that he heard someone talk about it? Or that it

---

[76] There are two papers of Van Dantzig on intuitionistic mathematics, [Dantzig 1947, Dantzig 1949].

'shows some similarity' with the set of natural numbers? In the latter case it is certainly not 'begrifflich fertig definiert'. [77] But where is the borderline?

With the species concept you leave the constructive domain and you bring a vague element into the intuitionistic theory of the *same* kind as the comprehension axiom does in the classical theory. This would not be the case if you would restrict yourself to *subspreads* of a spread by extending the sterilization rules for choice sequences. Making the notion 'mathematical property', applied to elements of a spread V, more precise, would in my opinion consist of reducing this notion to verifiable properties of the individual choices in the choice sequence, properties that are connected by universal and existence predicates, so *for example*: for each natural k and each natural n_k there is an m_k, such that the m_k-th choice possesses the verifiable property E (possibly dependent on previous choices). The scope of the 'property'-notion then depends for example on admitting finitely or infinitely many universal or existential predicates. I guess that you will prefer an as wide as possible definition (considerably wider than the example above). But *that* such a specification is necessary is, in my view, beyond doubt. It is true that the notion of 'property' of a specific choice still remains undefined, but in any case we would have made a great deal of progress.

With friendly greetings,

t.à.t.
D. van Dantzig

[Typescript copy – in Dantzig]

1948-04-02

To Board UvA — 2.IV.1948   **Amsterdam** [78]
                                                Mathematisch Instituut der UvA

With reference to the application by the Foundation for Applied Mathematics, that your Board on March 17 has submitted to the Senate of the University of Amsterdam, to be allowed to establish an extraordinary chair in applied mathematics at the University of Amsterdam; which request has been discussed by the Senate in its meeting of last March 28 (in the opinion

---

[77] defined as conceptually completed.   [78] Addressed: College van Curatoren van de Universiteit van Amsterdam.

of the undersigned without sufficient time for preparation) the undersigned
feels he must draw the attention of your Board to the following points:

1. Mathematics is an introvert science and as such coalesces with philosophy, theology and reflective psychology, but it is constructive in
   a higher degree than these. And the mathematical creative urge is
   directed not only to inner enlightenment but also to beauty, a beauty
   related to that of architecture and music, but more immaterial.

2. In connection with this the mathematical state of mind is as a rule
   indifferent to natural science and definitely rejects expanding the ex-
   ploitation of nature, and the technology that creates the possibilities
   for this.

3. This is not altered by the generally known fact that *and* technology, *and* natural science *and* many other extrovert sciences only have
   reached their present range because they 'calculated' (arithmetically
   or graphically), in other words operated mathematically in the mathematical systems that had been 'projected' upon their enterprise or
   their field of research.

4. Hence, although the technical sciences in the first place, but further
   also almost all other extrovert sciences belong more or less to 'applied mathematics', their essence is nonetheless fundamentally different from that of introvert mathematics.

5. Where applied mathematics has been amalgamated as an all pervad-
   ing accidental circumstance with the activity of the university, and
   is almost the proper substance of the activity of a technical university, [79] there is, precisely because of this ubiquity, in neither of the
   two institutions a place for a separate educational task called 'applied
   mathematics'. On the contrary, every extrovert science is interwoven
   with its own specific applied mathematics, and this should in teaching
   remain inseparable from it.

6. Only in a very special, and in general better avoided, case an academic
   teaching assignment in applied mathematics, acquires a reasonable
   content.

   For, if one calls the relatively simple mathematics which is an indispensable part of the initial instruction of mathematics students, as
   well as a supplier of methods of calculation for many other sciences

---

[79] At the time technology was the business of the "technische hogeschool", much like
the ETH or MIT. Only much later (1985) it acquired the name 'technische universiteit'.

'propaedeutic', then it may happen that a natural science which is represented at the university needs methods of calculation whose theoretical foundations on the one hand exceed the level of propaedeutic mathematics, and on the other hand are of too subordinate an interest to be incorporated in the regular university curriculum. If under these circumstances those calculation techniques, the importance of which is to be found mainly outside of mathematics, are nonetheless taught by a mathematician, who is willing to make the sacrifice (for example if the relevant natural science is understaffed), then in *this* special case it is an activity that reasonably can be given a place as an educational task in applied mathematics.

7. In the opinion of the undersigned, the educational task 'in propaedeutic and applied mathematics' of Dr. Bruins, in the manner described above, sprung from a certain need that was felt here. Moreover, apart and separately from fulfilling this need, the courses of Dr. Bruins open, for mathematics students who wish such, access to a mathematical job in industry.

8. The above makes it clear that the *factual* intention of the application of the Foundation for Applied Mathematics is the construction of a possibility to attach Professor Van der Waerden, — who is said to be declared unacceptable as a public teacher in Dutch university by a decision of the Crown — in spite of this decision, to the University of Amsterdam as a professor.

9. The preliminary advice, dated March 13, 1948, by the Faculty of Mathematics and Physics to the Senate on the application of the Foundation for Applied Mathematics, refers to the letter of the Faculty to the Secretary of the Senate dated December 31, 1947, which in its turn bases itself on the judgment of the First Section of the Faculty. [80] However no decisive authority can be attached to this judgment of the Faculty. For the three lecturers in mathematics, who in Amsterdam are in charge of major part of mathematical teaching, and who all are professorial [81] scholars and who all maintain in an excellent way the spirit that at the time contributed to the flourishing of Amsterdam mathematics, were not admitted to the relevant meeting of the First Section on December 18, 1947, where the mentioned judgment

---

[80] i.e. the section of the faculty that covers mathematics and physics.   [81] In The Netherlands an academic teacher or researcher is called 'professorabel' if he possesses the qualities for a professorship. The translation 'professorial' is chosen, lacking a better term.

of the Faculty was established, *after* having been invited to the relevant Faculty meeting of October 22, 1947, and to the relevant meeting of the First and Second Department on December 11, 1947. Subsequently the afore-mentioned lecturers, together with the undersigned, did make their view known to the Central Committee of the Faculty through the enclosed letter, dated December 27, 1947, which view was based on the discussion they had attended. So when this Committee assumed responsibility for the contents of the aforementioned letter dated December 31, 1947, it knew that these contents clashed in essential points with the opinions of the majority of the mathematics teachers.

10. In the above mentioned Faculty meeting dated December 18, 1947, the undersigned has not concealed, how much in his opinion a minority of mathematical teachers in his opinion used their position of power and thereby more and more precluded the possibility for the City of Amsterdam to fulfill its promises made to undersigned in 1920, trusting which, he remained in the Netherlands at the time.

11. However great the mathematical merits of Professor van der Waerden are, with respect to applied mathematics, that is, with respect to other sciences than mathematics, his record of service is certainly not so well known that it is superfluous to submit his record as part of a proposal to appoint him.

The Director of the Mathematical Institute
(w.g.) L.E.J. Brouwer

[Signed carbon copy of typescript – in Corput]

––––––––––––

Editorial supplement

J. Clay and J.G. van der Corput to Chancellor — 9.IV.1948 [82]

With reference to the copy we received of the letter of Colleague L.E.J. Brouwer intended for the Board of the University of Amsterdam, we have the honor to inform you of the following:

––––––––––––––––––––

[82] Addressed: Chancellor and the board (curatoren) of the University of Amsterdam.

On the meeting of April 22 ult., [83] Colleague L.E.J. Brouwer has insisted that the Senate should advise unanimously favorably about the establishment of a chair in applied mathematics intended for Prof. van der Waerden. However, he regretted that in the explanation given by the Faculty about the meaning of applied mathematics his views about this had not been taken into accounted. The Senate accepted his proposal, that he would submit, very soon and after consulting and in agreement with Van der Corput, a document to the Board of Rector and Assessors, in which he would expound his views concerning the meaning of applied mathematics.

The undersigned object to sending the letter of Colleague L.E.J. Brouwer through the Senate to the Board of the University. Neither the condition 'very soon' nor the one of 'consulting' has been fulfilled, and moreover the proposed letter goes much further than an exposition by Mr. L.E.J. Brouwer concerning the meaning of applied mathematics. Moreover the undersigned object to this letter because it creates the unjustified impression that each of the gentlemen A. Heyting, E.M. Bruins, J. de Groot and F. Loonstra agrees with an action that would have the consequence of making the appointment of Prof.Dr. B.L. van der Waerden at the Municipal University of Amsterdam more difficult. [84]

Van der Corput, chairman of the 1st section of the Faculty of Mathematics and Physics has convened this section for April 14 next. The four mentioned gentlemen will be invited for part of the meeting, so they can expound their views to the Section.

Van der Corput has written a letter to Colleague L.E.J. Brouwer, in which he gives him to consider to see if it is recommendable that they will discuss this matter *before* the Section meets.

Prof.Dr. J. Clay, Chairman of the Faculty for Mathematic and Physics.
Prof.Dr. J.G. van der Corput, Chairman 1st Section.

[Typescript copy – in Brouwer [85]]

---

[83] I.e. one year earlier.   [84] Note that Clay and Van der Corput seem to question Brouwer's integrity.   [85] Copy received on 12 April 1948 — according to a note in Brouwer's handwriting.

948-06-03

To G. Mannoury — 3.VI.1948   **Blaricum**

Dear Gerrit [Beste Gerrit]

Thanks for your call. And for the pond of thoughts in the garden of life, which Part I of your Handbook of Analytica Significs [86] is for me (and probably for many). How long it is already that we haven't seen each other!
Many cordial greetings from house to house

Bertus

[Signed autograph – in Mannoury]

49-03-10

To H. Hopf — 10.III.1949   **Lugano**
                                                          Kurhaus Cademario

Dear Hopf, [Lieber Hopf]

Since a few weeks I am finally once more in the delightful Tessin and seek there recovery from the bronchial asthma, the attacks of which bother me again and again since August last year. The foreign currency needed for this trip finally has been allocated to me after many months of patient waiting. Unfortunately they vanish considerably quicker than I expected, so I will have to end my trip about the twentieth of this month.

I have now written to Saxer to invite him to become your successor in the editorial board of Compositio Mathematica. Please be so kind as to plead for my request with him.

I greet you and your spouse most cordially in most pleasant recollection of our meeting in October.

Your Brouwer

[Signed autograph, postcard – in Hopf]

---

[86] [Mannoury 1947].

1949-07-10

To Eds. Compositio Mathematica — 10.VII.1949   Blaricum [87]

Dear colleagues, [Mes chers collègues]

When the Noordhoff company in Groningen,[3] which functioned from 1934 to 1940 as publisher, bookseller, and commercial agent of Compositio Mathematica, offered us in 1945 to resume its old function, there was no reason to refuse it to prove itself up to this task. Meanwhile having resumed that task, it has started to work rather poorly, either for lack of equipment, or for lack of zeal, or for lack of willingness, and now finally it is demanding that before its work can be continued the board of editors must be reorganized in a way that would completely change its character, especially the international character of our journal.

As the contracts with the publisher are a matter of the Administrative Committee according to the Editorial Statute, I ask you under these circumstances to authorize me to withdraw, in the name of the Administrative Committee, the commission of our journal from the Noordhoff company and give it to another publishing company. I have good hopes that I can find one of high repute, well managed, and well equipped, which will serve us better than the one that has deceived us.

By replacing the editors that we regrettably have lost, I ask you to authorize me to propose to the board as new editors Messrs. Hodge (Cambridge), Newman (Manchester), Kloosterman (Leyden), Bernays (Zürich) and Kleene (Madison). By the choice of the last two the board will in the future be enriched by two representatives of mathematical logic.

Sincerely yours, [88]
(signed) L.E.J. Brouwer

[Carbon copy of typescript – in Brouwer]

---

[3]Nowadays the company Noordhoff is represented for us by the son of the prewar representative, who passed away.

[87]Addressed: Aux MM. les membres du Comité d'Administration de Compositio Mathematica.   [88]*Agréez, mes chers collègues, l'expression de mes sentiments cordiaux.*

49-08-24

To D. van Dantzig — 24.VIII.1949   **Blaricum**

Dear Van Dantzig, [Waarde Van Dantzig]

Many thanks for sending me a first copy of your 'Comment's. [89] I am glad to see that these developments make the essentially negative properties [90] meaningful also to those who do not recognize the intuitionistic creating subject, because with respect to mathematics they hold either a psychologistic point of view, or in any case stick to the 'plurality of mind'. [91]

As I told you in conversation, my example in question is for fundamental intuitionism so much more unassailable than for those of a different persuasion, because the intuitionistic creating subject can certainly, and from the outset put restrictions (or prohibitions of restrictions) on a specific growing mathematical entity, but not on his own possibilities of creation.

My belief that psychological pictures of intuitionistic mathematics, however interesting they may be, never can be adequate, has, if possible, been even strengthened by your comment.

With friendly greetings

t.t.
L.E.J. Brouwer

[Signed typescript – in Dantzig]

49-10-28

To A. Heyting — 28.X.1949   **Blaricum**

Dear Heyting, [Waarde Heyting]

In my opinion a yet living author, being in a state of scientific responsibility, who now brings again into the light his earlier published work, is obliged to give an account for each of the items of his work of both the meaning and impact it had on the state of science of that time when it first

---

[89] [Dantzig 1949].   [90] Cf. [Brouwer 1948].   [91] Brouwer's English terminology. See also [Brouwer 1949].

appeared, as well as on the present state of science, as if it had appeared only now for the first time. And he must consider on the basis of this account, to what elucidation the reprinted text the present reader is entitled.

Therefore a new edition of my collected works would burden me with such an amount of work, that I will not have the time for that for several years to come.

To a lesser degree this objection also holds for the planned re-issue of my dissertation and connected publications in English. For the effort related to that as well, I will have no time, as long as not in the first place Compositio Mathematica is again permanently functioning, and subsequently my Cambridge lectures have appeared, and finally the manuscript of my intuitionistic theory of functions is completed. But that point in time is, I believe (if at least my energy is not totally paralyzed by the consolidation of the nazification of Dutch mathematics [92]) in a not too distant future, so I see no objections to continuation of the activities that Welter en De Loor [93] were so kind to take on. But on closer consideration it seems very premature already now to get a publisher involved in this work. There will be time enough for that if the evolving text has passed through all stages and a definitive manuscript is ready. Because I am convinced that in that case every competent publisher, be it in this country or in England or in America, will be eager to accept the book.

With friendly greetings

t.t.

L.E.J. Brouwer

[Signed typescript – in Heyting]

Editorial supplement

A. Heyting to P. Bernays — XI.1949   *Laren*

   Dear Colleague,

   Although Mr. Prof. Brouwer first agreed to the plan to publish his collected works, he reached a different point of view after consideration,

----

[92] This remark illustrates Brouwer's bitterness about his treatment after the war by certain colleagues.   [93] Two South African mathematicians, who volunteered to translate parts of Brouwer's Dutch texts.

as becomes clear from the following translation of a paragraph of a letter from him to me.

[followed by a German translation of the first paragraph of the above letter]

So it is necessary to postpone the publication of his collected works for an indefinite time.

I thank you again for your willingness to cooperate; when at a later moment the plan is taken up again, I hope that I can count again on your sympathy and support.

Sincerely,
A. Heyting

[Signed typescript – in Bernays]

Chapter 7

1950 – 1966

50-02-28

To S. Carathéodory jr — 28.II.1950   Blaricum [1]

Dear Mr. Carathéodory, [Sehr geehrter Herr Carathéodory]

The notice of the death of your father, which arrived only today, has deeply moved me. His friendship and the awareness of his great importance as a thinker and as a human personality, have been something absolutely essential for me for many decades. His death has made the world poorer for me. How much I have since 1945 looked forward to an opportunity to meet him again! It could not be, and for me it only remains to offer my most cordial condolences to you, the other family members and the other relatives for the severe loss you suffered, and to assure you that I will hold the memory of Constantin Carathéodory with the highest regard.

With respectful greetings

Sincerely yours [2]
(signed) L.E.J. Brouwer

[Signed carbon copy – in Brouwer]

---

[1] Reply to an obituary notice of Constantin Carathéodory, dated 1950, February 2 (in the Brouwer Archive): 'Unser lieber Vater, Geheimer Regierungsrat Univ.Prof.Dr. Constantin Carathéodory ist heute nach schwerer Erkrankung im Alter von 76 Jahren sanft entschlafen, Münster, den 2.II.1950.' [Our dear father, privy councillor Prof. Dr. Constantin Carathéodory, aged 76, today has passed away peacefully after a serious illness.]
[2] Mit hochachtungsvollem Gruss – Ihr ergebener.

D. van Dalen, *The Selected Correspondence of L.E.J. Brouwer,*   443
Sources and Studies in the History of Mathematics and Physical Sciences,
DOI 10.1007/978-0-85729-537-8_7, © Springer-Verlag London Limited 2011

1950-12-22

To W. van Haersolte [3] **— 22.XII.1950**   **Blaricum**

Dear Sir [Hoogwelgeboren Heer]

Your letter of the 16th of this month came into my hands. Far from being finished, the conflict with the company Noordhoff concerning Compositio Mathematica has considerably escalated since the beginning of this year, but it has also stalled in connection with the following two catastrophic circumstances

1. Shortly after I wrote to you the last time, I had to note to my bewilderment, that my foreign colleagues in the Committee of Administration, who were in July 1949 still without reservation on my side, had abandoned me as a result of information and promises that remained secret for me.

2. The shock that was delivered to me through this stunning observation has left me, after a heart attack, mentally and physically incapacitated to such an extent, that I am with respect to my defense against the aggression concerned out of action for a considerable period, and that even any abiding in the realm of thought of this conflict is forbidden to me for a considerable time.

That observing this instruction has become the cause that I have not managed earlier to inform you about the new stage into which the affair has entered, fills me with shame, and I offer you my sincere apologies for it.

In the meantime I remain, in spite of all forced inaction, clearly aware that the conflict with Noordhoff not only concerns my personal honor, but also the honor of my country, so that is it my sacred duty to resume this struggle as soon as I am able to, do so. I sincerely hope that you are willing to continue your support for me in this matter.

reiterating my apologies, I remain,

yours sincerely
(signed) L.E.J. Brouwer

[Signed carbon copy – in Brouwer]

---

[3] Brouwer's legal adviser in the Noordhoff conflict.

151-02-05

From R. Fraïssé – 5.II.1951   **Algers**
Attaché de recherches au Centre National
de la Recherche Scientifique, Algers
187 rue Laperlier

Professor, [Monsieur le Professeur]

I am at the moment writing a mathematical dissertation at the University of Algeria, under the direction of professor De Possel, with whom I am working since January 1948; my investigations deal with the theory of relations and they originate in problems of formal logic, more in particular of semantics.

Mr. de Possel thought that even a short contact with the Dutch intuitionist School would be most profitable for me. He has obtained a study grant for me from the Dutch government. I have asked provisionally for the months of March and April. Would it be possible for me to meet you during that period, and if not what would be the period that would suit you best? I direct a similar request to professors A. Heyting and E.W. Beth. [4]

I enclose in this letter reprints of my notes that have now appeared in the Comptes Rendus de l'Académie des Sciences.

Hoping to have the pleasure soon to become acquainted with you through direct conversation, and not only through the publications of your articles,

Sincerely yours, [5]
R. Fraïssé

[Signed autograph – in Brouwer]

---

[4] Fraïssé stayed in 1951 in Amsterdam for research in logic, semantics, set theoretic theory of relations and intuitionistic mathematics; see R. Fraïssé, 'Rapport sur le séjour en Hollande', dd. Amsterdam, le 21 mai 1951, in the Beth Archive. The supervision was in Heyting's hands; Brouwer was in Switzerland at the time. [5] *Je vous prie d'agréer, Monsieur le Professeur, l'expression de mes sentiments les plus respectueux.*

1951-04-18

From Mrs. van der Corput — 18.IV.1951   Stanford University

Dear Professor Brouwer,   [Hooggeachte Prof. Brouwer]

To start with, I should apologize for this typed letter, which, as I fear, will get full of errors. In fact I have yesterday burned my right-hand so terribly in boiling water, that there is no skin left on the five fingers. Writing is impossible, and the larger part of the day I is still stay in bed, but now that my husband is answering your letter, I would like to enclose a letter, typed with my left hand.

Some weeks ago I was in Holland because of a serious illness of both of my parents. And I heard at that occasion from at least three sides, that you blame my husband for the affair with Compositio Mathematica. According to these sources you consider him as the auctor intellectualis in the background. I cannot possibly decide, if these sources have correctly rendered your opinion, but if you should really think so, then I set great store by assuring you that my husband has vigorously opposed the action concerned, and that has caused him trouble with others. I happened to be there when these discussions about the affair were going on, and thus know this by my own experience. The same thing must appear from the correspondence exchanged with the company Noordhoff.

Of course it occasionally happens that my husband differs of opinion with somebody else, but it is his invariable rule to settle these differences with the persons concerned, and immediately. It is not his habit to hide behind others. But he disagreed with the Compositio-action, and he has refused to join.

With my warm greetings to your wife, and my best wishes for your health, which, as I heard, is a problem these days,

yours, [6]

[Carbon copy – in Corput]

---

[6] *gaarne Uw*

Editorial supplement

J. van der Corput to E.W. Beth, A. Heyting — 12.VI.1951   Stanford [7]

Dear Friends, [Waarde vrienden]

In connection with a serious illness of her father in Groningen, Jeannet [8] has been back and forth to Holland. As she heard from various sides, Brouwer claims that I am the real auctor intellectualis of the Compositio Mathematica affair, and that the others just carry out my instructions. Jeannet wrote in the middle of April to him that his accusations are neither here nor there; at the same time I declared the same thing in my letter. He does not even bother to reply.

More important is his attitude with respect to the two vacancies. [9] In fact I should not worry about what Brouwer does or doesn't do, but I find it dreadful all the same. I think it is disgraceful, that a man to whom mathematics should be dear, dares to put forward such a proposal. It would mean the destruction of much of what we have built up at great cost. It is clear that the faculty will reject unanimously his proposal, nonetheless his action carries serious dangers, of which he must be aware. I am looking forward to a period in which mathematical matters at the Amsterdam University can be dealt with in a businesslike manner, and can be deliberated among ourselves.

[. . . .]

With many warm greetings, also from Jeannet, and also for your respective spouses.

tt.

Jan

[Signed typescript – in Beth]

———————

———————

[7] Only the parts relevant to Brouwer are reproduced.   [8] Mrs. J. van der Corput.
[9] Vacancies Mathematical Institute UvA after departure of Van der Waerden and retirement of Brouwer.

1951-05-01

To Mathematics & Physics UvA — 1.V.1951   Blaricum [10]

Dear Chairwoman, [Hooggeachte Voorzitster]

On the agenda of the faculty meeting of tomorrow, which I am unable to attend, I find listed the following agenda point: 'vacancies in mathematics', a manner of phrasing that seems to me less fortuitous, for indeed professor Van der Waerden has left and I myself will leave soon; however the position of Van der Waerden was in my view created last year less for the matter than for the man, [11] and my own function has been eroded gradually since 1946 by the establishment and the operation of the Mathematisch Centrum, and under the present circumstances it has lost its reasons to exist.

As furthermore the academic education in mathematics can proceed without disruption by merely extending the teaching assignments of Dr. Bruins by adding analysis to it, which subject Dr. Bruins has taken temporarily care of in an excellent manner; whereas applied mathematics already is for some time part of his teaching duties, so there is no urgent public interest to appoint new mathematics teachers, hence the authorities, for whom, after the most recent Government declaration, every admissible economizing is obligatory, have at the moment the duty to desist from such an appointment.

With this extension of his teaching duties naturally the title of professor should not be withheld any longer from Dr. Bruins, because among the Dutch mathematicians of his generation he is at the top with respect to his versatility and originality, and also internationally he probably has most attracted attention.

As a clear token of the appreciation and admiration that has been shown to Bruins from beyond our borders, I allow myself to enclose a copy of a letter by Prof. Turnbull dated March 24, 1951.

With this reduction of the existing excessive staffing of the mathematical teaching body of the university, which impedes the education of the students, also a first perspective would be opened onto restoring the prewar mathematical school of Amsterdam and the international influence it had,

---

[10] Addressed: Voorzitster der Faculteit der Wis- en Natuurkunde der Universiteit van Amsterdam; Chairman of the Department of Mathematics and Science [i.e. Van Arkel].
[11] *minder ad rem dan ad hominem.*

where for the positions earlier held by Mannoury, Heyting, and Beth would now be the proper persons.

For the realization of this restoration it would furthermore be necessary:

1. to appoint a lecturer, so as to relieve Heyting en Bruins from teaching undergraduates [12]

2. To restrict the academic mathematics teaching staff gradually further to four professors and a lecturer (after a considerable increase of the prosperity of the country and the intellectual capacity of the students possibly to be extended to five professors and a lecturer);

3. to make the curriculum for the doctoral examination [13] more uniform, and more specifically, to require a compulsory examination of some depth for all candidates by both specialists in intuitionism and symbolic logic;

4. To cut all connecting arteries between mathematics at the university and the extra-curricular Mathematisch Centrum, through which since 1946 the lifeblood of the Amsterdam mathematics school has been drained. [14]

In connection with the gradual erosion of my personal academic function through the establishment and operation of the Mathematisch Centrum, mentioned in the beginning of this letter, I submit hereby a copy of the letter I wrote on October 8, 1946, to Mayor and Aldermen of Amsterdam, an answer to which was never received, and the contents of which were also ignored. The page of history mentioned at the end of that letter has meanwhile almost been written to its end. A conciliatory final paragraph might however be added to it, if the City Council of Amsterdam would at the eleventh hour recognize the hollowness of the arguments that at the time persuaded it to allocate the requested subsidy to the Mathematisch Centrum at the expense of the University of Amsterdam, and if it would, under a brief extension of my term of employment, hand back to me for a short time the leadership of the Amsterdam mathematical organization with the assignment to strip this organization of all expensive pretense and all unjustified privileges, and concentrate it once more on the branch of science that in the old days gave some significance in the world to Amsterdam mathematics, and that still is represented better in Amsterdam than

---

[12] *candidandi.*   [13] Comparable to M.Sc., and formally prerequisite to get a Ph.D.
[14] Brouwer's proposals to the Curators caused considerable upheaval among the resident mathematicians, see [Van Dalen 2005] p. 854 ff.

anywhere else in the world, and that still receives ever more international attention.

With collegial greetings

the faculty member
signed L.E.J.Brouwer.

[Typescript copy – in Corput]

1951-05-16c

From D.R. Pye — 16.V.1951^c   **London** [15]
                                                   University College London
                                                        Gower Street, W.C. 1

Dear Sir,

In 1937 the late Dr. A.T. Shearman bequeathed to this College the residue of his estate to found a course of lectures on Symbolic Logic and Methodology. I enclose a copy of the scheme for the Shearman Lectureship which was established in 1938. Unfortunately the war intervened before it was possible to hold the first of these lectures, but the scheme was inaugurated in 1946 with a course of lectures by Earl Russell [16] on 'Scientific Inference'. A second course of lectures was given in 1948 by Dr. Schrödinger on 'The Origin and Nature of Scientific Thought' and a third course in 1950 by Professor Alfred Tarski on 'Fundamental Ideas and Problems in Meta-Mathematics'.

I am writing now, on behalf of the Standing Committee appointed to advise in this matter, to ask you if you will honor them by giving the next series of lectures in the forthcoming session (1951–1952).

As you will see from the Scheme, the funds at our disposal are only £100. Though from our point of view, the most suitable times in the session at which to hold the lectures are, in order of preference, February/March or mid-November/mid-December, we should as far as possible, wish to fix the dates to suit your convenience.

---

[15] Sender: D.R. Pye, C.B., M.A., Sc.D., F.R.S. Provost.   [16] Bertrand Russell.

I very much hope that you may find yourself able to accept our invitation.

Yours very truly,
D.R. Pye

[Signed typescript – in Brouwer]

Addendum

[Rules of the Fund (added on reverse side of the sheet]

The College Committee, on the recommendation of the Professorial Board, in March 1938, resolved as follows:—

(i)   That the income of the Shearman Fund be used for a biennial course of lectures, the payment to the lecturer, including stipend and expenses, being £100.

(ii) That the number of lectures be normally not less than three nor more than six, at the discretion of the Lecturer.

(iii) That the Lecturer be invited to deal with some problem within the general field of Methodology and Symbolic Logic.

(iv) That a Standing Committee of the Professorial Board be appointed to advise on all matters pertaining to the Shearman Lectureship; that the following be members:-

The Provost
The Professor of the History and Philosophy of Science
The Grote Professor of the Philosophy of Mind and Logic
The Professor of Psychology
The Professor of Political Economy

52-03-04

To W. Radley — 4.III.1952   **Blaricum**

Dear Miss Radley,

I approve of the dates proposed in your letter of February 28th (RS/143/162). The title of my lectures will be 'Outline of intuitionism'.

As friends of mine in London to be invited to the tea party I could mention Professor K.R. Popper of the London School of Economics, Professor H. Dingle of University College, and outside of the university Mr. and Mrs. Haynal Conyi, 7 The Park, NW11, and Miss Winifred Gordon Fraser, 2 Nottingham Str., W1. Outside London, I presume the following scholars specially interested in my subject: Whitehead, Kneale and Waisman in Oxford, Steen (Christ's), Braithwaite and Routledge (King's) in Cambridge, Newman, Polanyi and Turing in Manchester. If some of them would have opportunity to attend my first lecture, I should be happy to see them at the tea party.

I should like to receive a dozen of lecture notices to be sent by me to addresses which might cross my mind. [17]

Yours sincerely
(signed) L.E.J. Brouwer

[Signed typescript – in Brouwer]

1953-00-00

To D. Coxeter — 1953   **Blaricum** [18]

Confidential for Donald Coxeter
LEJB

As far as I see, the present legislation and distribution of power in the Netherlands is such that who neither belongs to a political party, nor to a church, nor to a coterie, nor to a category, who moreover is neither dishonest, nor insincere, nor stupid, nor stonehearted, nor pecuniarily independent, nor below seventy years of age, needs all his strength to remain alive, and has to forsake any vocation.

L.E.J. Brouwer

[Signed autograph – in Brouwer]

---

[17] A copy of the notice can be found in the Brouwer archive.   [18] The note is undated; it should not be dated before 1951; the terms used, suggest that it was written in the early fifties. The note shows Brouwer's bitterness over his postwar treatment.

53-07-28a

To H.S.M. Coxeter — 28.VII.1953[a]   Blaricum

My dear Coxeter

My arrival at Montreal airport has been fixed on August 7th at 10.10, and a reservation has been made for me in Montreal at the Berkeley Hotel by Professor Williams. So, if nothing will come between (which I keenly hope), I shall now very soon have the pleasure to meet you.

As you probably know, the following list of papers which could be usefully examined by my audience previously to my lectures, has been sent by me to Professor Williams:

> *Weyl*, 'Über die neue Grundlagenkrise der Mathematik', Mathematische Zeitschrift, vol. 10 (1921)
>
> *Dresden*, 'Brouwer's contributions to the foundations of mathematics', Bulletin of the American Mathematical Society, vol. 30 (1924).
>
> *Wavre*, 'Y a-t-il une crise des mathématiques?', Revue de Métaphysique et de Morale, vol. 31 (1924)
>
> *Wavre*, 'Logique formelle et logique empiriste', Revue de Métaphysique et de Morale, vol. 33 (1926)
>
> *Lévy*, *Wavre* et *Borel*, 'Discussions', Revue de Métaphysique et de Morale, vol. 33/34 (1926/27).
>
> *Brouwer*, 'Wissenschaft, Mathematik und Sprache', Monatshefte für Mathematik und Physik, vol. 36 (1928)
>
> *Brouwer*, 'Consciousness, Philosophy and Mathematics', Proceedings of the Xth International Congress on Philosophy (Amsterdam, 1948)

Kindest regards from house to house

ever yours
(signed) L.E.J. Brouwer

[Signed autograph, copy – in Brouwer]

1953-09-27

To W.G. Fraser — 27.IX.1953   New York

Dear Miss Fraser and you all members of the New Europe Group! [19]

The crushing news of Mitrinovic's death reached me here. [20] It's beyond expression how deeply I am moved by the passing away of this survivor of an era of vision, this herald of an era of realization.

In friendship
(signed) L.E.J. Brouwer

[Signed handwritten draft – in Brouwer]

1953-11-28

To Mr. E., Mrs. L. Gutkind — 28.XI.1953   Davenport (Iowa)

Dearest Ekalucia [Allerliebstes Ekalucia [21]]

Because of the many social occasions that were connected with my lecture tour, it had a much slower course than I expected. But the larger part is behind me, and between today and my return to you still lay Urbana, LaFayette, Toronto and Ithaca as stages. I will probably be again in New York on 12 December. My address until 3 December is c/o Dept. of Mathematics, Purdue University, La Fayette, Indiana. Then until 9 December c/o Professor H.S.M. Coxeter, 67 Roxborough Drive, Toronto 5, Ontario.

You are embraced by [22]
your Bertus

[Signed autograph – in Brouwer]

---

[19] A group of idealists founded by Mitronovic. It has artistic, philosophical, and mystical roots in the early part of the twentieth century, in particular the years before and after World War I. See [Van Dalen 2005] p. 864, [Rigby 1984].   [20] Mitrinovic died on 28.VIII.1953, cf. [Rigby 1984] p. 185. In the Brouwer Archive there is an obituary notice with: '[...] death of the Founder, Dimitrije Mitrinović [...] Requiem Service [...] 7th October [...] September 1953.'   [21] *Allerliebstes Ekalucia* – contraction of Eka (Erich Gutkind) and Lucia (Gutkind's wife).   [22] *Es umarmt Euch.*

54-12-31

To F. van Anrooy [23] **— 31.XII.1954**   **Blaricum** [24]

Dear Freddy, [Beste Freddy]

Thus Peter has fought the battle of his life to the end. [25] A life that, guided by a great and indomitable talent and by a tempestuous wealth of thoughts and feelings, has found the predestined triumphs and conflicts on his path.

My deepest sympathy and my assurance that I, with the many others who are better qualified for it, will hold his memory very dear.

Your
Bertus Brouwer

[Signed autograph – in Heyting]

55-01-04

To H.C. Marston Morse — 4.I.1955   **Blaricum**

Dear Morse,

All good wishes for 1955 to you and Mrs. Morse. Thinking back with the greatest pleasure to my service in Princeton during October 1953. Playing with the idea of being called back to the Institute some day for a longer stay. The which might also be to the profit of science, my circumstances in the Netherlands being absolutely prohibitive for scientific research.

Please remember me to the other members of the Institute, in particular to Veblen, Einstein, Oppenheimer and Von Neumann.

A small pile of reprints is following by slower mail.

---

[23] Peter van Anrooy's widow.   [24] A copy was made available by Van Anrooy's daughter Fien, the second wife of Arend Heyting.   [25] Peter van Anrooy, 13.X.1879 – 31.XII.1954; well-known musician, composer (Piet Heyn Rhapsody), conductor (Residentie Orchestra).

Kindest regards to you and Mrs. Morse.

Faithfully yours
(signed) Egbertus Brouwer

[Signed autograph, copy – in Brouwer]

1955-05-03

To J. Kok — 3.V.1955   Blaricum

Dear Colleague, [Hooggeachte Collega]

To my great regret I am, because of an indisposition, unable to attend the meeting of tomorrow and to hear your memorial address. [26] In the meantime I would, in case you plan to give the memorial address not only a collective, but also an individual character, take the liberty to commemorate two fallen [members of the mathematical institute] who were close to me, and whose memory is dear to me. They are

The late J.F. Koppers, in life the porter [27] of the Mathematical Institute of the University, who during the first four years of the occupation with an untiring and almost superhuman diligence and perseverance succeeded in keeping many hundreds of fellow countrymen out of the hands of the occupation forces, and moreover usually knew how to ensure their livelihood. In June 1944 he was arrested, and at the end of 1944 he died in Neuengamme. Of his family that was left behind, not only his widow will have to be taken care of, but one of his three adult children is probably not able to secure its livelihood.

Furthermore, the late Dr. M.J. Belinfante, in life private docent at our university, and one of the most gifted researchers of our country. With an equally surprising and admirable resignation, he has during almost four years refused steadfastly each opportunity to evade the threatening dangers. In the spring of 1944 he was arrested, and he died later in Theresienstadt [28].

---

[26] For the victims of the Second World War.   [27] A function that resembles that of the college porter.   [28] Belinfante was deported to Theresienstadt; he died in Auschwitz.

Apologizing for the possibly superfluity of the above information, I remain with friendly greetings,

Sincerely
your
(w.g.) L.E.J. Brouwer

[Typescript – in Brouwer]

57-02-03

To H. Hopf — 3.II.1957   Blaricum

Dear Hopf, [Lieber Hopf]

In view of rumors that reached me that some scholars have the plan to call attention to the coming fiftieth anniversary of the appearance of my doctoral thesis, I would like to announce my *explicit wish* to all colleagues that are scientifically more or less close to me, that *no* attention will be paid to this anniversary and that also *any printed reference of the mere fact will be refrained from.* And since not all these colleagues can be reached by me, I would like at the same time ask the colleagues that I approach personally in this matter, to do all they can to ensure that in their circle of influence my express wish is respected to the largest possible extent.

My warmest thanks in advance for your kindness, and cordial greetings from house to house.

Your
L.E.J. Brouwer

[Signed autograph – in Hopf]

58-07-23

From A.S. Esenin-Volpin — 23.VII.1958   Moscow
1st Volkonski pereulok, 11, apt. 4

Dear Professor Brouwer!

I am one of those who continue your criticism of the classical point of view — but that continued criticism destroys one of your fundamental notions, namely that of the natural number series. In virtue, what is it? Why are we sure that such numbers as 10^{12} exist — or, more exactly, why are we sure that 10^{12} is representable in the form $1+1+\ldots+1$? Is the complete induction principle compatible with the existence of the operations '+' and '·'?

Instead of an absolute notion of a natural number I introduce here a relative notion of a feasible number. And the unfeasible objects may be regarded as constituting a model for the infinity which appears in the Zermelo-Fraenkel set theory (of course after the banishing the logical operators '∨' and '∃' from the latter).

It is a copy of a letter I sent also to some other scientists in that domain. It was written before the last two weeks, and now I have found a most exact version of the theory I consider. The elaborated exposition of that new version I hope to achieve in September-October and I shall write to you about it if you desire. Some remarks about it are in PPS.

PPS. I continue the study of the theory of feasible objects. Now nearly everyday some new ideas come to me and I don't want to postpone the sending of this letter.

Concerning the question of the last paragraph on the page 20 I say:

The thoughts of a person considered there, if relating to ordinary objects, are accessible to the traditional 'omnipotent' subject and therefore the thinking of that person about these objects must be consistent, but it leads only to true conclusions if the first premises are true (in an intuitive sense). And it is evident also that the introducing of the new variables for the feasible objects cannot destroy the consistency unless a postulate concerning the existence of some unfeasible object is added. But this postulate is a true one.

Such are the reasons that the thinking of our imagined person must be consistent if the thinking of the traditional person is. And concerning the last — we can consider the natural number theory as the theory of the decimal (or dyadic) notations — in this theory the addition and the multiplication are easier. Of course, even here we have a postulate that for each notation there is a corresponding natural number. It is a main postulate. The second — and I hope essentially the last for the foundations of Z^- — is the postulate on the existence of some unfeasible number relative to '+'.

The methods of reasoning must be such that each reasoning is to be feasible for the considered person — in particular, each reasoning justifying the application of some induction principle. So, we cannot always apply the principle $(I3')$ — because if for $F(m)$ and $F(n)$ the reasoning is feasible, for $F(m+n)$, according to $F(m)\&F(n) \supset F(m+n)$, it may be feasible only relative to '+'. Sometimes nevertheless such reasonings are admissible — namely, if they don't lead to a contradiction of an unfeasible object which is to be regarded as a feasible one (i.e. to be substituted for a variable corresponding to the feasible numbers). The proof of the induction step is always to be examined in this connection. It seems to me that everywhere it is necessary for my purpose of establishing the truth of the axioms of Z^- in the model M_Z (or axioms of ZF^- in the model N_Z). That condition is satisfied. (An example where it is not satisfied we obtain if we try to prove the existence of the last letter in every word — and that existence leads to a contradiction if regarded in connection with the set of all a_n with feasible n — or if we try to obtain the paradox of the heap here).

Sincerely yours,
A.S. Esenin-Volpin

[Signed autograph – in Brouwer]

59-04-25

To KNAW — 25.IV.1959   Amsterdam

W. Sierpiński in Warsaw [29]

When about the turn of the century the physiology of the real functions had drawn the general attention, a new vast field of new problems had arisen, the treatment of which required a radical deepening of epistemology, and to face questions like those concerning the rationale and scope of the axiom of choice and of the notion of continuity. The group of researchers thus stimulated has from the beginning experienced the powerful guidance of Sierpínski, whose mental power and originality obtained results that mark him as a Grandmaster, and whose inspiration gave birth to the Polish mathematical school, which found its expression in the renown journal *Fundamenta Mathematica*.

During his entire life so far Sierpínski has widened, deepened, renewed, and juvenated his realm of thoughts. Not long ago his researches carried

---

[29] Brouwer's text in support of Sierpiński's foreign membership of the KNAW. The proposal was not universally applauded, see [Van Dalen 2005] p. 897 ff.

him into number theory [1]; on this topic he opened with zeal a seminar.

Apart from articles in journals, counting more than six hundred, Sierpínski has written a number of comprehensive texts: *Leçons sur les nombres transfinis* (1928), *Hypothèse du continu* (1934), *Les ensembles projectifs et analytiques* (1950), *Algèbre des ensembles* (1951), *General topology* (1952). In particular in the *Algèbre des ensembles* light is shed, in addition to the treatment of the subject expressed in the title, on the mutual relation of almost all fields of research on mathematical-epistemological subjects that were laid open in the first half of the twentieth century, including those that were so far practiced by the Polish mathematical school.

With Chopin, Paderewski, and Madame Curie, Sierpínski belongs to the admirable persons that through the ages, Poland has given to the world.

[Carbon copy – in Brouwer]

1959-08-07

From B.N. Moyls — 7.VIII.1959   **Vancouver**
                                       The University of British Columbia
                                             Department of Mathematics

Dear Professor Brouwer:

We have heard that you might be interested in a sessional appointment at a university on this continent. Would you please let us know if this report is true. In particular, would you be interested in such a position in Canada.

We remember with pleasure your visit to Vancouver a few years ago. Please accept the best wishes of the members of the Department of Mathematics here.

Yours sincerely,
B.N. Moyls, Acting Head,
Department of Mathematics.

[Signed typescript – in Brouwer]

---

[1][Handwritten:] A Schinzel et W. Sierpiński, 'Sur certains hypothèses concernant les nombres premiers', *Annales Analytiques* 4 (1958), p. 185–208; W. Sierpiński, 'Sur les nombres premiers ayant des chiffres initiaux et finals donnés', *Annales Analytiques* 5 (1959) p. 205–206.

60-04-30

To KNAW — 30.IV.1960   Amsterdam

Improvised words I uttered at the extraordinary meeting of 30 April 1960, between the first and the second vote on the nomination of a foreign member. [30]

When the content of the memorandum of recommendation of Sierpínski for foreign member of the mathematics section was disputed by a member of this section in the extraordinary meeting of 26 March last, I have, in view of the late hour, referred, as sole refutation, to what I said about the significance of Sierpínski in the extraordinary meeting of 25 April 1959.

Now that however another member of the mathematics section has portrayed Sierpínski as a very old man, deep into his eighties, whose foreign membership of the section would probably be granted a span of perhaps only a year, [31] In order to avert the dreaded consequences, I am indeed obliged, notwithstanding the late hour, to ask for a few minutes in order to add to the content of the memorandum concerned, some elaborations, elaborations that should have been superfluous under the existing international opinion.

The direct and indirect influence of the work done by Sierpínski (born 1882) in the first decennia of this century. has at the time impregnated and enveloped the thinking of the practitioners of mathematics and epistemology, who have entered the field after him, to such an extent, that for these collaborators it is well-nigh impossible to take a sufficient distance from Sierpínski, in order to objectivize him critically, while those who nonetheless try this, may perhaps shed new light on themselves, but not on Sierpínski.

During his entire life so far Sierpínski has widened, deepened, renewed, and juvenated his realm of thoughts. Not long ago his researches carried him into number theory. His leadership in the field opened up by him has remained undisputed, however famous, perspicacious and original some of his collaborators may be. Among them I mention, without aiming at completeness, Borel, Baire, Lebesgue, Hausdorff, Young, Hobson, Alexandroff,

---

[30] The topic of this letter is the nomination of Sierpínski as a foreign member of the KNAW.   [31] In a preliminary draft Brouwer wrote: " '... portrayed as old gentleman who, had already for a long time become senile, who would probably be dead 'anyway' within a year', I am indeed obliged to avert the threatening unfortunate consequences of this insulting qualification."

Kuratowski, Tarski and Carnap. Among these Sierpínski shines as a star of the first magnitude in the epistemological firmament.

If in an academy that has considered Sierpínski, but nevertheless has not associated him with it, a later generation might become aware of this oversight, there would in my opinion nothing unjustified, if, following an illustrious example, this academy had Sierpínski's bust placed in its most characteristic room with the caption:

nothing was missing from his glory; he was missing from ours [32]

Finally I may point to article 8. I of the rules of the Academy which charges the section to see to it that also among the foreign members all relevant subjects are represented as far as possible; thus, if possible, to have also foreign members associated to it who represent the epistemological foundations of pure and applied mathematics.

[Typescript copy – in Brouwer]

1960-07-26

To Mrs. Whitehead — 26.VII.1960   **London**
                                                              7, The Park

Dear Mrs. Whitehead,

Through Miss Cartwright I learned the recent sudden decease of your husband whom I loved and highly appreciated. I'm sending my heartfelt condolences to you and your children. May the luster of Henry's memory give you fortitude to bear this blow of fate.

Bertus Brouwer

[Signed autograph, copy – in Brouwer]

---

[32] On the bust of Molière, placed, in 1778, in the Académie Française

61-05-26

From L. Hardenberg — 26.V.1961   **Amsterdam**
                                                   Advocaat en Procureur

Dear Sir, [Hooggeleerde Heer]

Estate Brouwer-de Holl.

Yesterday I had an extensive telephone conversation with notary Van
der Ploeg about various aspects of this matter, in particular the conflicts
with your stepdaughter. [33]

The latter has indeed announced several times, via her lawyer, that she
is on her side not disinclined to cooperate in such a settlement that appears
most desirable to the other legatees; but on the other hand she has added
that she wants nonetheless to dispose of a certain sum.

Furthermore we still have the matter of the interpretation of the will. [34]

[....]

If I don't hear otherwise, I'll be expecting you,

Sincerely yours, [35]
L. Hardenberg

[Signed typescript – in Brouwer]

52-05-17

To KNAW, physics section — 17.V.1962   **Blaricum** [36]

The attention that you paid to my golden anniversary on the 15th of
this month gave me great pleasure. During this half century I have had the
Section (and also the Academy as a whole) continuously at heart, with the

---

[33] A.L.E.(Louise) Peijpers   [34] To be discussed with notary Van der Ploeg.
[35] *Hoogachtend, – Uw dw.*   [36] Addressed: Bestuur der Afdeeling Natuurkunde der
Koninklijke Nederlandse Akademie van Wetenschappen, Amsterdam.

consequence that I can look back now on my membership as an over the years developing, lively and as yet unfinished adventure.

Thanking you with cordial feelings, I remain

Your retired fellow-member
(signed) L.E.J. Brouwer

[Signed autograph, copy – in Brouwer]

1965-02-18

To Mrs. J.A.L. van Lakwijk-Najoan — 18.II.1965   Blaricum [37]

Dear Johanna [38],

Concerning our future banking and Giro [39] account numbers, the only line of conduct which seems to offer me security of fortifying in a tangible way my precarious chances of survival, is that all banking and Giro account numbers connected with pharmacy Brouwer-de Holl will be closed and replaced by others. (Apart from the account of Brouwer-de Holl you may perhaps have requested also other banking and Giro account numbers for other accounts you manage.)

Every other line of conduct perpetuates the situation in which I can die any moment, *suddenly* without having been able to take any measure for further winding up the estate Brouwer-de Holl, for the recovery of my good name in international science, or even for my funeral.

By the way, the tangible strengthening of my chances of survival by the mentioned line of conduct is of such importance for my co-heirs of the Brouwer-de Holl estate, that I consider it completely justified to these co-heirs that I offer f 10,000.- for the acceptance of the mentioned line of conduct and your cooperation to implement energetically the mentioned line of con-

---

[37] Addressed: Comeniusstraat 195 IV, Amsterdam. The letter deals with the sale of the pharmacy to Mrs. Van Lakwijk. The pharmacy had played an important role in the lives of Brouwer and his wife, Lize. He was very much attached to it, and one can see that he postponed parting with it till almost the last moment. [38] English in original.
[39] Postal banking system.

duct, which hence amounts to a further decrease of the sales price of the business by this amount.

With cordial greetings

affectionately yours [40]

L.E.J. Brouwer

[Signed typescript – in Lakwijk]

66-07-06

From J. Myhill — 6.VII.1966   **Buffalo (New York)**

Dear Professor Brouwer,

I have been interested for a year now in the arguments you use in your later papers to provide counter-examples to classical theorems. I have been trying to formalize them. While I now have several formalisms [41] in which I believe I can obtain your results, the methods are not quite the same as yours; I have difficulties in achieving this.

I give you one example which I would very much like you to comment on; the proof that if it is impossible for a real number α to be 0, then one cannot necessarily conclude that α is separated from 0, i.e. one cannot necessarily exhibit a rational number between α and 0.

In your version, if I understand it correctly, the proof runs as follows. For each real number $\alpha \in [0, 1]$, the real number $\phi(\alpha)$ is defined as follows: as long as the creating subject has not judged the proposition 'α is rational', let $[\phi(\alpha)](n) = \frac{1}{2^n}$; if at the kth step (after k choices for α) he decides that α is rational or irrational, let $[\phi(\alpha)](k+q) = \frac{1}{2^k}$ for all q. Then $\phi(\alpha)$ cannot be 0, for if it were, α could be neither rational nor not rational. All this is quite clear. The difficulty lies in the second half of the proof.

Here we have to show that we cannot find, for every α in $[0, 1]$, a number β (rational) separating $\phi(\alpha)$ from 0. If we could, we could find a number n such that the proposition 'α is rational' would be judged after n choices for α. Now in my formalism this is immediately contradictory, because it implies that the species of all real numbers in $[0, 1]$ would be split up into the rational and the irrationals, q.e.d.

------

[40] English in original.   [41] See [Myhill 1966, Myhill 1968].

However you proceed differently, reasoning as follows: if for every α in $[0, 1]$, we could find such an n, then by the fan theorem there would be a bound on the n's, say n_0. Now take a real number which up to the n_0 stage is completely unrestricted (except to belong to $[0, 1]$); it is absurd that we could decide at the n_0 stage whether it is rational or not.

It is the application of the fan theorem which I question here. As I understood it, the fan theorem applies only to those cases in which to every free choice sequence α belonging to a finitary spread F we can assign a natural number n_α *using only the values* $\alpha(0), \alpha(1), \alpha(2) \dots$. The proof of the fan theorem, it seems to me, depends essentially on this condition (which is met in the usual mathematical cases: for instance in the theorem, that I used above, that $[0, 1]$ has no detachable subspecies.) But it is not met in the situation to which you apply the fan-theorem here, because in computing the n from the α one is allowed to use also the values of $\phi(\alpha)$, which may depend not only on α but also on what restrictions have been placed on α, and on what properties of $\phi(\alpha)$ the creating subject may have inferred from these.

Any comments you may have on these and related questions, or any reprints since 1929, either mathematical or philosophical, would be very much appreciated.

Very sincerely
John Myhill

P.S. My return address: Department of Mathematics, State University of New York, Buffalo, N.Y.

[Brouwer's note on the envelope:]
Suggereert de mogelijkheid van een wijziging in de argumentatie van 'Points and Spaces' [42]

[Signed autograph – in Brouwer]

---

[42] Suggests the possibility of a change in the argument of 'Points and Spaces'.

Chapter 8

Appendices

List of Enclosures, Editorial Comments and Editorial Supplements

1910-09-00 Draft Brouwer to Korteweg, ed. suppl. see *To D.J. Korteweg — late summer 1910*

1910-10-27 O. Blumenthal to D. Hilbert, ed. suppl. see *Brouwer, Note on Lebesgue III.1911*

1911-03-14 O. Blumenthal to D. Hilbert, ed. suppl. see *Brouwer, Note on Lebesgue III.1911*

1911-05 L.E.J. Brouwer, *Bemerkung zu den Invarianzbeweisen des Herrn Lebesgue* (note on Lebesgue), ed. suppl. see *Brouwer to Blumenthal 9.V.1911*

1911-07-08 L.E.J. Brouwer, Further remarks on Lebesgue's proof, ed. suppl. see *Brouwer to Blumenthal 8.VII.1911*

1911-11-05 L.E.J. Brouwer Note related to the letter *Brouwer to Baire 5.XI.1911*, ed. suppl.

1911-12-22 L.E.J. Brouwer, some remarks, ed. comm. on various versions, see *Brouwer to Fricke 22.XII.1911*

1913-08-16 L.E.J. Brouwer, some remarks, ed. suppl. see *Brouwer to Schoenflies 16.VIII.1913*

1915-06-09 H.A. Lorentz to P. Ehrenfest, ed. suppl. see *Lorentz to Brouwer 11.VI.1915*

1917-03-17 M. Buber to H. Borel, ed. suppl. see *Brouwer to Buber 4.II.1918*

D. van Dalen, *The Selected Correspondence of L.E.J. Brouwer*,
Sources and Studies in the History of Mathematics and Physical Sciences,
DOI 10.1007/978-0-85729-537-8_8, © Springer-Verlag London Limited 2011

1918-02-16 L.E.J. Brouwer, remarks on the Scientific Comm. for Advice and Investigation (photogrammetry), ed. suppl. see *Brouwer to Lorentz 16.II.1918*

1920-05-16 D. Hilbert to H. Weyl, ed. suppl. see *Weyl to Brouwer 6.V.1920*

1921-11-26 Algemeen Handelsblad, Report KNAW, ed. suppl. 1 see *Brouwer to Algemeen Handelsblad 27.XI.1921*

1921-11-26 L.E.J. Brouwer, Note KNAW, ed. suppl. 2 see *Brouwer to Algemeen Handelsblad 27.XI.1921*

1923-11-29b Freudenthal's comments on Brouwer's slip of the pen (dimension definition), see *Brouwer to Urysohn 29.XI.1923b*

1924-06-21 C. Jongejan to L.E.J. Brouwer, Note, ed. suppl. see *Urysohn to Brouwer 21.VI.1924*

1928-09-27 H. Härlen to I. Gawehn, ed. suppl. see *Härlen to Brouwer 27.IX.1928*

1928-12-19 Carathéodory re Hilbert's motives. ed. suppl. see *Carathéodory to Courant — 19.XII.1928*

1929-07-11 Ed. comment on dimension discussion, see *Brouwer to Hahn 11.VII.1929*

1929-08-09 L.E.J. Brouwer, Note Lennes - connectedness, encl. see *Brouwer to Hahn 9.VIII.1929*

1931-04-24 R. Carnap, CV and Bibliography, encl. see *Carnap to Brouwer 24.IV.1931*

1948-04-09 J. Clay, J.G. van der Corput to Chancellor UvA. ed. suppl. see *Brouwer to Board UvA 2.IV.1948*

1949-11 A. Heyting to P. Bernays, ed. suppl. see *Brouwer to Heyting 28.X.1949*

1951-06-12 J. van der Corput to A. Heyting and E.W. Beth, ed. suppl. see *Corput (mrs) to Brouwer 18.IV.1951*

Biographical Information

[The selection below contains only items relevant to the translated letters. The online version of the untranslated correspondence contains more biographical items.]

Adama van Scheltema, Carel Steven 1877–1924. Leading Dutch socialist-poet. Was Brouwer's closest friend since his student days. Gave up his medicine study to turn to poetry and literature. Greatly influenced Brouwer's position with respect to society and the world.

Alexandrov, Pavel (Paul) Sergeevich 1896–1982. 1913 - Study at Moscow University. Teachers Egorov and Luzin. Turned to topology with his friend Urysohn (1920); Göttingen 1921 and 1923. Supported by Noether, Hilbert and Courant. 1924 - visits Hausdorff in Bonn and Brouwer in Blaricum. 1925 - took part in the "Amsterdam topological school"; 1927/28 at Princeton together with Hopf. Alexandrov-Hopf's "Topologie" became the prime source for the subject (1934). 1929 - full professor in Moscow. Close friend of Kolmogorov.

Anrooy, Peter van 1879–1954. Dutch composer, director of the Residentie Orchestra in The Hague. Influential pedagogue.

Arkel, Cornelia G. van 1902–1980. Professor of pharmacology; in 1949 chairman of the Faculty of Mathematics and Physics at the UvA.

Aten, Adriaan Hendrik Willem 1877–1950. 1896–1904 - Study at the UvA; 1904 - PhD with Bakhuis Roozeboom. 1919–1949 - Chemistry professor at the UvA.

Baire, René-Louis 1874–1932. Student of the École Normale Supérieure. As a student created the Baire hierarchy. 1899 - PhD, thesis on discontinuous functions. Suffered from poor health and published little. 1901 - 'Maître de conférence' at Montpellier. 1905 - at Faculty of Science at Dijon; 1907 - professor of analysis. 1922 - Member of Académie des Sciences. Published textbooks on irrational numbers and on analysis. He moved away from the intuitive idea of continuity of functions, accepting that a theory of infinite sets was fundamental for rigorous real analysis.

Belinfante, Maurits Joost 1896–1944. Student of Brouwer; 1921 - graduation, Brouwer's assistant. 1923 - PhD; dissertation *On infinite series*, Privaatdocent at the UvA. Dismissed as Jew in 1940; deported to Theresienstadt, died in Auschwitz (1944).

Bernstein, Felix 1878–1956. Studied with Cantor at Halle and with Hilbert and Klein at Göttingen; Göttingen PhD on set theory. 1907–1934 - taught at Göttingen; 1921 professor of mathematics, founder of the Institute of Mathematical Statistics. Dismissed in 1934 while visiting the USA He stayed in the USA until his return to Göttingen in 1948. Bernstein is best known for his contributions to set theory (in particular the Cantor-Bernstein theorem). Published also on statistics, mathematical biology.

Beth, H.J.E. 1880 –1952. Studied mathematics at the UvA. 1910 - PhD with Korteweg. Taught mathematics at high schools; he was the director of various high schools. He was a prolific author of a large number of books and papers. His books on the history of mathematics were generally admired.

Beth, Evert Willem 1908 –1964. Son of H.J.E. Beth. Initially studied pharmacy at Utrecht University, then switched to mathematics. He graduated in 1934 and continued to study at Leyden and Brussels and graduated once more at Utrecht in philosophy and psychology. 1935 - PhD with J.C. Franken *Reason and intuition in mathematics*. Subsequently he was a high school teacher of mathematics and physics; 1946 - extraordinary professor of logic and its history and of philosophy of science at the UvA; 1948 - full professor. 1952 - visiting professor at Berkeley (1952), 1957 at Johns Hopkins. 1964 - honorary doctorate at the University of Gent. He also participated in the 'Signific Circle'. Together with Brouwer and Heyting, he was an editor of the series Studies in Logic and the Foundations of Mathematics. Known for his semantic tableaux and Beth models. His definability theorem was an important contribution to model theory.

Bieberbach, Ludwig G.E.M. 1886–1982. Study at Heidelberg and Göttingen. 1910 - PhD with Klein–*Zur Theorie der automorphen Funktionen*, 1911 - Habilitation: groups of Euclidean motions - an important step towards the solution of Hilbert's eighteenth problem. He was a professor of mathematics in Basel, Frankfurt am Main and Berlin (1921). Bieberbach contributed to geometry and function theory and he was well known as an excellent teacher and PhD supervisor. Together with Schur he published in 1928 *Über die Minkowskische Reduktionstheorie der positiven quadratischen Formen*. He has been severely criticized for his active participation in the Nazi movement before and during World War II.

Birkhoff, Georg David 1884–1944. American mathematician. Birkhoff studied at Chicago and at Harvard; 1907 - PhD–thesis *Asymptotic Properties of Certain Ordinary Differential Equations with Applications to Boundary Value and Expansion Problems*. 1911 - professor at Princeton, 1912 - professor at Harvard. His main contributions are in the field of mechanics and ergodic theory, but he also did important work on pure mathematics (e.g. on the four color theorem), on the foundations of relativity and quantum theory and on 'philosophy and science'. Solved Poincaré's last problem.

Blaschke, Wilhelm 1885–1962. Study in Graz; PhD at the university of Vienna in 1908, after which he spent semesters in Pisa with Bianchi and in Göttingen with Klein, Hilbert and Runge. 1910 - Habilitation with Study, Bonn. Next he worked with Engel at Greifswald. He held positions in Prague (1913), in Leipzig (1915) where he published *Kreis und Kugel*; 1917 - full professor in Königsberg (now Kaliningrad). After a brief spell at Tübingen, he accepted a position at Hamburg in 1919. Visiting professor at John Hopkins University, Chicago, Istanbul and at the Humboldt University of Berlin. Prominent in the German Mathematical Society. Main research was on various aspects of geometry. He published *Vorlesungen über Differentialgeometrie*. Also active in topology.

Blumenthal, Otto 1876–1944. Studied mathematics in Göttingen from 1894 until 1898; 1901 - Habilitation with Hilbert. Privatdocent at Göttingen until 1905. He was a function theorist, with a strong penchant for applications. 1905 - professor at the Technische Hochschule in Aachen and editor of the Mathematische Annalen; 1924 - editor of the Jahresberichte der Deutschen

Mathematiker-Vereinigung. 1933 - discharged. Blumenthal moved to the Netherlands in 1939; was eventually deported and died in the Theresienstadt.

Bohr, Harald 1887–1951. Younger brother of the physicist Niels Bohr. 1904 - Study at University of Copenhagen. He was an excellent soccer player, member of the Danish national team. 1915 - professor of mathematics at the Polytechnic Institute in Copenhagen, 1930 - professor of mathematics at the University of Copenhagen. Research: Dirichlet series and applications to the theory of numbers. Collaboration with Landau on the Riemann zeta function. From 1923 to 1926 research on 'almost periodic functions', connected with his interest in representation by Dirichlet series.

Bolk, Louis 1866–1930. Dutch biologist and anatomist. 1898 - professor of medicine at the UvA. Founder of the New Dutch Anatomic School. He also did anthropological research. Around 1923 he was General Secretary of the Dutch 'Royal Academy of Science'.

Borel, Félix Édouard Justin Émile 1871–1956. Study: École Normale Supérieure. 1893 - professor at the University of Lille; 1896 - professor at the École Normale Supérieure; 1909, a personal special chair at the Sorbonne. 1910 - Director of the École Normale Supérieure; 1921 - member of Académie des Sciences, president in 1934. Borel created a theory of measure for point sets, along with Baire and Lebesgue. He published on game theory, on relativity, on function theory and, in 1946, on the paradoxes of infinity. Was of constructive inclination (pre- or semi-intuitionist). He was also politically active.

Brouwer, Egbertus Luitzen 1854–1947. Father of L.E.J. Brouwer, headmaster in Overschie (now Rotterdam), Medemblik and Haarlem.

Brouwer, Hendrik Albertus (Aldert) 1886–1973. Brother of L.E.J. Brouwer; studied geology at Delft technical university. 1911–1917 - research in Dutch Indies (now Indonesia). 1917 - Professor in Delft, 1928 - geology professor at UvA.

Brouwer, Luitzen Egbertus Jan 1881–1966. 1897 - Study at the University of Amsterdam. 1907 - PhD with Korteweg; dissertation *On the Foundations of Mathematics* (Over de grondslagen der wiskunde). As a student he published on four dimensional rotations and potential theory. From 1909 research in topology. Breakthrough in 1910, 1911- invariance of dimension, domain. Father of the new topology. 1912 - Extraordinary professor at the UvA; 1914 - full professor. Inaugural address *Intuitionism and Formalism*; 1918 mature intuitionistic program with choice sequences. 1924- fan theorem, continuity theorem. During the 1920s he was involved in the foundational conflict with Hilbert, briefly joined by Hermann Weyl.

Brouwer-de Holl, Reinharda Bernardina Frederica Elizabeth (Lize) 1870–1959. L.E.J. Brouwer's wife (married 1904). Studied pharmacy, was in charge of the pharmacy in Amsterdam.

Bruins, Evert Marie 1909–1990. Mathematician and physicist, 1928 - PhD with Clay for a thesis on cosmic radiation. Succeeded Freudenthal in 1941, after the Jewish staff members were dismissed. 1943 - lecturer of mathematics (real analysis). His refusal to make place for Freudenthal after the war, left bad feelings. Became historian of mathematics (specialist Babylonian mathematics).

Bruijns-Oosterbaan, Cor Wife of the mathematician E.M. Bruijns.

Buber, Martin 1878–1965. Austrian born Jewish philosopher of religion. He studied at Vienna, Leipzig, Berlin and Zurich. He defended his own version of Zionism, which was not directly aimed at the foundation of a Jewish state and which promoted solidarity with the Arabs in general and those in Palestine in particular. 1938 Buber - professor at Hebrew University of Jerusalem. He translated chassidic texts as well as the Old Testament.

Carathéodory, Constantin 1873–1950. Carathéodory was of Greek descent. He started his studies in Belgium. He worked as a military engineer in Egypt, subsequently entered in 1900 the University of Berlin. In 1902 he moved to Göttingen where he worked on the calculus of variations. 1904 PhD with Minkowski; 1905 - Habilitation *Über die starken Maxima und Minima bei einfachen Integralen* at Göttingen; subsequently Privatdocent there. After a year at Bonn, professor at Hanover Technical University and in 1910 in Breslau. 1913 - professor at Göttingen. 1916 - Professor at Berlin university; 1920 - at Athens. He helped to found the Greek University at Smyrna. Returned to Athens after the Turkish invasion of Smyrna (Izmir). 1924 - professor at Munich University. He taught in the United States (1928). His main contributions are in the field of calculus of variations, measure theory of point sets, theory of functions of a real variable.

Carnap, Rudolf 1891–1970. Studied physics, mathematics and philosophy at Jena (with Frege) and Freiburg. Studied relativity with Einstein in Berlin. 1921 - PhD with Bruno Bauch *Der Raum* (Space). 1923 - Carnap joined the Vienna Circle. 1925 - Assistant professor at the University of Vienna. 1928 - publication *The Logical Structure of the World* and *Pseudo Problems in Philosophy*. In 1929 Carnap met Tarski, moved towards mathematical logic. 1931 - professor of natural philosophy in Prague; publication of *The Logical Syntax of Language* (1934). 1935 - emigration to the US. 1936–1952 - professor at the University of Chicago; visiting professor at Harvard. 1954 - professor at UCLA, after a visit at the Institute for Advanced Study in Princeton.

Cauwelaert, Frans van 1880–1961. Belgian Catholic statesman, studied philosophy at Leuven, Leipzig and Munich. From 1907 to 1910 professor of psychology at Fribourg University. After that he studied law and became lawyer in Antwerp. As a politician he was also active in the Flemish movement and in making the University of Gent Flemish. He became mayor of Antwerp and later Minister for Economic Affairs of Belgium.

Claparède, Édouard 1873–1940. Swiss psychologist; studied originally biology and zoology, later switched to medicine; doctorate in 1897. 1904 - director of a laboratory of psychology at the University of Geneva and professor of psychology, specializing in infantile psychology and teaching. Founded in 1912 the Institute J.J. Rousseau, a school for the science of education.

Clay, Jacob 1882–1955. Dutch physicist and philosopher, known for his research on cosmic rays. He started his career in philosophy as a Bolland adept. Study at Leyden University with Kamerlingh Onnes and Lorentz. 1908 - PhD on low temperature physics. He published on philosophy and physics. Clay was a teacher in Leyden and Delft, lecturer at Delft Technical University. Professor at the Technical University of Bandung (then Dutch East Indies, now Indonesia). 1929 - professor of physics at the UvA. After World War II he served as Dean of the Faculty of Mathematics and Physics and President of the Physics Section of the Royal Dutch Academy of Science.

Corput, Johannes Gualtherus van der 1890–1975. 1908–1914 - Study at Leyden University. Taught mathematics at secondary schools. 1919 - PhD at Leyden. Worked at Göttingen with Landau; 1920–1922 - Denjoy's assistant at Utrecht University. 1922–1923 - professor at Fribourg (Switzerland). 1923 - professor of mathematics at Groningen University, 1946–1953 at the UvA. After 1953 he held various positions in the United States, to return to the Netherlands after retirement. Best known for his number theory. Van der Corput was the first director of the Mathematical Center in Amsterdam (1946–1953).

Courant, Richard 1888–1972. 1905 - Study at University of Breslau. 1907 - moved via Zurich to Göttingen to continue his studies with Hilbert and Minkowski. 1908 - Hilbert's assistant; 1910 - doctorate with Hilbert. Lecturer at Göttingen after Habilitation (1912). Served in the army in the first World War, returned to Göttingen as a Privatdocent. Professor in Münster (1920) subsequently in Göttingen. He founded the Mathematics Institute (1922). With Hilbert he wrote the influential *Methoden der mathematischen Physik* 1924 (Methods of Mathematical Physics, 1953); his *Vorlesungen über Differential- und Integralrechnung* 1928 (Differential and Integral Calculus) became a standard text on the subject. 1933 - emigration, ousted by the nazis, first to Cambridge, subsequently to New York. He founded an applied mathematics research center, based on the Göttingen model, now the Courant Institute.

Coxeter, Harold Scott MacDonald 1907–2003. Study at University of Cambridge; 1931 - PhD with H.F. Baker. He spent two years at Princeton with Veblen (1932); 1936 - professor at the University of Toronto. Coxeter remained in Toronto for the rest of his life. He was geometry incarnate, with a great sense of beauty. Author of a large number of wonderful books, e.g. *The real projective plane* (1955), *Introduction to geometry* (1961) and *Non-Euclidean geometry* (1965).

Dantzig, David van 1900–1959. Van Dantzig wrote his first mathematical paper at the age of thirteen, when in secondary school. Shortly after commencing to study chemistry, he was forced to give up the academic studies and to find a job in order to support his family. He taught himself enough mathematics to pass the required state examinations. After his master's degree Schouten appointed him as an assistant in 1927 at Delft Technical University. 1932 - PhD with Van der Waerden at Groningen University, thesis *Studies on topological algebra*, and appointed lecturer at Delft. 1938 - Extraordinary professor; 1940 - full professor. After the Germans occupied the Netherlands, Van Dantzig, being Jewish, was dismissed. 1946 - professor at the UvA. He was cofounder of the Mathematisch Centrum. Van Dantzig studied differential geometry, electromagnetism and thermodynamics; his main topic of research was topological algebra. He worked on metrization of groups, rings and fields. After the war his main topics were statistics and probability.

Denjoy, Arnaud 1884–1974. Study at the École Normale Supérieure with Borel, Painlevé and Picard, 1902. 1909 - PhD thesis, entitled *Sur les Produits canoniques de l'ordre infini*. 1910 - appointed 'maître de conférence' at the university of Montpellier. 1917 - professor at Utrecht University; 1922 - professor in Paris. In his work Denjoy used topological and metrical methods to attack problems in real analysis. He was a member of the Académie des Sciences and vice-president of the International Mathematical Union.

Dijksterhuis, Eduard Jan 1892–1965. Dutch historian of science; studied at Utrecht University. 1918 - doctorate. 1953 - extraordinary professor, 1960 - full professor in history of mathematics and science at Utrecht University. 1955 to 1960 - also extraordinary professor at Leyden University. Active in the pedagogy of mathematics. After 1950 he rose to international fame, with his *The mechanization of the world picture* (Dutch, 1950, English 1961). He initiated the publication of Simon Stevin's Selected works (1955).

Dingle, H. 1890–1978. Astronomer, 1938 professor of Natural Philosophy at Imperial College; 1946 - professor of history and philosophy of science at University College, London. Known for his opposition to Albert Einstein's special theory of relativity and the subsequent controversy.

Dingler, Hugo Albert Emil Hermann 1881–1954. Studied mathematics and physics at Erlangen, Munich and Göttingen. 1906 - PhD, 1912 - Habilitation at the University of Munich. 1920–1932 professor at Munich. 1932–1934 at Darmstadt. 1934 - return to Munich. Scientific interest mainly in the field of the foundations of science. He propagated the Erlanger constructivism.

Doetsch, Gustav 1892–1977. 1911–1914 - Study mathematics, physics and philosophy at Göttingen, Munich and Berlin. 1920 - PhD with Landau. 1921 - Habilitation at the Technical University of Hanover. He taught at Halle and at Stuttgart. 1922 publication of a famous paper about the applicability of mathematics to natural science. He collaborated with Felix Bernstein on a modern version of Laplace transforms. 1924 joined the peace movement. 1931 - professor at Freiburg and until 1936 he supported the Nazi policy towards Jewish mathematicians. After 1936 he led a rather isolated life in Freiburg. During World War II he joined the German Air Force; after the war he returned to Freiburg, where he was suspended until 1951; then taught until his retirement in 1961.

Donder, Théophile Ernest de 1872–1957. Belgian mathematician and physicist; 1899 - PhD in physics at the University of Brussels. 1911 - professor of mathematical physics at that university. In the 1920s De Donder supported the relativistic ideas of Einstein. His research included relativity, electromagnetism, wave mechanics, thermodynamics and calculus of variations. 1929 - member of the Académie Royale de Belgique.

Dresden, Arnold 1882–1954. American mathematician of Dutch origin. Studied in Amsterdam and Chicago. 1909 - PhD at the University of Chicago with Oskar Bolza. 1909–1927 Assistant professor at the University of Wisconsin. 1927 professor of mathematics at Swarthmore College (Pa). He translated into English Brouwer's inaugural lecture (1912) - published in the Bulletin of the American Mathematical Society (1914).

Dubislav, Walter 1895–1937. Studied mathematics (with Hilbert) and philosophy. 1928 - 'Privatdocent' of philosophy of mathematics at the Technical University of Berlin; 1931 - extraordinary professor. He was co-founder of the 'Berliner Gesellschaft für Empirische Philosophie'. 1936 moved to Prague. Dubislav's main work was on the logical and scientific foundation of mathematics and physics.

Eeden, Frederik Willem van 1860–1932. Dutch poet, essayist and playwright. Founder of the community 'Walden'. Active in the 'Signific Movement' with Brouwer, Mannoury and Jacob Israel de Haan. Also PhD in medicine in Amsterdam; first psychiatrist in Holland.

Ehrenfest, Paul 1880–1933. 1890–1899 attended the Akademisches Gymnasium in Vienna. Followed by study at the Technische Hochschule in Vienna with Boltzmann. 1901 - study with Hilbert, Klein and Zermelo. 1904 - PhD in Vienna with Boltzmann for a thesis on classical mechanics. 1907 - moved with his Russian wife Tatyana Alexeyevna Afanasyeva to St Petersburg. In 1911 he published a paper on quantum theory in the Annalen der Physik. 1912 - professor at Leyden. He published many important papers on quantum theory and had long discussions about quantum theory both with Bohr and Einstein. His took his own life.

Ehrenhaft, Felix 1879–1952. Austrian physicist. 1903 - PhD at the University of Vienna. He worked on observations and measurements of the Brownian motion. He came into conflict with Millikan about the measurement of the elementary electric charge. He was a fearless independent thinker. He diverged more and more from the mainstream of physics. 1920 - professor of experimental physics at Vienna. 1938 - emigration to England and USA; 1946 - return to Vienna.

Einstein, Albert 1879–1955. 1896–1900 - Study at the ETH in Zürich. Failing to obtain a position at the ETH, he taught at schools; 1902–1909 - position

at the patent office in Bern. 1906 - PhD with Grossmann at the University of
Zürich. 1908 - Privatdocent in Bern. 1911 - assistant professor in Zürich and
full professor at the German University in Prague. 1912 - professor at the
ETH. 1915 - director of the Kaiser Wilhelm Institute for Physics at Berlin.
There were unsuccessful attempts to get Einstein to Holland. 1920–1946
extraordinary professor in Leyden. He visited Holland regularly. During
his visit in 1918 to the KNAW Brouwer met him for the first time. His
paper on special relativity of 1905 (together with his other papers) made
him instantaneously famous. 1921 - Nobel Prize for Physics for his work on
photoelectric effect. From 1933 on Einstein was at the Institute for Advanced
Study at Princeton.

Engel, Friedrich 1861–1941. 1879 - Study at the universities of Leipzig and
Berlin. 1883 - PhD in Leipzig with Adolph Mayer. Studied then with Klein,
and at Klein's suggestion he worked between 1884 and 1885 with Lie in
Christiania. 1885 - Habilitation (Leipzig) and lecturer at Leipzig. The next
year Lie was appointed at Leipzig and the collaboration between the two
continued. 1889 - assistant professor. 1904 - professor at Greifswald; 1913
at Giessen where he remained for the rest of his career. His collaboration
with Lie led to the publication, between 1888 and 1893, of the *Theorie der
Transformationsgruppen*. Later Engel published Lie's collected works.

Esenin-Volpin, Alexander Sergeyevich 1924– . Russian-American mathematician and, during his Russian years, a dissident, political prisoner and poet.
1946 graduation from Moscow State University. After that he was imprisoned
and exiled several times for psychiatric and other reasons. In 1953, after the
death of Stalin, he was granted amnesty. He became known in mathematical
circles for his ultra-finitism and ultra-intuitionism. In 1972 he emigrated to
the United States and worked at Boston University.

Feigl, Georg 1890–1945. 1909 - Study mathematics and physics at the University of Jena. 1919 - PhD (Conformal mappings) with Koebe in Jena. 1919
- assistant of Erhard Schmidt in Berlin, turned to topology under his influence. 1925 - editor of the *Jahrbuch über Fortschritte der Mathematik*;
1933-35 extraordinary professor at Breslau University, 1935 - full professor
of mathematics. 1941 Feigl member of Executive Committee of the German
Mathematical Society, but strongly opposed the Nazi's. He did not survive
the war due to lack of medication during the final stage of the war. His work
was mainly on geometry, in particular its foundations, and on topology, style
Brouwer-Hopf. Through him the modern approach of Klein and Hilbert was
introduced into universities and even into secondary schools.

Fraenkel, Adolph Abraham Halevi 1891–1965. Studied at Munich, Marburg,
Berlin and Breslau. 1916 - lecturer at Marburg, 1922 - professor. 1928
professor at the University of Kiel. 1929 professor at the Hebrew University
of Jerusalem. He spent the rest of his career in Jerusalem. Fraenkel became
Dean of the Faculty and eventually Rector of the University. He is best
known for his work on set theory; his first major work on that topic was

Einleitung in die Mengenlehre (1919). He made two attempts to put set theory into an axiomatic setting to avoid the paradoxes. He improved the Zermelo axiom system, and showed the independence of the Axiom of Choice within Zermelo's system with ur-elements. In 1922 his axiom system was modified by Skolem to what is now known ZFC. Fraenkel also published on the history of mathematics.

Fraïssé, Roland 1920–2008. Studied mathematics at the University of Algiers during the 1950s. 1953 - doctorate (Paris) Known for the Ehrenfeucht - Fraïssé games.

Fraser, Winnifred G. Member of the New Atlantis group, connected with Mitrinovic.

Fréchet, Maurice René 1878–1973. 1906 - PhD thesis on metric spaces with Hadamard. 1910–1919 professor in Poitiers; 1920–1927 in Strasbourg; 1928–1949 in Paris. 1956 member of Académie des Sciences. His major contributions lie in the field of topology of point sets, functional analysis, and probability theory.

Freudenthal, Hans 1905–1990. 1923 - Study at University of Berlin. 1931 - PhD with Hopf on the theory of ends (topology). When 1927 Brouwer lectured in Berlin Freudenthal impressed him as a knowledgeable and original student. As a result Freudenthal was invited to become Brouwer's assistant. With Hurewicz he developed homotopy theory. He is known for his Suspension Theorem. He published on a remarkably wide range of topics, not only in mathematics. In 1940, when Holland was overrun he, as a Jew, was dismissed. His marriage to a non-jewish wife saved him from deportation. 1946 - professor of pure and applied mathematics at Utrecht University. He retired in 1975. In 1971 Freudenthal was the first director of the Institute for the Development of Education in Utrecht, which was founded by him. In 1991 it was renamed the 'Freudenthal Institute'. He was in the true sense of the word a universal mathematician and scholar.

Fricke, Karl Emmanuel Robert 1861–1930. Function theorist, known for his work on automorphic functions. 1894–1930 - professor at Technische Hochschule in Braunschweig. He published with Klein the famous *Vorlesungen über die Theorie der automorphen Functionen.*

Gawehn, Irmgard 20.2.1900–? Studied in Heidelberg. 1925 - PhD with Rosenthal. Studied philosophy and mathematics in Berlin. Assistant of Brouwer 1927-30. In psychiatric clinic from end thirties. Died after Second World War.

Ginneken, Jacobus Joannes Antonius van 1877–1945. Dutch catholic linguist and philologist. 1907 - PhD at Leyden university. 1923 - professor at the Catholic University of Nijmegen. He applied the foundations of psychology to linguistics. He played a major role in the emancipation of the Catholics in the Netherlands. Together with Brouwer, Mannoury and Van Eeden he founded the Signific Circle. Under his influence Van Eeden and Brouwer's stepdaughter Louise converted to Catholicism.

Gonseth, Ferdinand 1890–1975. Swiss mathematician, studied at the University of Bern and at the École Polytechnique Fédérale (ETH) in Zurich; 1915 - privat docent; 1919–1929 mathematics teacher at the University of Bern. 1929–1960 - professor at ETH (foundations of geometry and philosophy of mathematics). Invited Brouwer in 1934 for the Geneva Lectures.

Griss, George François Cornelis 1898–1953. Dutch mathematician and philosopher, studied at the UvA. 1925 - dissertation on the theory of invariants with Weitzenböck in 1925. Mathematics teacher. Initiated negationless intuitionistic mathematics.

Groot, Johannes de 1914–1972. Study at Groningen University. 1942 - PhD with Schaake, *Topological Studies*. Taught at high schools; 1946 - Mathematical Center in Amsterdam. 1947 - lecturer at the UvA, 1948 - professor at the Technical University of Delft, 1952 - professor at the UvA. De Groot mainly worked on topology and on group theory. He is best known for his set-theoretic topology; he was the father of the school of that branch of topology in Holland.

Gross, Wilhelm 1886–1918. Austrian mathematician. Topology and differential geometry.

Gutkind, Erich 1877–1965. Deeply religious Jewish thinker from Berlin; author (under pseudonym Volker) of *Siderische Geburt–Seraphische Wanderung vom Tode der Welt zu Taufe der Tat.* (1910). He published with Van Eeden in 1911 *Welt-Eroberung durch Heldenliebe*. He was part of the failed initiative to form an international group of leading intellectuals, located at Forte dei Marmi (Italy). Emigrated to the USA in 1933.

Haalmeyer, B.P. Brouwer's first PhD student (1917) - *Contributions to the theory of elementary surfaces*. Mathematics teacher. Authored with J.H. Schogt the first Dutch textbook on set theory.

Hadamard, Jaques Salomon 1865–1963. 1884 - Study at the École Normale Supérieure; 1892 - doctorate. 1893–1896 - professor in Bordeaux; 1897 - return to Paris. 1906 - president of the French mathematical society. 1909 - professor at the Collège de France; 1912 - professor of analysis at the École Polytechnique; 1916 - member of the Académie des Sciences. He spent the war years (of WW II) in the USA and in the UK. A versatile mathematician, published on a wide variety of topics.

Haersolte, Baron W. van Dutch lawyer, who acted in the Compositio conflict between Brouwer and the publisher Noordhoff.

Hahn, Hans 1879–1934. Studied at the Technische Hochschule Vienna. Also in Strasbourg, Munich and Göttingen. 1921 - professor of mathematics in Vienna. Hahn pioneered in the field of set theory, real functions, and functional analysis. He is best remembered for the Hahn-Banach theorem. Also active in the calculus of variations, following Weierstrass. During the 1920s Hahn was a member, together with Frank and Von Mises, of the Vienna Circle.

Hamel, Georg Karl Wilhelm 1877–1954. Studied at the Rheinisch-Westphälische Hochschule in Aachen. 1897 - Study at Berlin University where he attended lectures by Planck. 1900 - Study with Klein and Hilbert. 1901 - PhD with Hilbert *Über die Geometrien, in denen die Geraden die Kürzesten sind,* re Hilbert's fourth problem. 1903 - Habilitation at the Technical University of Karlsruhe. 1905 - professor of mechanics at the German Technical University of Brün. 1912 - professor in Aachen; 1919 - at the Technical University of Charlottenburg in Berlin. Specialization: function theory, on mechanics (fluid dynamics) and on the foundations of mathematics.

Hardy Godfrey Harold 1877–1947. 1896 - study – Trinity College, Cambridge. 1900 - Fellow of Trinity; 1911 - start collaboration with J.E. Littlewood, which lasted for 35 years. Recognised genius of Ramanujan. 1919 - Savilian professor of geometry in Oxford. 1928/9 - at Princeton. 1931 - return to Cambridge. Hardy's main interests were in the area of Diophantine analysis, Riemann zeta functions and the distribution of primes. His only passion besides mathematics was cricket. Hardy was known for his eccentricities, famous for his book *A mathematicians apology.* Opposed the exclusion of Germany and Austria from the scientific community.

Härlen, Hasso 1903–1989. German mathematician and 'Privatgelehrter'.

Hasse, Helmut 1898–1979. Prominent number theorist. In 1917, while on navy duty in Kiel, Hasse attended lectures of Toeplitz at Kiel; 1918 - study in Göttingen. Took courses of Landau, Noether, Hilbert and Hecke. 1921 - PhD and Habilitation. 1922 - lecturer at the University of Kiel; 1925 - professor at Halle; 1930 at Marburg; 1934 - Weyl's successor in Göttingen. There Hasse clashed with hard line Nazi functionaries within the Mathematics Institute. Application for membership of the Nazi party was refused. During the Second World War Hasse was in the navy; after the war he was at first barred from teaching. 1949 - professor at the Humboldt University in East-Berlin; 1950 - professor at the Hamburg University.

Hecke, Erich 1887–1947. Born in Buk, Posen, now Poznan. Studied in Breslau, Berlin and in Göttingen. 1910 - doctorate with Hilbert. Subsequently assistant to Klein and Hilbert. 1912 - Habilitation. 1915 - extraordinary professor in Basel, full professor 1916. 1918 - professor in Göttingen; 1919 in Hamburg. Hecke contributed to a number of topics, such as Hilbert modular functions, Riemann zeta functions, algebraic functions, etc., but his best work was in analytic number theory where he continued work of Riemann, Dedekind and Weber.

Herzberg, Lily Journalist for science of the Berliner Tageblatt.

Heyer, Rud. Deputy notary in Amsterdam.

Heyting, Arend 1898–1980. 1916–1922 - Study at the UvA; 1925 - PhD thesis with Brouwer on *Intuitionistic axiomatization of projective geometry.* 1928 - wrote a prize winning essay on axioms for intuitionistic logic for the Dutch Mathematical Association. 1930 - attended the Erkenntnis Symposium at

Königsberg (now Kaliningrad), representing intuitionism. Heyting also corresponded with Kolmogorov on intuitionistic topics. 1934 publication *Intuitionism and Proof Theory*. 1936 - privaatdocent at the UvA, 1937 - lecturer, 1948 - full professor. Heyting published on intuitionistic mathematics. His *Intuitionism: an Introduction* (1956) was very influential.

Hilbert, David 1862–1943. Gymnasium and university study in his hometown Königsberg (now Kaliningrad), 1885 - PhD with Lindemann. 1886 - privat docent in Königsberg, professor in 1892. 1895 - professor in Göttingen. Hilbert published on a wide variety of topics. 1900 - Hilbert's 23 mathematical problems, including the continuum hypothesis, consistency of arithmetic, Goldbach conjecture and the Riemann hypothesis. Epochal monograph: *Grundlagen der Geometrie* (1899) On the foundational side: the 'Heidelberg lecture' *Über die Grundlagen der Logik und der Arithmetik* (1904), *Axiomatisches Denken* (1918), *Neubegründung der Mathematik (Erste Mitteilung)* (1922), *Die Logische Grundlagen der Mathematik* (1923) *Über das Unendliche* (1925), the 'Hamburg lecture' *Die Grundlagen der Mathematik* (1928). With P. Bernays *Die Grundlagen der Mathematik* (2 vols.) 1934, 1939. Father of *proof theory*, and proponent of *Formalism*.

Hölder, Otto Ludwig 1859–1937. Study engineering at the Polytechnic in Stuttgart - 1876; 1877 - mathematics at the University of Berlin; lectures by Weierstrass, Kronecker and Kummer. 1882 - PhD at the University of Tübingen. 1884 - lecturer at Göttingen, research on the convergence of Fourier series; discovery of the Hölder-inequality. Through Von Dyck and Klein he became interested in group theory. 1890 - professor in Tübingen, contributed to Galois theory. From 1900 interested in philosophical questions.

Hopf, Heinz 1894–1971. 1913 - Study at the Friedrich Wilhelms University in Breslau. Volunteer in World War One. During a fortnight's leave in 1917 Hopf attended a class by Schmidt on set theory which had great influence on him. 1919 - study in Breslau and Heidelberg. 1920 - study in Berlin. 1925 - PhD on the topology of manifolds, with Schmidt and Bieberbach. 1925 - met Emmy Noether and Alexandrov in Göttingen. 1926 - Habilitation in Göttingen. 1928 Hopf and Alexandrov in Princeton. 1935 - publication of 'Alexandrov-Hopf'. 1930 - successor of Weyl in Zürich. Most of his work is on algebraic topology, in the tradition of Brouwer.

Hurewicz, Witold 1904–1956. Born in Poland; studied at the University of Vienna with Hahn and Menger, PhD in 1926. 1927/28 - in Amsterdam with a Rockefeller stipend; 1928–1936 assistant of Brouwer; he worked on higher homotopy groups. Moved to the United States - Institute for Advanced Studies in Princeton, University of North Carolina. 1945 at M.I.T.

Hurwitz, Adolf 1859–1919. 1877 - Study with Klein in Munich, 1878 - Berlin with Kummer, Weierstrass and Kronecker. 1881 - PhD on elliptic modular functions with Klein at the University of Leipzig. 1882- Habilitation in Göttingen, Privatdocent. 1884 - extraordinary professor at Königsberg (now Kaliningrad), where Hilbert and Minkowski were among his students. 1892 -

professor at the Eidgenössische Polytechnikum Zürich. His work was strongly influenced by Klein. He published on a wide variety of mathematical topics, such as the genus of Riemann surfaces, complex function theory, Fourier series and algebraic number theory.

Jaeger, Frans Maurits 1877–1945. Dutch chemist; 1902 - Privatdocent, 1908 - lecturer and 1909–1943 professor at Groningen University.

Jahnke, Paul Rudolf Eugen 1863–1921. German mathematician, professor at Berlin and founder of the Berlin Mathematical Society.

Jongejan, Cor 1893–1968. Classmate of Brouwer's stepdaughter Louise Peijpers. Cor Jongejan moved in with the Brouwer family and eventually became his secretary. Was a licensed assistant in the pharmacy.

Juel, Sophus Christian 1855–1935. Danish mathematician. Studied first at the Technical University, then at the University of Copenhagen. 1885 - PhD on geometry. He taught at the Polytechnic Institute of Copenhagen, 1897 - full professor. His main contributions to mathematics are in the field of projective geometry.

Julia, Gaston Maurice 1893–1978. French mathematician, severely wounded during the first world war. In 1918, at the age of 25, he published his masterwork *Mémoire sur l'itération des functions rationelles*. Later he became professor of mathematics at the Sorbonne.

Kagan, Benjamin Fedorovich 1869–1953. 1887 - Study at the Odessa University, expelled in 1889 for participating in the Democratic Students Movement. In 1892 he received a degree from Kiev University and in 1895 a master's degree from St Petersburg University. 1897–1922 at Novorossysky University, 1917 - full professor. He held more teaching positions and he edited the Journal of Experimental Physics and Elementary Mathematics from 1902 until 1917. 1922 - head of the newly founded Department of Differential Geometry of Moscow State University. He also did research in the field of vector and tensor analysis and he worked on the foundations of geometry and on Lobachewski's geometry. In 1902 he proposed definitions and axioms very different from Hilbert. He wrote a history of non-Euclidean geometry, also a biography of Lobachewski, whose complete works he edited.

Kapteyn, Willem 1849–1927. Professor of mathematics at Utrecht University (1916), brother of the astronomer J.C. Kapteyn.

Kerékjártó, Béla von 1889–1946. 1920 - PhD from Budapest University. 1922 - privatdocent Szeged; 1925 - full professor at Szeged. 1938 professor at Budapest. He worked and published on topology. Known for his book *Vorlesungen über Topologie* (Lectures on Topology) 1923.

Kerkhof, Karl 1877–1945. 'Regierungsrat' during the interbellum, head of the 'Reichszentrale für naturwissenschaftliche Berichterstattung', which had to counteract the boycott of Germany by the international scientific community.

Klaarenbeek, J.J. 1884–? Klaarenbeek was mayor of the village of Blaricum (Brouwer's home town) from 1922 to 1946.

Kleene, Stephen Cole 1909–1994. Studied at Amhurst College; 1934 - PhD with Church. *A Theory of Positive Integers in Formal Logic.* Kleene taught at Princeton; 1935 - moved to University of Wisconsin; 1948 - full professor in 1948. He shaped the theory of algorithms and recursive functions as we know it now. With Turing, Gödel and Church he belongs to founding fathers of the subject. He also contributed to intuitionism; he introduced the realizability interpretation. His *Introduction to Metamathematics* (1952) became *the* textbook on the subject.

Klein, Felix Christian 1849–1925. 1865 - Study mathematics and physics at Bonn university. 1868 - PhD on line geometry and its application to mechanics with Plücker. 1871 - lecturer at Göttingen; 1872 - professor at Erlangen at the age of 23. 1875 - professor at the Technische Hochschule at Munich. Among his students were Von Dyck, Hurwitz, Runge and Planck. 1880 - professor of geometry at Leipzig. 1886 - professor at Göttingen where he taught until his retirement in 1913. He is best known for his work in non-Euclidean geometry, for his work on the connection between geometry and group theory, and for results in function theory. His Erlanger Programm from 1872 studies the properties of spaces that are invariant under a given group of transformations; this program gives a unified approach to geometry. He turned Göttingen into a first rate mathematical research centre, and made the Mathematische Annalen the leading mathematical journal.

Kluyver, Jan Cornelis 1860–1932. Studied at Delft; 1892–1930: professor of algebra and analysis at Leyden University. Received an honorary doctorate from Groningen University. Kluyver participated with Korteweg and Schoute in the modernization of mathematics in the Netherlands.

Kneser, Hellmuth 1898–1973. Son and father of well-know mathematicians. 1916 - Study in Breslau, where his father was professor of mathematics. Subsequently in Göttingen. 1921 - dissertation with Hilbert, on the mathematics of quantum mechanics: *Untersuchungen zur Quantentheorie.* 1925 - professor in Greifswald. 1937 - professor in Tübingen. Published on topology, analytic functions, groups, non-Euclidean geometry, differential geometry of manifolds.

Koebe, Paul 1882–1945. 1900 - enrolled in Kiel, then in Berlin. PhD with Hermann Schwartz. 1907 - Habilitation at Göttingen. 1910 - extraordinary professor at Leipzig University, 1920 - professor at Jena; 1926 back in Leipzig. His mathematical oeuvre consists mainly of (lengthy) papers on complex function theory. He was famous for his proof of the uniformization theorem.

Kohnstamm, Philip Abraham 1875–1958. Dutch physicist, obtained his PhD with Van der Waals. After the retirement of Van der Waals he became full professor at the UvA. Later his interest shifted towards philosophy and pedagogy. Kohnstamm was an editor of the *Tijdschrift voor Wijsbegeerte* (Journal of Philosophy).

Kok, J. One-time Rector Magnificus of the UvA.

Koppers, J.F. ?–1944. Janitor of the Mathematical Institute of the UvA. He organized from the mathematical institute activities of the resistance during the Second World War. He was arrested and taken to a German concentration camp; he died in Neuengamme.

Korteweg, Diederik Johannes 1848–1941. Studied in Delft. 1869–1881 math. teacher. 1878 - PhD at UvA with Van der Waals. His work in mainly in the field of applied mathematics. Edited the Collected works of Huygens. He is remembered for the 'Korteweg-de Vries equation' on solitary waves. He was the father of a new generation of Dutch mathematicians, including L.E.J. Brouwer.

Lakwijk-Najoan, Johanna A.L. van Pharmacist, bought Brouwer's pharmacy in Amsterdam in 1965.

Landau, Edmund Georg Hermann 1877–1938. Studied at the University of Berlin; 1899 - PhD with Frobenius for a dissertation on number theory. 1901 - Habilitation on Dirichlet series. 1899–1909 Privatdocent at the University of Berlin. 1909 - full professor at Göttingen as successor of Minkowski. 1933 - Landau, as a Jewish mathematician, was excluded from all university duties, but he was allowed to teach at Groningen University in the Netherlands until his official retirement in 1934. He returned to Germany where, in 1938, he died of a heart attack. His main work was in analytic number theory and the distribution of primes.

Lebesgue, Henri 1875–1941. Study at the École Normale Supérieure from 1894 to 1897. 1899–1902 teacher at the Lycée centrale at Nancy. In 1901 he formulated the theory of measure and defined the Lebesgue integral, thus generalizing the Riemann integral. 1902 - PhD at the Faculty of Science in Paris, *Intégrale, Longeur, Aire* (Integral, Length, Area). Subsequently 'maître de conférence' in mathematics at the Faculty of Science in Rennes. In 1910 - same position at the Sorbonne, 1918 - full professor. 1921 - professor of mathematics at the Collège de France.

Lévy, Paul 1886–1971. Studied at the École Polytechnique in Paris. After graduating, he continued studies at the École des Mines and attended courses at the Sorbonne of Darboux, Picard, and Hadamard. 1912 - thesis on functional analysis with Picard, Poincaré and Hadamard. 1913 - professor at the École des Mines; 1920 - professor of analysis at the École Polytechnique in Paris. His main work was in functional analysis, probability theory, but also on partial differential equations, series, Laplace transforms and geometry.

Loor, Barend de 1900–1962. Born in the Netherlands. Studied in Pretoria, South Africa. 1924 - PhD with Brouwer. He published, together with Brouwer, an intuitionistic proof of the fundamental theorem of algebra. 1925 - teacher at Pretoria university, 1939 - full professor.

Lorentz, Hendrik Antoon 1853–1928. Dutch physicist, studied at Leyden University. 1875 - PhD for a dissertation on a refinement of Maxwell's theory of electromagnetism. 1878–1912 - full professor at Leyden; subsequently honorary professor and director of research at the Teyler Institute in Haarlem.

1902 - Nobel Prize for his mathematical theory of the electron, together with his student Pieter Zeeman. He is also famous for the relativistic so-called FitzGerald-Lorentz contraction.

Mannoury, Gerrit 1867–1956. Largely autodidact in mathematics; 1903 - privaat docent at the UvA; 1917 - extraordinary professor; 1918 - full professor. He early lectures were on the foundations of mathematics, as a professor he taught a mixture of topics, including mechanics. 1946 - honorary doctorate at UvA. In 1906 appeared his *Methodologisches und Philosophisches zur Elementarmathematik*. He was the driving force behind *Significs*.

Mauve, Rudolf 1878–1963. Fellow student and friend of L.E.J. Brouwer. Studied at the UvA and later moved to the Technical University of Delft to study architecture. He was the son of the famous painter Anton Mauve (1838–1888). Designed Brouwer's hut.

Menger, Karl 1902–1985. 1920 - Study at University of Vienna, switched from physics to mathematics. 1924 - doctorate on dimension theory with Hahn. Introduced and studied the notions of 'curve' and 'dimension' independently of Brouwer and Urysohn. 1925 - Brouwer's assistant. 1927 - professor of geometry at the University of Vienna, member of the Vienna Circle. Left Austria in 1938; became professor at Notre Dame University, where he attempted to set up a Mathematical Colloquium as influential as the Vienna Circle. 1948 - professor at the Illinois Institute of Technology. In America he broadened his interest in mathematics to the fields of hyperbolic geometry, probabilistic geometry and the algebra of functions.

Michels, A.M.J.F. 1889–1969. Professor at the UvA in Physics. Specialist in thermodynamics. Founder of the Van de Waals Laboratory at Amsterdam.

Minnaert, Marcel Gilles Jozef 1893–1970. Belgian-Dutch astronomer, studied biology at Gent University; 1914 - doctorate. Was active in the Flemish University at Gent during the war. Had to flee in 1918. Further study at Utrecht University; 1925 - doctorate in physics. 1937–1963 –professor of astronomy at Utrecht University and director of the astronomical observatory. He became famous for his pioneering work on photometric analysis of spectral lines. One of his other merits was the popularization of physics and astronomy. He is the author of the books *De natuurkunde van het vrije veld* (The Nature of Light and Color in the Open Air).

Mises, Richard von 1883–1953. Born in Lvov, then Austria. 1901 - Study at the Technische Hochschule in Vienna; 1907 - Hamel's assistant; 1908 - PhD at Vienna; 1908 - Habilitation at Brno. 1909–1918 - professor of applied mathematics at Strasbourg. 1913 - lectured on aircraft design. Joined the Austrian-Hungarian air-force as a test pilot and instructor. 1918 - professor of hydrodynamics and aerodynamics at the Dresden. 1919 - University of Berlin, director of the novel Institute of Applied Mathematics. In 1921 he founded the journal *Zeitschrift für Angewandte Mathematik und Mechanik* and edited it. In 1933 he left Germany for the University of Istanbul; 1939

- emigration to the United States, professor at Harvard. His work remained mainly in the field of applied mathematics, but he was also interested in philosophy. Published *Kleines Lehrbuch des Positivismus. Einführung in die empiristische Wissenschaftsauffassung* 1939. Introduced the notion of 'collective' in probability theory.

Mittag-Leffler, Magnus Gösta 1846–1927. 1865 - Study at the University of Uppsala; 1872 - PhD. 1873 in Paris, heard lectures by Hermite on elliptic functions. 1875 - visit to Berlin to attend lectures by Weierstrass, which inspired Mittag-Leffler's research. 1876 - professor at the university of Helsinki; 1881 - professor at Stockholm University. Founded the international journal *Acta Mathematica*. Active in mathematical analysis, analytic geometry and probability theory. Best known for the analytic representation of a one-valued function, a study resulting from Weierstrass' lectures. He supported Cantor's set theory and he proposed some general topological notions on infinite point sets based on Cantor's work.

Marston Morse, H.C. 1892–1977. Study at Harvard; 1917 - PhD with Birkhoff, for his thesis *Certain Types of Geodesic Motion of a Surface of Negative Curvature*. After serving in France during World War I he worked at Harvard; 1920–1925 Cornell University; 1925–1926 Brown University; 1935–1962 - Institute of Advanced Study at Princeton. Morse developed variational theory and its applications in physics, now known as Morse theory. He did major publications on variational theory and topology. He served on various commissions and councils to improve mathematics in the United States and he did important work for the International Congress of Mathematicians.

Myhill, John R. 1923–1987. American mathematician and logician. 1949 - PhD at Harvard University with Quine and Loomis for his dissertation *A Semantically Complete Foundation for Logic and Mathematics*. Specialist in recursion theory and intuitionistic logic. Professor at Buffalo - 1966. Worked on choice sequences, in particular the creating subject and Kripke's Schema.

Nagy, Julius (Gyulia) von Sz. Hungarian mathematician, topologist.

Neumann, John von 1903–1957. 1921 - Study chemistry at the University of Berlin, 1923 - at Zurich, 1926 - diploma chemical engineering. On the side he became an excellent mathematician, and took the final exams at Budapest. 1926 - PhD on set theory at the University of Budapest. 1926–1927 - study at Göttingen with Hilbert. He lectured at Berlin and Hamburg from 1926 to 1930. 1930 professor at Princeton University, on the invitation of Veblen. 1933 - one of the six mathematics professors (together with Alexander, Einstein, Morse, Veblen and Weyl) at the newly founded Institute for Advanced Study in Princeton. 1933 - editor of the *Annals of Mathematics*; 1935 of the *Compositio Mathematica*. 1932 his famous *Mathematische Grundlagen der Quantummechanik*. Von Neumann is known for the wide variety of different scientific topics on which he published in several journals. Famous for Von Neumann set theory (von Neumann hierarchy, ordinals). The first mathematician to grasp Gödel's incompleteness theorem, and to see the generalisation the second Gödel theorem. Among many other things, he put game theory on the map of mathematics, took part in the development of logical design for computers, introduced, what later was called, Von Neumann algebras, participated in the research for atomic weapons.

Newman, Maxwell Herman Alexander 1897–1984. 1915 - Study – St John's college, Cambridge. 1922–23 in Vienna; 1923 - Fellow of St John's College. 1927 - lecturer at Cambridge. 1928–29 and 1937–38 - visit to Princeton. 1939 - Fellow of the Royal Society in recognition for his contributions to combinatory topology, Boolean algebra and mathematical logic. During World War II he worked, in cooperation with Turing, at the Government Code and Cipher School. After the war he was appointed to the Fielden Chair at Manchester University. His main work was in the field of combinatorial topology on which topic he published a series of papers. In computer science he is known for Newman's Lemma on reductions and the Church-Rosser property.

Nielsen, Jakob 1890–1959. Danish mathematician; born on the island of Alsen in Schleswig, German at the time, now Danish. 1908 - study at the University of Kiel. Dehn introduced him to the new mathematical subjects topology and group theory; 1913 - dissertation. During World War I Nielsen served in the military navy and army. 1919 - study in Göttingen; 1920 he followed Hecke to Hamburg as an assistant; 1921 - professor at Breslau University. In 1921 Nielsen became a Danish citizen. 1925 - professor of theoretical mechanics at Copenhagen. 1951 - successor of Harald Bohr as professor of mathematics. He is known for his work on group theory in the combination with topology; but he also contributed to many other fields of mathematics.

Noether, Emmy Amalie 1882–1935. 1900–1902 - mathematics study in Erlangen; 1903–1904 at Göttingen. 1907 - doctorate with Gordan. The Habilitation was at the time denied to females. She worked as an assistant to her father, a mathematics professor at Erlangen, at the same time turning towards abstract mathematics à la Hilbert. 1909 - joined the 'Deutsche Mathematiker Vereinigung'. 1915 - invited to Göttingen by Hilbert and Klein. 1919 - Habilitation by special permission. There she proved an important relation between symmetries in physics and conservation principles. After 1919 she published on modern algebra (1921: theory of ideals and rings). 1924 - Van der Waerden studied with Noether; 1927 collaboration with Hasse and Brauer. Being Jewish, she left for America in 1933. Visiting professor at Bryn Mawr College, USA. She had a tremendous influence on the development of algebra in the twentieth century.

Noordhoff, J. Dutch publisher, strong reputation in mathematics. Firm founded in 1858 in Groningen.

Ornstein, Leonard Salomon 1880–1941. Dutch physicist; 1908 PhD - with Lorentz. 1914 - full professor at Utrecht University. He is the founder of the Netherlands' Physical Society. Briefly associated with the Signific Circle.

Ostrowski, Alexander 1893–1986. Ostrowski had a highly irregular education, he attended the School of Commerce in his hometown Kiev, studying mathematics at the side. 1912 - mathematics study in Marburg with Hensel. During World War I he was interned as a hostile foreigner, but he could continue his studies while in captivity. 1918 - study at Göttingen; 1920 - PhD at Göttingen, (Hilbert's eighth problem) with Hilbert and Landau. 1922 - habilitation in Hamburg. 1923 - lecturer at Göttingen. 1925–1926 - visit to Britain. Subsequently professor at the University of Basel. In the 1960's he travelled to the USA as a visiting lecturer. Ostrowski published around 275 papers on algebra, number theory, topology. numerical analysis and many other subjects. His contemporaries considered him as one of the last universal mathematicians.

Pannekoek, Antonie 1873–1960. Dutch astronomer (and prominent Marxist). Studied at Leyden University. 1902 - PhD with H.G. van de Sande Bakhuyzen. 1918 - lecturer at the UvA; 1925 - extraordinary professor, 1932 - full professor; 1942 suspended by the German authorities. He was very well up to date on the development of the new quantum theory, which he applied in astronomy. Pannekoek was one of the founders of the theory of stellar spectra, he is also the father of Dutch astrophysics.

Planck, Max Karl Ernst Ludwig 1858–1947. 1874 - Study in Munich. 1877 - move to the University of Berlin to study with Weierstrass, Helmholtz and Kirchhoff. He read and studied Clausius, being very much impressed by the absolute nature of the second law of thermodynamics. 1879 - PhD *On the Second Law of the Mechanical Theory of Heat*. 1880 - Habilitation on entropy; privatdocent at the University of Munich. 1885 - extraordinary professor of theoretical physics at Kiel and the publications on thermodynamics. 1888 - extraordinary professor at the University of Berlin, 1892 - full professor, and director of the Institute for Theoretical Physics. 1900 - radiation formula, thereby introducing the quanta of energy and rejecting his belief in the absoluteness of the second law of thermodynamics and accepting instead Boltzmann's interpretation that it was a statistical law. 1918 - Nobel Prize for Physics in for his achievements. He also was Secretary of the Science section of the 'Prussian Academy of Sciences' and he was active in the Kaiser Wilhelm Gesellschaft.

Poincaré, Henri 1854–1912. 1873–1875 - Study at the École Polytechnique. Worked as mining engineer while completing his doctoral thesis. 1879 - PhD with Charles Hermite; his thesis was on differential equations. 1879–1881 taught at the University of Caen; 1881 - University of Paris; 1886 - Sorbonne. He is considered as one of the greatest geniuses of all time and he published on many topics, physical (celestial mechanics, fluid mechanics, special relativity, quantum theory), mathematical (automorphic functions, topology, analytic functions of several complex variables) and philosophical. He also published for a greater public wonderful expositions of contemporary advances, such as *La Science et l'Hypothèse, Science et Méthode, La Valeur de la Science*.

He had strong opinions on foundational matters like the role of logic and the role of intuition in mathematics, which influenced Brouwer. He was critical of set theory and the axiomatic method.

Popper, K.R. 1902–1994. Austrian-born British philosopher of science. Popper was professor at the London School of Economics. Study at the University of Vienna, 1928 - PhD in philosophy. 1930–1936 - school teacher. 1934 - *Logik der Forschung* (The Logic of Scientific Discovery) 1937 - emigration to New Zealand - Canterbury University College. 1946 he moved to London; 1949 professor at University of London. 1965 - knighthood; 1976 Fellow of the Royal Society in 1976.

Pye, D.R. Provost at University College, London.

Radley Winifred Assistant secretary at the University College in London.

Reymond, Arnold 1874–1958. Swiss philosopher. Reymond studied theology at Lausanne. 1908 - PhD at the University of Geneva. 1912 - professor at the University of Neuchâtel; 1925–1944 - professor of philosophy at Lausanne. He published on subjectivism and the problem of knowledge (especially of religious knowledge), and on 'spiritual philosophy', but also on the philosophy of mathematics and logic: *Logique et mathématique, Essai historique et critique sur le nombre infini* in the *Revue de Métaphysique et de Morale* (1911). In 1932 appeared his *Les principes de la logique et la critique contemporaine*.

Riesz, Frigyes 1880–1956. Studied at Budapest, Göttingen and Zürich. 1902 - PhD at the University of Budapest. Specialist in functional analysis, measure theory. 1911 - professor at Kolozsvàr, which became Rumanian after 1920; the University moved to Szeged, where in 1922 he set up, together with Haar, the Bolyai Mathematical Institute. 1945 - professor at the University of Budapest.

Rosenthal, Arthur 1887–1959. Studied at the University of Munich; 1909 - PhD. 1912 - Habilitation and Privatdocent; 1920 - Professor at Munich. 1922 - professor at Heidelberg until his forced retirement in 1935. Was put in the concentration camp Dachau (1938). In 1939 he emigrated via the Netherlands to America. Was professor in Michigan, New Mexico, Purdue.

Saxer, Walter 1896–1974. Swiss mathematician, studied at the ETH in Zurich, PhD in 1923. From 1927 to 1966 he worked as professor of mathematics at the ETH, teaching geometry and analysis. Well-known for his books on insurance mathematics.

Schmidt, Erhard 1876–1959. Studied at the local university in his hometown Dorpat (Estonia) before moving to the University of Berlin. 1905 - PhD with Hilbert on integral equations. 1906 - Habilitation in Bonn; after that he held positions in Zurich, Erlangen and Breslau. 1917 - professor at the University of Berlin as successor of Schwartz. Carathéodory was appointed in 1918 as the second full professor, but left after one year; Schmidt nominated Brouwer, Weyl, and Herglotz (in that order) as Carathéodory's successor. The position was eventually filled in 1921 by Bieberbach. Schmidt promoted the founding

of an Institute of Applied Mathematics. Richard von Mises accepted the
two positions of Director of the Institute and of its associated chair. After
World War II Schmidt was appointed Director of the Mathematics Research
Institute of the German Academy of Science and he became the first editor
of the *Mathematische Nachrichten*, a journal that he co-founded in 1948. His
main interest was in integral equations and Hilbert space. His ideas were to
lead to the geometry of Hilbert spaces.

Schoenflies, Arthur Moritz 1853–1928. Studied at the University of Berlin,
1877 - PhD in 1884 - Habilitation. 1892 - professor in applied mathematics at
Göttingen; 1899 - professor at Königsberg (now Kaliningrad); 1911 professor
at Frankfurt. Being at first active in the field of geometry and kinematics,
he became known for his work on group theory (applied to crystallography).
He specialized in set theory and topology. His two volumes *Die Entwickelung
der Lehre von den Punktmannigfaltigkeiten I* (1900) and *II* (1908) were very
influential. These books (Bericht) contained omissions and errors, which were
discovered and corrected for the 1914 edition by Brouwer.

Scholz, Heinrich 1884–1956. Studied theology and philosophy at the universities
of Berlin and Erlangen. 1910 - Habilitation at Berlin. 1917 - professor
at Breslau, teaching philosophy of religion. 1924 to 1928 - study of exact
science and logic. Scholz started a school for logic and the foundations of
mathematics at the University of Münster. His chair became the first in
Germany on this topic. Scholz was a Platonist.

Schouten, Jan Arnoldus 1883–1971. Studied electrical engineering at the Technical
University Delft; was an electrical engineer for some years, then moved
to Leyden University. 1914 - PhD thesis (on tensor analysis) and professor
of mathematics at Delft. From 1948 Schouten was professor of mathematics
at the UvA without, however, teaching there. He was director of the Mathematical
Center at Amsterdam for five years. In his own research work he
applied tensor analysis to Lie groups, relativity, unified field theory and systems
of differential equations. He also made, independently of Levi-Civita,
the discovery of connections in Riemannian manifolds.

Severi, Francesco 1879–1961. Born in Arezzo, Severi studied at the University
of Turin; 1900 - PhD on enumerative geometry with Corrado Segre. 1904 -
professor of Projective and Descriptive Geometry at Parma; 1905 - professor
at Padua. During World War I he served in the artillery. 1922 - professor at
the University of Rome. His most important contributions are to algebraic
geometry, rigorizing the subject. After working on enumerative geometry,
Severi turned to birational geometry of surfaces. He introduced many new
concepts in geometry, like the notion of algebraic equivalence.

Sierpínski, Waclaw 1882–1969. 1899 - Study at the University of Warsaw. 1903
- gold medal for a paper on number theory. After working as a teacher
for some time, he went to Krakow; 1908 - PhD and appointment at the
University of Lvov. 1909 - Sierpínski moved to set theory. At the beginning
of World War I he was interned, but by intervention of Luzin and Egorov he

was released and spent the remaining war years in Moscow. After the war he returned to Lvov. 1919 - professor of mathematics at the University of Warsaw. 1920 - founded the journal *Fundamenta Mathematicae*. From then on Sierpínski worked mainly on set theory (in which he made contributions to the continuum hypothesis and to the axiom of choice), on point set topology and on functions of a real variable. During World War II he continued to work in the Underground Warsaw University; in 1944 his house, his library and his personal letters were burnt. He retired in 1960 as professor, after the publication of more than 720 papers and some 50 books.

Snijders, Cornelis Jacobus 1882–1939.   Supreme commander of the Dutch armed forces during the First World War.

Sommerfeld, Arnold Johannes Wilhelm 1868–1951. 1886 - Study at University of Königsberg (now Kaliningrad). 1891 PhD with Lindemann. 1893 - move to Göttingen; 1894 - Klein's assistant, who at that time was involved in applying the theory of functions of a complex variable to a range of physical topics. In 1895 - Habilitation with Klein at the University of Göttingen, Privatdocent in mathematics. 1897–1910 - Klein-Sommerfeld *Die Theorie des Kreisels*. 1897 - professor at the mining academy of Clausthal. 1898–1926 - Sommerfeld edited volume V of the *Encyklopädie der mathematischen Wissenschaften*. 1900 - professor of mechanics at the Technische Hochschule of Aachen; 1906 - professor of theoretical physics at the University of Munich with Debye, Ewald, Pauli, Heisenberg and Bethe as students. In Munich Sommerfeld worked on atomic spectra and quantum theory, improving Bohr's theory.

Springer, Ferdinand 1881–1965. Head of the Springer Verlag. Managed the medical section of the firm. In 1924 the exact sciences were incorporated. In spite of the nazi policies, the firm could survive under the supervision of a caretaker, T. Lange. In 1942 Springer had to withdraw from the firm. After surviving in hiding, he could take up his position again in 1945.

Springer, Julius 1817–1877. German publisher, founder of the 'Springer Verlag' (1842). His grandson Julius took over the branch of the exact sciences after World War II.

Stomps, Th J. 1885–1973. 1903 - Study in Amsterdam; PhD - 1910; extraordinary professor of biology (botanics) at the UvA; 1920 full professor and director of the Hortus Botanicus in Amsterdam. After the war criticized for his alleged wartime conduct; 1946 honorable discharge. 1956 rehabilitation.

Study, Eduard 1862–1930. 1880 - Mathematics study at the universities of Jena, Strasbourg, Leipzig and Munich. 1884 - PhD at Munich. 1885 - lecturer at the University of Leipzig; 1888 - appointment at University of Marburg; 1893–1894 - visit to the United States where he taught mainly at John Hopkins University in Baltimore. 1894 - extraordinary professor at Göttingen; 1897 professor at Greifswald; 1904 - professor at Bonn University. He became a leader in the geometry of complex numbers and he worked in invariant theory. He was also an amateur biologist.

Tarski, Alfred 1902–1983. Polish-American mathematician and logician of Jewish descent. 1918 - study at the University of Warsaw; switched from biology
to mathematics under the influence of Sierpínski, Łukasiewicz and others.
1924 - PhD with Leśniewski. Tarski turned to set theory, partly in cooperation with Banach (the Banach-Tarski paradoxical decomposition of the
sphere). He taught at the University of Warsaw; 1930 visited the University
of Vienna where he met Menger and Gödel. 1933 - *the concept of truth in
formalized languages*. 1939 - visited Harvard University. After several temporary positions, he obtained in 1942 a permanent post at Berkely, becoming
full professor there in 1949. Tarski moved around a great deal. He made major contributions to logic, set theory, topology and geometry; he was one of
the fathers of modern model theory.

Teubner, B.G. German publisher of scientific books and journals.

Tornier, W.H. Erhard 1894–1982. Study at Breslau, Berlin, Marburg. 1922 -
PhD with Hensel (Marburg). German Nazi, who, in 1934 after the expelling
of all Jewish scientists, took over control of the Faculty of Mathematics of
the University of Göttingen for a short period. 1936–39 - professor in Berlin.
He tried to persuade Brouwer to accept a chair at Göttingen.

Turnbull, Herbert Western 1885–1961. Turnbull studied at Trinity College,
Cambridge. 1909 - teacher at St Catharine's College, Cambridge; 1910 - lecturer at the University of Liverpool; 1911–1915 teacher at St Stephen's College
in Hong Kong. 1919–1926 - Fellow at St John's College, Oxford; 1921 - Regius
Professor of mathematics at the University of St Andrews. He published on
the theory of invariants; also active in the history of mathematics. After his
retirement he published two volumes of the *Correspondence of Isaac Newton*.

Urysohn, Pavel (Paul) Samuilovich 1898–1924. 1915 - Study at the University of Moscow, where he soon switched from physics to mathematics. Graduation in 1919, after which he continued his studies for his doctorate. 1921 -
Habilitation on a thesis on integral equations. 1922 - Assistant professor at
the University of Moscow. Inspired by Egorov he turned to topology. 1922
- found (independent of Brouwer 1913) an intrinsic definition of dimension.
Visiting Göttingen in 1923, he noted an error in Brouwer's paper. 1924 -
Urysohn and Alexandrov again visited Göttingen, and Hausdorff in Bonn.
That summer they visited Brouwer in Blaricum. From there they went on to
France, where they stayed in Brittany. Urysohn lost his life, when going for
a swim in rough weather. Brouwer and Alexandrov supervised the editing of
his posthumous papers.

Uven, M.J. van 1878–1959. Professor in mathematics at Wageningen University.

Vahlen, Karl Theodor 1869–1945. Study at the University of Berlin; 1893 -
PhD on additive number theory; 1897 - Privatdocent at the University of
Königsberg (now Kaliningrad); 1904 - professor. 1911 - professor of mathematics at the University of Greifswald, but he was discharged for anti-
republican actions. After the Nazi's came to power in 1933 he returned to
that university (after a short professorship at Vienna). 1934 - appointed at

the University of Berlin; served as 'kommissarischer Präsident der Akademie'. Was *Ministerialdirektor* of the ministry for education (mathematics). He was an active Nazi. He published on applied matters, in particular ballistics; his *Abstrakte Geometrie* was well-known.

Veblen, Oswald 1880–1960. American mathematician of Norwegian descent. Veblen studied at the universities of Iowa, Harvard and Chicago. 1903 - PhD with Eliakim Moore at the University of Chicago - *A System of Axioms for Geometry*. 1905 - publication *Theory on Plane Curves in non-metrical analysis situs*. 1905–1932 - professor at Princeton University. He taught at Oxford as part of an exchange with G.H. Hardy and in 1932 he spent time in Germany, lecturing at Göttingen, Berlin and Hamburg and in the same year he helped organize the Institute for Advanced Studies in Princeton, choosing Alexander, Einstein, Von Neumann and Weyl as original institute members. His main interest was in all areas of geometry, especially its foundations; he made Princeton one of the leading centers for topology research. After the publication of Einstein's general theory of relativity, he turned his attention to differential geometry, applying it to relativity theory.

Vietoris, Leopold 1891–2002. Austrian topologist. Studied at the Technical University of Vienna. Wounded and Italian P.O.W. during World War I. 1919 - PhD with Escherich and Wirtinger at the University of Vienna. 1921 - Habilitation, Vienna. He spent three semesters with L.E.J. Brouwer in Amsterdam, together with Aleksandrov and Menger, where he started working on algebraic topology–e.g. Vietoris homology, the Mayer-Vietoris sequence. 1927 - associate professor at Innsbruck; then full professor at Vienna. In 1930 full professor in Innsbruck.

von Dyck, Walther Franz Anton 1856–1934. Von Dyck studied at Munich, meanwhile spending time at the universities of Berlin and Leipzig. 1879 - PhD with Klein. 1880 - Klein's assistant in Leipzig; 1882 - Habilitation. 1884 - Professor at the Munich Polytechnikum (later Technische Hochschule); 1900 - director. He also played an important role in the creation of the German Museum of Natural Science and Technology. Furthermore he was active in the publication of the complete works of Kepler. His contributions to function theory, group theory, topology and potential theory were recognized as innovative.

Vries, Hendrik (Hk.) de 1867–1954. De Vries studied with Korteweg at the UvA. 1901 - PhD with Korteweg. 1905 - professor in Delft; 1906 - professor of mathematics at the UvA. Emigrated before the Second World War to Israel. A gifted teacher, wrote a number of text books, published also on the history of mathematics.

Waals jr, Johannes Diderik van der 1873–1971. Dutch physicist and son of the Nobel laureate in physics J.D. van der Waals; 1903 - professor of physics at Groningen University; 1908 - successor of his father as professor of theoretical physics at the UvA. In later years his interest shifted towards the foundations of physics and other philosophical topics.

Waerden, Bartel Leendert van der 1903–1996. 1919–1926 - Study at the universities of Amsterdam and Göttingen (at Göttingen with Emmy Noether). Strongly influenced by Emmy Noether's abstract algebra. 1926 - PhD with Hk. De Vries at UvA. 1927 - assistant/privat docent in Göttingen, 1928 - Habilitation; professor in Groningen. 1929 - visiting professor in Göttingen, 1931 - professor in Leipzig. 1948 - visiting professor at Johns Hopkins, 1948 - professor at UvA, 1951 - professor at ETH Zürich. He was the author of the extremely influential book, *Moderne Algebra*. Made important contributions to algebraic geometry. Later publications on the history of mathematics.

Weitzenböck, Roland 1885–1955. Austrian mathematician. 1910 - PhD Vienna. During 1911/1912 he studied at Göttingen and habilitations at Vienna. He taught mathematics in Graz and after World War I he was appointed extraordinary professor at Prague, where he subsequently was promoted to ordinary professor. 1921 - professor at UvA. During World War II he gave up his Dutch nationality, and took the German one. Was dismissed after the war. Renown specialist in the theory of invariants.

Welter, Kees Former student of Brouwer, who emigrated to South-Africa.

Went, F.A.F.C. 1863–1935. Dutch biologist; studied at the UvA; 1886 - PhD. He became professor of biology at Utrecht University where he initiated a study of plant hormones. He was director of the Botanical Gardens.

Weyl, Hermann Klaus Hugo 1885–1955. Study: Munich and Göttingen; 1908 - PhD at Göttingen with Hilbert. Privatdocent at Göttingen; philosophically influenced by Husserl. 1913 - professor at ETH Zürich. 1930–1933 - professor at Göttingen; 1933–1952 appointment at the Institute of Advanced Study at Princeton. Renown for his *Die Idee der Riemannschen Fläche*. Important contributions to topology, topological groups, constructive mathematics; significant contributions to the field of mathematical physics. He attempted to unite the geometry of general relativity with electromagnetism; he also applied group theory to quantum mechanics. Famous are his *Space, Time, Matter* and *The Theory of Groups and Quantum Mechanics*. It is, in our context, worth mentioning that he was involved in the foundational debate during the 1920s and that he, in that period, agreed with Brouwer's intuitionistic view. He published a monograph *Das Kontinuum* (1918), and a much praised book *Philosophy of Mathematics and Natural Science* (1928, 1949 (English)).

Whitehead, John Henry Constantine 1904–1960. British mathematician, nephew of Alfred North Whitehead. He studied at Oxford. 1920 - followed Veblen to Princeton, and worked there on differential geometry and on topology. 1928 - return to Oxford. 1930 - PhD at Princeton. His joint work with Veblen led to the classic *The Foundations of Differential Geometry*. Towards the end of his three year stay at Princeton he worked with Lefschetz on topology. After his Princeton period Whitehead returned to Oxford and during the war he worked for the Admiralty in London and at Bletchley Park with the codebreakers. 1947 - professor at Oxford. He was instrumental in developing homotopy theory.

Whittaker, John Macnaughten 1905–1984. Born in Cambridge as the son of the mathematician Edmund Taylor Whittaker. 1920 - study at Edinburgh University; 1923 - Trinity College. PhD in the late 20s from Edinburgh University, where he worked as lecturer and where his father was professor of mathematics. 1933 - professor at the University of Liverpool. Served in Egypt during World War II. 1945 - return to Liverpool. In 1953 he became Vice-Chancellor of Sheffield University. His important work in mathematics was on complex analysis, extending results of his father. He wrote three important books on complex functions.

Wijdenes, Pieter 1872–1972. Self-made Dutch mathematician, high school teacher and author of acclaimed Dutch textbooks on mathematics.

Williams, W.L.G. (Lloyd Williams) 1888–1976. 1921 - PhD from the University of Chicago for a thesis on modular forms. 1924 professor at the McGill University, where he remained until his retirement in 1954. In 1936 Williams and G. de B. Robinson founded the Canadian Mathematical Society. Williams was instrumental in founding the Canadian Mathematical Congress; he was its first Treasurer from 1945 to 1965.

Wilson, Wilfrid 1928 - PhD at the UvA *Afbeeldingen van Ruimten*, (Mappings of Spaces), with L.E.J. Brouwer. Around 1950 he was teacher of mathematics in the USA.

Wirtinger, Wilhelm 1865–1945. Studied at the University of Vienna 1887 - PhD; 1889 - Habilitation. During his Vienna years he also studied at Berlin and Göttingen, where he was strongly influenced by Klein. 1895 - professor at Vienna, 1896 - at the University of Innsbruck. 1905 - return to the University of Vienna. Known for his work on the general theta function. He wrote a great number of important papers on function theory, geometry, algebra, number theory, and the theory of invariants. He published also on Einstein's theory of relativity and even on rainbows.

Wolff, Julius 1882–1945. Fellow student of L.E.J. Brouwer; 1908 - PhD with Korteweg; *Dynamen van duale vectoren*. (Dynams, considered as dual vectors). 1917 - professor of mathematics at Groningen University. 1922 - professor at Utrecht. Being a Jew he was deported to the concentration camp Bergen-Belsen, where he died.

Woude, W. van der 1876–1974. Dutch mathematician. Study in Groningen. 1908 - PhD with P.H. Schoute. 1916 - professor at Leyden University.

Zeeman, Pieter 1865–1943. 1885 - Study at Leyden University, a student of Kamerlingh Onnes and Lorentz. 1890 - assistant to Lorentz. 1893 - PhD, followed by a stay in Strasbourg; 1894 - return to Leyden as Privatdocent. 1896 - discovery of the 'Zeeman-effect', the splitting of spectral lines in an inhomogeneous magnetic field. 1897 - lecturer at the UvA; 1900 - extraordinary professor; 1908 - full professor after the retirement of Van der Waals. At the same time he became director of the Physics Laboratory. A new laboratory was set up for him in 1923, named Zeeman laboratory in 1940, now Van der Waals - Zeeman laboratory. His main theme of investigation was optical phenomena and the influence of magnetism on the nature of light radiation. He received several honorary doctorates; 1902 - Nobel Prize in 1902, together with H.A. Lorentz.

List of Letters

1901-12-15 from - Scheltema

1903-05-23 to - Scheltema

1903-08-09 to - Scheltema

1903-11-15a to - Scheltema

1904-01-18 to - Scheltema

1904-07-04 to - Scheltema

1905-05-13 from - Korteweg

1906-09-07a to - Korteweg

1906-09-07b to - Scheltema

1906-10-16 to - Korteweg

1906-11-05a to - Korteweg

1906-11-06 to - Korteweg

1906-11-11 from - Korteweg

1907-01-10a from - Korteweg

1907-01-10b to - Korteweg

1907-01-11 to - Korteweg

1907-01-18 to - Korteweg

1907-01-23 to - Korteweg

1908-02-21 to - Scheltema

1908-05-00b to - Korteweg

1908-06-08 to - Korteweg

1908-06-24 to - Scheltema

1908-11-08 from - Korteweg

1909-03-01 to - Scheltema

1909-03-16 to - Korteweg

1909-05-27 from - Schoenflies

1909-06-00 to - Korteweg

1909-06-18 to - Korteweg

1909-07-26 to - Hilbert

1909-08-08 from - Schoenflies

1909-11-09a to - Scheltema

1909-12-15 from - Lorentz

1909-12-19 from - Schoenflies

1909-12-24a from - Hadamard

1909-12-24b to - Korteweg

1909-12-24c to - Hadamard

1910-01-01 to - Hilbert

1910-01-04 to - Hadamard

1910-03-18 to - Hilbert

1910-09-00 to - Korteweg

1911-00-00 to - Blumenthal

1911-03-00a re Lebesgue Proof

1911-03-00b Lebesgue to - Blumenthal

1911-03-25 from - Blumenthal

1911-03-27 to - Blumenthal

1911-03-31 to - Hilbert

1911-05-09 to - Blumenthal

1911-06-11 to - Blumenthal

1911-06-14 from - Blumenthal

1911-06-16a from - Blumenthal

1911-06-19a to - Blumenthal

1911-06-20 to - Blumenthal

1911-06-22 from - Blumenthal

1911-07-02 to - Blumenthal

1911-07-08 to - Blumenthal

1911-07-14b to - Hilbert

1911-08-19a to - Blumenthal

1911-08-19b to - Scheltema

1911-08-26 from - Blumenthal

1911-09-14 to - Korteweg

1911-10-08 from - Blumenthal

1911-10-12 from - Blumenthal

1911-10-28 from - Baire

1911-11-02 from - Baire

1911-11-05 to - Baire

1911-11-07a to - Scheltema

1911-11-21 to - Blumenthal

1911-12-05 from - Baire

1911-12-10a to - Poincaré

1911-12-10b from - Poincaré

1911-12-21 to - Schoenflies

1911-12-22 to - Fricke

1911-12-30b to - Hurwitz

1912-01-04a Poincaré

1912-01-13 to - Klein
1912-01-16 from - Blumenthal
1912-01-21 to - Engel
1912-01-28 from - Engel
1912-02-03 from - Blumenthal
1912-02-04b from - Engel
1912-02-12a from - Blumenthal
1912-02-12b from - Koebe
1912-02-14 to - Koebe
1912-02-24 to - Hilbert
1912-02-27 to - Hilbert
1912-03-06a to - Engel
1912-03-06b from - Koebe
1912-03-07 to - Hilbert
1912-03-09b to - Hilbert
1912-03-26 from - Engel
1912-03-29 to - Engel
1912-05-16Weyl-Klein
1912-05-22 from - Bieberbach
1912-05-31 to - Hilbert
1912-11-07 from - Bernstein
1913-02-06 from - Bieberbach
1913-04-16 to - Hilbert
1913-06-16 to - Hilbert
1913-07-04 to - Hilbert
1913-08-16 to - Schoenflies
1913-11-08 from - Borel
1914-06-04 from - Korteweg
1914-06-20 to - Hamel
1914-07-13 from - Korteweg
1915-06-11 from - Lorentz
1915-06-19 to - Zeeman
1915-09-18 to - Snijders
1915-10-12 from - Snijders
1915-11-04 from - Blaschke
1915-11-19 to - Blaschke
1916-02-07 to - Ehrenfest
1916-05-06 to - Ehrenfest
1916-09-16 to - Belgian Government
1917-04-16 from - Schoenflies
1917-06-09 to - Mannoury
1917-10-01 from - Jaeger
1918-01-09 to - Schouten
1918-02-04a to - Buber
1918-02-15 from - Lorentz

1918-02-16 to - Lorentz
1918-05-23 from - Carathéodory
1918-11-25a to - Hilbert
1918-11-28 from - Denjoy
1919-02-16 from - Noordhoff
1919-02-26 from - Jaeger
1919-06-10 to - Hurwitz
1919-06-28 to - Hilbert
1919-09-08 from - Klein
1919-09-19 to - Klein
1919-10-18 from - Nielsen
1919-10-21b to - Klein
1919-11-09 from - Klein
1919-11-10 from - Schoenflies
1919-12-04a from - Teubner
1919-12-29 to - Schoenflies
1920-01-25 from - Lorentz
1920-02-04 to - Mayor Amsterdam
1920-02-12 from - Mayor Amsterdam
1920-02-21 to - Mayor Amsterdam
1920-03-00 to - KNAW
1920-03-25b from - Schouten
1920-04-01 from - Blumenthal
1920-05-06a from - Weyl
1920-05-06b to -Weyl
1920-07-26 from - Dingler
1920-08-07 to - Klein
1920-08-20 from - Mittag-Leffler
1920-08-28 from - Wolff
1920-09-07 to - Weyl
1920-10-04 Denjoy to - Blumenthal
1920-10-17 to - Denjoy
1920-10-20 from - Denjoy
1920-10-27 to - Denjoy
1920-10-29 from - Denjoy
1921-01-01 to - Weyl
1921-01-17b to - Schoenflies
1921-01-31 from - Weitzenböck
1921-02-01a from - Kneser
1921-02-14 from - Schoenflies
1921-04-10 from - Fraenkel
1921-04-11 to - Mauve
1921-11-27 to - Handelsblad
1922-00-00 to - Weitzenböck
1922-04-21 from - Ehrenfest-Afanassjewa

1922-04-26 to - Ehrenfest
1922-10-10a from - Dresden
1922-11-24 to - Mannoury
1922-11-25 to - Went
1922-12-16 to - Mannoury
1923-04-00 to - Wisk. Genootschap
1923-04-18 from - Fraenkel
1923-08-25 from - Ehrenfest-Afanassjewa
1923-09-01 to - Schoenflies
1923-10-24 from - Urysohn
1923-11-29a to - Klein
1923-11-29b to - Urysohn
1924-01-16 from - Dubislav
1924-01-22 to - Urysohn
1924-03-12 from - Menger
1924-03-25a to - Mannoury
1924-04-06 to - Menger
1924-04-09 to - Urysohn
1924-06-14b to - Urysohn
1924-06-21 from - Urysohn
1924-06-24 to - Urysohn
1924-06-27b Urysohn-Sierpiński
1924-07-09b to - Zeeman
1924-07-29 from - Urysohn
1924-08-21 to - Kneser
1924-08-31 to - Alexandrov
1924-09-07 from - Alexandrov's mother
1924-10-13 to - Alexandrov
1924-10-20b to - Alexandrov
1924-10-21 to - Kneser
1924-11-02 to - Menger
1924-11-06 from - Carathéodory
1924-11-13 from - Menger
1924-12-21 to - Alexandrov
1925-02-11 from - Menger
1925-06-22b from - Kagan
1925-07-03 from - Menger
1925-07-08 to - Menger
1925-12-15 to - Von Dyck
1925-12-21a to - Hopf
1926-04-10a from - Hahn
1926-04-10b from - Menger
1926-05-11 to - Heyting
1926-07-23 from - Planck
1926-08-08 from - Planck

1926-08-19 from - Menger
1926-08-20 from - Scholz
1926-12-13 to - Hopf
1926-12-21 to - Fraenkel
1927-01-12 to - Fraenkel
1927-01-28 to - Fraenkel
1927-02-03 to - Alexandrov
1927-03-08 to - Hopf
1927-04-09 to - Hopf
1927-09-07 from - Scholz
1927-11-08 from - Herzberg
1927-11-16 from - Scholz
1928-01-17 from - Menger
1928-01-20 from - Bieberbach
1928-01-23 to - Bieberbach
1928-02-16b to - Weyl
1928-03-24 from - Sommerfeld
1928-04-12b to - Mises
1928-07-03 from - Bohr
1928-07-17 to - Heyting
1928-09-27 from - Haerlen
1928-10-25a from - Hilbert
1928-11-02a to - Blumenthal
1928-11-02b to - Carathéodory
1928-11-05a to - eds. Math.Ann.
1928-11-06b Blumenthal - eds. Math.Ann.
1928-11-16a Blumenthal - eds. Math.Ann.
1928-12-22 Hilbert/Springer - eds. Math. Ann.
1928-12-23b Courant to - Carathéodory
1929-01-23 to - eds. Math.Ann.
1929-04-30 to - eds. Math.Ann.
1929-07-11 to - Hahn
1929-08-09a to - Hahn
1929-10-07 from - Heyting
1929-10-26 from - Donder
1930-01-10 to - Hahn
1930-01-11a to - Hahn
1930-06-07 to - Monatshefte
1930-06-10a to - Hopf
1930-08-03 to - Freudenthal
1930-09-20 to - Heyting
1930-10-09 to - Donder
1931-04-24 from - Carnap
1931-05-20 from - Feigl
1931-09-11 from - Belinfante

1931-09-20 from - Belinfante
1932-08-13 from - Freudenthal
1932-10-20a to - Alexandrov
1932-12-01 from - Courant
1933-04-11 from - Van der Corput
1933-05-14 from - Ehrenfest
1933-11-23 to - Van der Corput
1933-12-10 to - Gutkind
1934-04-13 to - Heyting
1934-06-19 from - Tornier
1934-08-03 from - Hasse
1934-08-13 to - Hasse
1934-10-12 from - Hasse
1934-11-07 from - Gonseth
1935-01-08 from - Bieberbach
1935-02-05 to - Hasse
1935-03-20 to - Doetsch
1936-03-20 to - Alexandrov
1936-03-30 to - Heyting
1936-04-10 to - Heyting
1939-10-04a to - Freudenthal
1939-10-09 from - Freudenthal
1940-05-08 from - Freudenthal
1940-08-10 from - Freudenthal
1940-11-30 from - Freudenthal
1941-01-14 to - Freudenthal
1941-04-19 from - Griss
1942-05-26 from - Freudenthal
1942-05-29 to - Freudenthal
1942-05-30 to - Freudenthal
1942-05-31 from - Freudenthal
1944-05-20 from - Dijksterhuis
1945-12-01 from - De Vries, Hk.
1945-12-03 to - Restauration Comm.
1945-12-06a to - Van der Corput
1945-12-12 to - Restauration Comm.
1946-01-07 to - Mayor & Aldermen

1946-01-10b to - Van der Corput
1946-01-23 to - Clay
1946-05-01b from - Minnaert
1946-08-03b to - Dresden
1946-10-08a to - Mayor & Aldermen
1947-08-21 to - Van Dantzig
1947-09-03 from - Van Dantzig
1948-04-02 to - Curators UvA
1948-06-03 to - Mannoury
1949-03-10 to - Hopf
1949-07-10 to - Compositio
1949-08-24 to - Van Dantzig
1949-10-28 to - Heyting
1950-02-28 to - Carathéodory
1950-12-22 to - Haersolte
1951-02-05 from - Fraisse
1951-04-18 from - mrs. Van der Corput
1951-05-01 to - Math. and Physics UvA
1951-05-16c from - Pye
1952-03-04 to - Radley
1953-00-00 to - Coxeter
1953-07-28a to - Coxeter
1953-09-27 to - Fraser
1953-11-28 to - Gutkind
1954-12-31 to - Anrooy
1955-01-04 to - Morse
1955-05-03 to - Kok
1957-02-03 to - Hopf
1958-07-23 from - Esenin Volpin
1959-04-25 to - KNAW
1959-08-07 from - Moyls
1960-04-30 to - KNAW
1960-07-26 to - mrs. Whitehead
1961-05-26 from - Hardenberg
1962-05-17 to - KNAW
1965-02-18 to - Van Lakwijk
1966-07-06 from - Myhill

Abbreviations

	in full	language	translation
a.	an	German	to
a.S.	an Seite	German	on page
a.s.	aanstaande	Dutch	next
apt.	apartement	Dutch	apartement
art.	artikel	Dutch	article/paper
asst.	assistent	German	assistant
B en W	Burgemeester en Wethouders	Dutch	Mayor and Aldermen
Bd.	Band	German	volume
Bde.	Bände	German	volumes
bezw.	beziehungsweise	German	respectively
bijv.	bijvoorbeeld	Dutch	for example
bijz.	bijzonder	Dutch	particular
blz.	bladzijde	Dutch	page
bzw.	beziehungsweise	German	respectively
c/o	care of		
CWI	Centrum voor Wiskunde en Informatica	Dutch	Centre for mathematics and informatics
d.d.	de dato	Dutch	
Dept.	Departement	Dutch	
d.h.	das heisst	German	that is (i.e.)
d.J.	dieses Jahres	German	of this year
d.M.	dieses Monats	German	of this month
Dr.	Doctor		
dw	dienstwillige	Dutch	obedient
d.w.z.	dat wil zeggen	Dutch	that is
enz.	enzovoorts	Dutch	etc.
ETH	Eidgenössische Technische Hochschule	German	
Ex.	Exemplar	German	copy
i.v.m.	in verband met	Dutch	in connection with
id.	idem	Dutch	
i.h.b.	in het bijzonder	Dutch	in particular
inkl.	inklusive	German	including
inst.	instituut	Dutch	institute
iur. docts.	iuris doctorandus		
j.l.	jongstleden	Dutch	last
Jul.	Julius		
H. Mis	Heilige Mis	Dutch	Mass
Kr.	Kronor	Swedish	Crones

Lab.	Laboratorium		laboratory
M.	Monsieur	French	Mr.
M.	Mark	German	
Me.	Maître	French	Master
M. E.	Meines Erachtens	German	in my opinion
Mevr.	Mejuffrouw	Dutch	miss
Mej.	Mevrouw	Dutch	misses
m.i.	mijns inziens	Dutch	in my opinion
MM.	Messieurs	French	Gentlemen
Mme.	Madame	French	mrs.
M.	Monat	German	
Mr.	Mister		
Ms.	Manuscript		manuscript
Mss.	Manuscripts		manuscripts
n.	nächsten	German	next
n.l.	namelijk	Dutch	namely
n.b.	nota bene		
no.	numero		
Nos.	numeros		
Nov.	November		
o.a.	onder andere	Dutch	among other things
p.	pagina, pagina's	Dutch	page, pages
p/a	per adres	Dutch	c.o.
pag.	pagina	Dutch	page
p.o.	per omgaand	Dutch	by return mail
Prof.	Professor		
r.	regel	Dutch	line
resp.	respectievelijk	Dutch	respectively
s.	sur	French	on
S.	Seite	German	page
s.	siehe	German	see
s. Z.	seiner Zeit	German	at the time
Str.	Strasse	German	street
t.a.p.	ter aangehaalde plaatse	Dutch	loc. cit.
t.t.	totus tuus –	Latin	totally yours
	geheel de Uwe	Dutch	totally yours
t/m	tot en met	Dutch	from . . to
u.	und	German	and
usw.	und so weiter	German	etc
u.a.	unter Anderem	German	among other things
v.	van	Dutch	of
v.	vorigen	German	preceding
v.b.	van boven	Dutch	from the top

vgl.	vergleich	German	compare
v.o.	van onderen	Dutch	from the bottom
v.h.t.h.	van huis tot huis	Dutch	from hpous to house
v.M.	vorigen Monats	German	last month
v.v.	vice versa		
w.g.	was getekend	Dutch	was signed
w.o.	waaronder	Dutch	among which
WG	Wiskundig Genootschap	Dutch	Mathematical Society
Z.	Zeile	German	line
z.B.	zum Beispiel	German	for example
z.Zt.	zur Zeit	German	at the time

Organizations and Journals

AIPS – Académie Internationale de Philosophie des Sciences [Dockx institution]

AISC – Association Internationale de Collaboration Scientifique [Dockx institution]

Annalen – Mathematische Annalen

ANP – Algemeen Nederlands Persbureau

ASBL – association sans but lucratif

ASL – Association for Symbolic Logic

CELS – Centre d'Étude de Logique Symbolique de l'Université de Paris

CNRS – Centre National de Recherches Scientifiques

Compositio – Compositio Mathematica

Conseil – Conseil Internationale des Recherches

C.R. – Comptes rendus de l'Académie des sciences

Crelle – Journal für die reine und angewandte Mathematik

CW – Collected Works Brouwer

CWI – Centrum voor Wiskunde en Informatica

DMV – Deutsche Mathematiker-Vereinigung

ETH – Eidgenössische Technische Hochschule (Zürich)

FISP – Fédération Internatonale des Sociéteés de Philosophie

Forum – Forum (Zürich) [Gonseth institution]

Fundamenta – Fundamenta Mathematicae

GGA – Gemeentelijk Archief Amsterdam, now Stadsarchief Amsterdam

ICHS – International Council of Humanistic Sciences [= CIPHS, Conseil International de Philosophie et des Sciences Humaines]

ICSU – International Council of Scientific Unions [= Conseil International des Unions Scientifiques]

IIP – Institut International de(Collaboration) Philosophi(qu)e

IIST – Institut International des Sciences Théoriques [Dockx institution]

IMU – International Mathematical Union

Indagationes – Indagationes Mathematicae

ISS – Institut de Synthèse Scientifique [Dockx institution]

KNAW – Koninklijke Nederlandse Akademie van Wetenschappen

KWG – Koninklijk Wiskundig Genootschap [= WG]

Math. Annalen – Mathematische Annalen [= MA]

Math. & Physics – Faculteit Wis- en Natuurkunde UvA

MC – Mathematisch Centrum [now CWI]

Monatshefte – Monatshefte für Mathematik und Physik

NHU[M –] Noord-Hollandsche Uitgevers Maatschappij (Studies in Logic)

NWO – Nederlandse Organisatie voor Wetenschappelijk Onderzoek

OK & W – Ministerie van Onderwijs, Kunsten en Wetenschappen

Proceedings – KNAW Proceedings

SILPS – Société Internationale de Logique et de Philosophie des Sciences

UIHS – Union Internationale d'Histoire des Sciences [= IUHS, International
      Union of History of Science]

UIHPS – Union Internationale d'Histoire et de Philosophie des Sciences
      [= IUHSP, International Union of History and Philosophy of Science]

UIPS – Union Internationale de Philosophie des Sciences [= IUPS, International
      Union of Philosophy of Science; = DMLPS, Division of Mathematical Logic
      and Philosophy of Science]

UNESCO – United Nations Educational Scientific and Cultural Organisations

Union – Union internationale de Mathématique

UvA – Universiteit van Amsterdam [formerly Gemeente Universiteit]

Verslagen – Verslagen van de KNAW

WG – Wiskundig Genootschap [now KWG]

ZWO – Nederlandse Organisatie voor Zuiver-Wetenschappelijk Onderzoek [now
      NWO]

Bibliography

[Adama van Scheltema 1903] C.S. Adama van Scheltema. *Levende Steden. Dusseldorp of de ontmoeting van Petrus Cordatus. Een satirisch-dramatisch gedicht.* S. van Looy, Amsterdam, 1903.

[Adama van Scheltema 1907] C.S. Adama van Scheltema. *De Grondslagen eener nieuwe Poëzie. Proeve tot een maatschappelijke kunstleer tegenover het naturalisme en anarchisme, de tachtigers en hun decadenten.* Brusse, Rotterdam, 1907.

[Adama van Scheltema 1911] C.S. Adama van Scheltema. *Goethe's Faust. Eerste deel. In Nederlandsche verzen, vertaald en toegelicht door C.S. Adama van Scheltema.* Wereld bibliotheek, Rotterdam, 1911.

[Adama van Scheltema 1914] C.S. Adama van Scheltema. *Italië. Indrukken en Gedachten. Een Causerie.* Brusse, Rotterdam, 1914.

[Alexandroff 1932] P. Alexandroff. *Einfachste Grundbegriffe der Topologie.* Springer, Berlin, 1932.

[Alexandroff and Hopf 1935] P. Alexandroff and H. Hopf. *Topologie I.* Springer Verlag, Berlin, 1935.

[Alexandrov 1969] P.S. Alexandrov. Die Topologie in und um Holland in den Jahren 1920–1930. *Nieuw Arch Wiskunde*, 17:109–127, 1969.

[Alexandrov 1972] P.S. Alexandrov. Poincaré and Topology. *Russian Mathematical Surveys*, 27:157–166, 1972.

[Barzin, M 1929] M. Barzin et A. Errera. Sur le principe de tiers exclu. *Arch. Soc. Belge Phil.*, 1:26, 1929.

[Belinfante 1931] M.J. Belinfante. Über die Elemente der Funktionentheorie und die Picardschen Sätze in der intuitionistischen Mathematik. *Nederl Ak Wetensch Proc*, 34:1395–1397, 1931.

[Bieberbach 1910] L. Bieberbach. *Zur Theorie der automorphen Funktionen.* PhD thesis, Göttingen, 1910.

[Bohl 1916] P. Bohl. Über die hinsichtlich der unabhängigen und abhängigen Variabeln periodische diffential Gleichung erster Ordnung. *Acta Mathematica*, 40:321–336, 1916.

D. van Dalen, *The Selected Correspondence of L.E.J. Brouwer,*   503
Sources and Studies in the History of Mathematics and Physical Sciences,
DOI 10.1007/978-0-85729-537-8, © Springer-Verlag London Limited 2011

[Borel 1912] É. Borel. Quelques remarques sur les principes de la théorie des ensembles. *Mathematische Annalen*, 60:194–195, 1912.

[Brouwer 1905A] L.E.J. Brouwer. *Leven, Kunst en Mystiek*. Waltman, Delft, 1905. Translation by W.P. van Stigt in *Notre Dame J. of formal Logic* 37, (1966), 381–429.

[Brouwer 1906A2] L.E.J. Brouwer. Polydimensional vector distributions. *Koninklijke Nederlandse Akademie van Wetenschappen. Proceedings of the Section Sciences*, 9:66–78, 1906a.

[Brouwer 1906b] L.E.J. Brouwer. The force field of the non-Euclidean spaces with negative curvature. *Koninklijke Nederlandse Akademie van Wetenschappen. Proceedings of the Section Sciences*, 9:116–133, 1906b.

[Brouwer 1906c] L.E.J. Brouwer. The force field of the non-Euclidean spaces with positive curvature. *Koninklijke Nederlandse Akademie van Wetenschappen. Proceedings of the Section Sciences*, 9:250–266, 1906c.

[Brouwer 1908a] L.E.J. Brouwer. About difference quotients and differential quotients. *Koninklijke Nederlandse Akademie van Wetenschappen. Proceedings of the Section Sciences*, 11:59–66, 1908.

[Brouwer 1908b] L.E.J. Brouwer. De onbetrouwbaarheid der logische principes. *Tijdschrift voor Wijsbegeerte*, 2:152–158, 1908.

[Brouwer 1909a] L.E.J. Brouwer. Continuous one-one transformations of surfaces in themselves. *Koninklijke Nederlandse Akademie van Wetenschappen. Proceedings of the Section Sciences*, 11:788–798, 1909a.

[Brouwer 1909b] L.E.J. Brouwer. Die Theorie der endlichen kontinuierlichen Gruppen, unabhängig von den Axiomen von Lie, I. *Mathematische Annalen*, 67:246–267, 1909b.

[Brouwer 1909c] L.E.J. Brouwer. *Het wezen der meetkunde*. Clausen, Amsterdam, 1909c. Inaugural address privaat docent, 12.10.1909. Also in 1919e.

[Brouwer 1909d] L.E.J. Brouwer. On continuous vector distributions on surfaces. *Koninklijke Nederlandse Akademie van Wetenschappen. Proceedings of the Section Sciences*, 11:850–858, 1909d. Corr. in 1910e.

[Brouwer 1909e] L.E.J. Brouwer. Over continue vectordistributies op oppervlakken. *Koninklijke Nederlandse Akademie van Wetenschappen. Verslagen van de Gewone Vergadering der Afdeling Natuurkunde*, 17:896–904, 1909e.

[Brouwer 1909f] L.E.J. Brouwer. Over de niet-Euclidische meetkunde. *Nieuw Archief voor Wiskunde*, 9, 1909f.

[Brouwer 1909g] L.E.J. Brouwer. Over één-éénduidige, continue transformaties van oppervlakken in zichzelf. *Koninklijke Nederlandse Akademie van Wetenschappen. Verslagen van de Gewone Vergadering der Afdeling Natuurkunde*, 17:741–752, 1909g.

[Brouwer 1910a] L.E.J. Brouwer. Berichtigung. *Mathematische Annalen*, 69:180, 1910a. re 1909 C.

[Brouwer 1910b] L.E.J. Brouwer. Beweis des Jordanschen Kurvensatzes. *Mathematische Annalen*, 69:169–175, 1910b.

[Brouwer 1910c] L.E.J. Brouwer. Die Theorie der endlichen kontinuierlichen Gruppen, unabhängig von den Axiomen von Lie, II. *Mathematische Annalen*, 69:181–203, 1910c.

[Brouwer 1910d] L.E.J. Brouwer. On continuous vectordistributions on surfaces, II. *Koninklijke Nederlandse Akademie van Wetenschappen. Proceedings of the Section Sciences*, 12:716–734, 1910d.

[Brouwer 1910e] L.E.J. Brouwer. On continuous vectordistributions on surfaces, III. *Koninklijke Nederlandse Akademie van Wetenschappen. Proceedings of the Section Sciences*, 12:171–186, 1910e.

[Brouwer 1910f] L.E.J. Brouwer. On the structure of perfect sets of points. *Koninklijke Nederlandse Akademie van Wetenschappen. Proceedings of the Section Sciences*, 12:785–794, 1910f.

[Brouwer 1910g] L.E.J. Brouwer. Über eineindeutige, stetige Transformationen von Flächen in sich. *Mathematische Annalen*, 69:176–180, 1910g.

[Brouwer 1910h] L.E.J. Brouwer. Zur Analysis Situs. *Mathematische Annalen*, 68:422–434, 1910h.

[Brouwer 1911a] L.E.J. Brouwer. Bemerkung. *Mathematische Annalen*, 71:598, 1911a.

[Brouwer 1911b] L.E.J. Brouwer. Berichtigung. *Mathematische Annalen*, 71:598, 1911b.

[Brouwer 1911c] L.E.J. Brouwer. Beweis der Invarianz des n-dimensionalen Gebiets. *Mathematische Annalen*, 71:305–313, 1911c.

[Brouwer 1911d] L.E.J. Brouwer. Beweis des Jordanschen Satzes für den n-dimensionalen Raum. *Mathematische Annalen*, 71:314–319, 1911d.

[Brouwer 1911e] L.E.J. Brouwer. Über Jordansche Mannigfaltigkeiten. *Mathematische Annalen*, 71:320–327, 1911e.

[Brouwer 1912a] L.E.J. Brouwer. *Intuïtionisme en Formalisme*. Clausen, Amsterdam, 1912a. Inaugural address, professor.

[Brouwer 1912b] L.E.J. Brouwer. Über den Kontinuitätsbeweis für das Fundamentaltheorem der automorphen Funktionen im Grenzkreisfall. *Jahresbericht der Deutschen Mathematiker-Vereinigung*, 21:154–157, 1912b.

[Brouwer 1912c] L.E.J. Brouwer. Über die Singularitätenfreiheit der Modulmannigfaltigkeit. *Nachrichten der Akademie der Wissenschenschaften in Göttingen. II. Mathematisch–Physikalische Klasse*, 803–806, 1912c.

[Brouwer 1912d] L.E.J. Brouwer. Über die topologischen Schwierigkeiten des Kontinuitätsbeweises der Existenztheoreme eindeutig umkehrbarer polymorpher Funktionen auf Riemannschen Flächen (Auszug aus einem Brief an R. Fricke). *Nachrichten der Akademie der Wissenschenschaften in Göttingen. II. Mathematisch–Physikalische Klasse*, 603–606, 1912d.

[Brouwer 1913a] L.E.J. Brouwer. Eenige opmerkingen over het samenhangstype η. *Koninklijke Nederlandse Akademie van Wetenschappen. Proceedings of the Section Sciences*, 15:1256–1263, 1913a.

[Brouwer 1913b] L.E.J. Brouwer. Intuitionism and Formalism. *Bulletin of the American Mathematical Society*, 20:81–96, 1913b.

[Brouwer 1913c] L.E.J. Brouwer. Some remarks on the coherence type η. *Koninklijke Nederlandse Akademie van Wetenschappen. Proceedings of the Section Sciences*, 15:1256–1263, 1913c.

[Brouwer 1913d] L.E.J. Brouwer. Über den natürlichen Dimensionsbegriff. *Journal f. die reine und angewandte Mathematik*, 142:146–152, 1913d. Corr. in 1924b.

[Brouwer 1914] L.E.J. Brouwer. A. Schoenflies und H. Hahn. Die Entwickelung der Mengenlehre und ihrer Anwendungen, Leipzig und Berlin 1913. *Jahresbericht der Deutschen Mathematiker-Vereinigung*, 23:78–83, 1914.

[Brouwer 1917a] L.E.J. Brouwer. Addenda en corrigenda over de grondslagen der wiskunde. *Koninklijke Nederlandse Akademie van Wetenschappen. Verslagen van de Gewone Vergadering der Afdeling Natuurkunde*, 25:1418–1423, 1917a. Separate sheet with corrections inserted.

[Brouwer 1917b] L.E.J. Brouwer. Addenda en corrigenda over de grondslagen der wiskunde. *Nieuw Archief voor Wiskunde*, 12:439–445, 1917b. Corrected w.r.t. Brouwer17a.

[Brouwer 1918a] L.E.J. Brouwer. Begründung der Mengenlehre unabhängig vom logischen Satz vom ausgeschlossenen Dritten. Erster Teil, Allgemeine Mengenlehre. *Verhandelingen der Koninklijke Akademie van Wetenschappen te Amsterdam*, 5:1–43, 1918a.

[Brouwer 1918b] L.E.J. Brouwer. Lebesguesches Mass und Analysis Situs. *Mathematische Annalen*, 79:212–222., 1918b.

[Brouwer 1918c] L.E.J. Brouwer. Schreiben der Herrn Dr. Martin Buber in Heppenheim mit Beantwortung. *Mededelingen Int Inst Wijsbegeerte*, 1:28–30, 1918c. Letter of Buber + reply of Brouwer.

[Brouwer 1919a] L.E.J. Brouwer. Begründung der Mengenlehre unabhängig vom logischen Satz vom ausgeschlossenen Dritten. Zweiter Teil, Theorie der Punktmengen. *Verhandelingen der Koninklijke Akademie van Wetenschappen te Amsterdam*, 7:1–33, 1919a.

[Brouwer 1919b] L.E.J. Brouwer. Énumération des groupes finis de transformations topologiques du tore. *Comptes Rendus*, 168:845–848, 1168, 1919b.

[Brouwer 1919c] L.E.J. Brouwer. Énumération des surfaces de Riemann regulières de genre un. *Comptes Rendus*, 168:677–678, 832, 1919c.

[Brouwer 1919d] L.E.J. Brouwer. Intuitionistische Mengenlehre. *Jahresbericht der Deutschen Mathematiker-Vereinigung*, 28:203–208, 1919d. Appeared in 1920.

[Brouwer 1919e] L.E.J. Brouwer. *Wiskunde, Waarheid, Werkelijkheid*. Noordhoff, Groningen, 1919e.

[Brouwer 1923a] L.E.J. Brouwer. Über den natürlichen Dimensionsbegriff. *Koninklijke Nederlandse Akademie van Wetenschappen. Proceedings of the Section Sciences*, 26:795–800, 1923a.

[Brouwer 1923b] L.E.J. Brouwer. Über die Bedeutung des Satzes vom ausgeschlossenen Dritten in der Mathematik insbesondere in der Funktionentheorie. *Journal f. die reine und angewandte Mathematik*, 154:1–8, 1923b.

[Brouwer 1924a] L.E.J. Brouwer. Bemerkungen zum natürlichen Dimensionsbegriff. *Koninklijke Nederlandse Akademie van Wetenschappen. Proceedings of the Section Sciences*, 27:635–638, 1924a.

[Brouwer 1924b] L.E.J. Brouwer. Berichtigung. *Journal f. die reine und angewandte Mathematik*, 153:253, 1924b. re 1913d.

[Brouwer 1924c] L.E.J. Brouwer. Bewijs van de onafhankelijkheid van de onttrekkingsrelatie van de versmeltingsrelatie. *Koninklijke Nederlandse Akademie van Wetenschappen. Verslagen van de Gewone Vergadering der Afdeling Natuurkunde*, 33:479–480, 1924c.

[Brouwer 1924d] L.E.J. Brouwer. Zum natürlichen Dimensionsbegriff. *Mathematische Zeitschrift*, 21:312–314, 1924d.

[Brouwer 1925] L.E.J. Brouwer. Zur Begründung der intuitionistischen Mathematik I. *Mathematische Annalen*, 93:244–257, 1925.

[Brouwer 1926] L.E.J. Brouwer. Intuitionistische Einführung des Dimensionsbegriffes. *Koninklijke Nederlandse Akademie van Wetenschappen. Proceedings of the Section Sciences*, 29:855–873, 1926.

[Brouwer 1927] L.E.J. Brouwer. Über Definitionsbereiche von Funktionen. *Mathematische Annalen*, 97:60–75, 1927.

[Brouwer 1928a] L.E.J. Brouwer. Intuitionistische Betrachtungen über den Formalismus. *Koninklijke Nederlandse Akademie van Wetenschappen. Verslagen van de Gewone Vergadering der Afdeling Natuurkunde*, 36:1189, 1928a.

[Brouwer 1928b] L.E.J. Brouwer. Intuitionistische Betrachtungen über den Formalismus. *Koninklijke Nederlandse Akademie van Wetenschappen. Proceedings of the Section Sciences*, 31:374–379, 1928b.

[Brouwer 1928c] L.E.J. Brouwer. Intuitionistische Betrachtungen über den Formalismus. *Die Preussische Akademie der Wissenschaften. Sitzungsberichte. Physikalisch–Mathematische Klasse*, pages 48–52, 1928c.

[Brouwer 1928d] L.E.J. Brouwer. Zur Geschichtsschreibung der Dimensionstheorie. *Koninklijke Nederlandse Akademie van Wetenschappen. Proceedings of the Section Sciences*, 31:953–957, 1928d. Corr. in KNAW Proc. 32, p. 1022.

[Brouwer 1942a] L.E.J. Brouwer. Die repräsentierende Menge der stetigen Funktionen des Einheitskontinuums. *Indagationes Mathematicae*, 4:154, 1942a.

[Brouwer 1942b] L.E.J. Brouwer. Zum freien Werden von Mengen und Funktionen. *Indagationes Mathematicae*, 4:107–108, 1942b.

[Brouwer 1948] L.E.J. Brouwer. Essentieel negatieve eigenschappen. *Indagationes Mathematicae*, 10:322–323, 1948. transl. "Essentially negative properties" in CW 1, p. 478.

[Brouwer 1949] L.E.J. Brouwer. Consciousness, Philosophy and Mathematics. *Proceedings of the 10th International Congress of Philosophy, Amsterdam 1948*, 3:1235–1249, 1949.

[Brouwer 1975] L.E.J. Brouwer. *Collected works 1. Philosophy and Foundations of Mathematics. (ed. A. Heyting)*. North-Holland Publ. Co., Amsterdam, 1975.

[Brouwer 1976] L.E.J. Brouwer. *Collected works 2. Geometry, Analysis Topology and Mechanics. (ed. H. Freudenthal)*. North-Holland Publ. Co., Amsterdam, 1976.

[Brouwer 1981] L.E.J. Brouwer. *Brouwer's Cambridge Lectures on Intuitionism* (Ed. D. van Dalen). Cambridge University Press, Cambridge, 1981.

[Brouwer 1992] L.E.J. Brouwer. *Intuitionismus* (Ed. D. van Dalen). Bibliographisches Institut, Wissenschaftsverlag, Mannheim, 1992.

[Brouwer 1937] L.E.J. Brouwer et al. Signifische dialogen. *Synthese*, 2:168–174, 261–268, 316–324, 1937. Also published by Bijleveld (Utrecht) in 1939.

[Dantzig 1947] D. van Dantzig. On the principles of intuitionistic and affirmative mathematics. *Indagationes Mathematicae*, 9:429–440, 506–517, 1947.

[Dantzig 1949] D. van Dantzig. Comments on Brouwer's theorem on essentially-negative predicates. *Indagationes Mathematicae*, 11:347–355, 1949.

[Dresden 1924] A. Dresden. Brouwer's contributions to the foundations of mathematics. *Bull. Am. Math. Soc.*, 30:31–40, 1924.

[Engel 1913] F. Engel. Review of Brouwer 1910c. *Jahrbuch über die Fortschritte der Mathematik*, 41:181–182, 1913.

[Fedorchuk and Van Mill 2000] J. van Mill and V.V. Fedorchuk. Dimensionsgrad for locally connected Polish spaces. *Fundamenta Mathematicae*, 163:77–82, 2000.

[Finsterwalder 1899] S. Finsterwalder. Die geometrischen Grundlagen der Photogrammetrie. Mit 19 Fig. im Text. *Jahresbericht der Deutsche Mathematiker.-Vereinigung*, 6:1–41, 1899.

[Fraenkel 1920] A. Fraenkel. *Zahlbegriff und Algebra bei Gauss. Mit einem Anhang von A. Ostrowski. Zum ersten und vierten Gaussschen Beweis des Fundamentalsatzes der Algebra. (Materialien für eine wissenschaftliche Biographie von Gauss. Gesammelt von F. Klein, L. Schlesinger und M. Brendel, Heft VIII.)*, Nachrichten der Akademie der Wissenschenschaften in Göttingen. Beiheft, 61. Teubner, Leipzig, 1920.

[Fraenkel 1921] A. Fraenkel. Über einfache Erweiterungen zerlegbarer Ringe. *Journal für die Reine und Angewandte Mathematik*, 151:121–167, 1921.

[Fraenkel 1922] A. Fraenkel. Axiomatische begründung der transfiniten Kardinalzahlen. I. *Mathematische Zeitschrift*, 13:153–188, 1922.

[Fraenkel 1923] A. Fraenkel. *Einleitung in die Mengenlehre*. Springer Verlag, Berlin, 1923. second edition.

[Fraenkel 1927] A. Fraenkel. *Zehn Vorlesungen über die Grundlegung der Mengenlehre*. Teubner, Leipzig, 1927. Reprinted by the Wissenschaftliche Buchgesellschaft Darmstadt, 1972.

[Freudenthal 1936] H. Freudenthal. Zur intuitionistischen Deutung logischer Formeln. *Compositio Mathematica*, 4:112–116, 1936.

[Fricke 1897] R. Fricke, F. Klein. *Vorlesungen über die Theorie der automorphen Functionen I*. Teubner, Leipzig, 1897.

[Fricke 1912] R. Fricke, F. Klein. *Vorlesungen über die Theorie der automorphen Functionen II*. Teubner, Leipzig, 1912.

[Gawehn 1928] I. Gawehn. Über unberandete 2-dimensionale Mannigfaltigkeiten. *Mathematische Annalen*, 98:321–354, 1928.

[Griss 1944] G.F.C. Griss. Negatieloze intuïtionistische wiskunde. *Kon Ned Ak Wet Verslagen*, 53:261–268, 1944.

[Griss 1946] G.F.C. Griss. Negationless intuitionistic mathematics I. *Indagationes Mathematicae*, 8:675–681, 1946.

[Griss 1950] G.F.C. Griss. Negationless intuitionistic mathematics II. *Indagationes Mathematicae*, 12:108–115, 1950.

[Griss 1951] G.F.C. Griss. Negationless intuitionistic mathematics III, IV. *Indagationes Mathematicae*, 13:193–200, 452–471, 1951.

[De Groot 1941] J. de Groot. Sätze über topologische Erweiterung von Abbildungen. *Nederl Ak Wetensch Proc*, 44:934–938, 1941.

[Hadamard 1910] J. Hadamard. Sur quelques applications de l'indice de Kronecker. In J. Tannéry, editor, *Introduction à la théorie des fonctions, 2e ed.*, volume 2, pages 437–477, 1910.

[Heyting 1927a] A. Heyting. Die Theorie der linearen Gleichungen in einer Zahlenspezies mit nichtkommutativer Multiplikation. *Mathematische Annalen*, 98:465–490, 1927a.

[Heyting 1927b] A. Heyting. Zur Axiomatik der projektiven Geometrie. *Mathematische Annalen*, 98:491–538, 1927b.

[Heyting 1930a] A. Heyting. Die formalen Regeln der intuitionistischen Mathematik III. *Die Preussische Akademie der Wissenschaften. Sitzungsberichte. Physikalische-Mathematische Klasse*, 158–169, 1930a.

[Heyting 1930b] A. Heyting. Sur la logique intuitionniste. *Ac. Royale de Belgique. Bull. de la Classe des Sciences*, 5:957–963, 1930b.

[Heyting 1931a] A. Heyting. Die Intuitionistische Grundlegung der Mathematik. *Erkenntnis*, 2:106–115, 1931a.

[Heyting 1931b] A. Heyting. Die Intuitionistische Mathematik. *Forschung und Fortschritte*, 7:38–39, 1931b.

[Heyting 1936] A. Heyting. Bemerkung zu dem Aufsatz von Hern Freudenthal "Zur intuitionistischen Deutung logischer Formeln". *Compositio Mathematica*, 4:117–118, 1936.

[Hilbert 1902] D. Hilbert. Ueber die Grundlagen der Geometrie. *Mathematische Annalen*, 56:381–422, 1902.

[Hilbert 1922] D. Hilbert. Neubegründung der Mathematik (Erste Mitteilung). *Abh Math Sem Univ Hamburg*, 1:157–177, 1922.

[Hopf 1926] H. Hopf. Vektorfelder in n-dimensionalen Mannigfaltigkeiten. *Mathematische Annalen*, 96:225–250, 1926.

[Johnson 1981] Dale M. Johnson. The Problem of the Invariance of Dimension in the Growth of Modern Topology, Part II. *Archive for History of Exact Sciences*, 25:85–267, 1981.

[Karo 1926] G. Karo. Der geistige Krieg gegen Deutschland. *Zeitschr. f. Völkerpsychologie und Soziologie.*, 2, 1926.

[Kaufmann 1930] F. Kaufmann. *Das Unendliche in der Mathematik und seine Ausschaltung*. Franz Deuticke, Leipzig-Vienna, 1930.

[Kerékjártó 1919] B.v. Kerékjártó. Über die Brouwerschen Fixpunktsätze. *Mathematische Annalen*, 80:29–32, 1919.

[Klein 1882] F. Klein. Neue Beitrage zur Riemann'schen Functionentheorie. *Mathematische Annalen*, 21:630–710, 1882.

[Klein 1892] F. Klein. *Riemannsche Flächen*. Teubner, Leipzig, 1892. Autographed text; second print 1906.

[Kneser 1921] H. Kneser. Kurvenscharen auf geschlossenen Flächen. *Jahresbericht der Deutschen Mathematiker-Vereinigung*, 30:83–85, 1921.

[Kneser 1924] H. Kneser. Ein topologischer Zerlegungssatz. *Koninklijke Nederlandse Akademie van Wetenschappen. Proceedings*, 27:601–616, 1924.

[Koebe 1910] P. Koebe. Über die Uniformisierung der algebraischen Kurven. II. *Mathematische Annalen*, 69, 1910.

[Koebe 1912] P. Koebe. Begründung der Kontinuitätsmethode im Gebiete der konformen Abbildung und Uniformisierung (Voranzeige). *Nachrichten der Akademie der Wissenschaften in Göttingen*, 879–886, 1912.

[Lebesgue 1911a] H. Lebesgue. Sur la non–applicabilité de deux domaines appartenant respectivement des espaces à n et $n + p$ dimensions (Extrait d'une lettre à M.O. Blumenthal). *Mathematische Annalen*, 70:166–168, 1911a.

[Lebesgue 1911b] H. Lebesgue. Sur l'invariance du nombre de dimensions d'un espace et sur le théorème de M. Jordan relatif aux variété fermées. *Comptes Rendus*, 152:841–843, 1911b.

[Lebesgue 1921] H. Lebesgue. Sur les correspondances entre les points de deux espaces. *Fundamenta Mathematicae*, 2:256–285, 1921.

[Lennes 1911] N.J. Lennes. Curves in non-metrical analysis situs with an application in the calculus of variations. *American Journal of Mathematics*, 33:287–326, 1911.

[Mannoury 1947] G. Mannoury. *Handboek der Analytische Significa. Geschiedenis der Begripskritiek*. Kroonder, Bussum, 1947.

[Menger 1924a] K. Menger. Über die Dimension von Punktmengen. *Koninklijke Nederlandse Akademie van Wetenschappen. Proceedings*, 27:639–643, 1924a.

[Menger 1924b] K. Menger. Über die Dimension von Punktmengen. II. *Monatshefte für Mathematik und Physik*, 34:137–161, 1924b.

[Menger 1925a] K. Menger. Grundzüge einer Theorie der kurven. *Koninklijke Nederlandse Akademie van Wetenschappen. Proceedings*, 28:67–71, 1925a.

[Menger 1925b] K. Menger. Grundzüge einer Theorie der Kurven. *Mathematische Annalen*, 95:67–71, 1925b.

[Menger 1928a] K. Menger. Bemerkungen zu Grundlagenfragen. *Jahresbericht der Deutschen Mathematiker-Vereinigung*, 37:213–226, 1928a. (On the analogy Spreads-Analytic sets).

[Menger 1928b] K. Menger. *Dimensionstheorie*. Teubner, Leipzig, 1928b.

[Menger 1929] K. Menger. Allgemeine Räume und Cartesische Räume. III: Beweis des Fundamentalsatzes. *Koninklijke Nederlandse Akademie van Wetenschappen. Proceedings*, 32:330–340, 1929.

[Myhill 1966] J. Myhill. Notes towards an axiomatization of intuitionistic analysis. *Logique et Analyse*, 9:280–297, 1966.

[Myhill 1968] J. Myhill. Formal Systems of intuitionistic analysis I. In B. van Rootselaar, J.F. Staal (eds.), *Logic, Methodology and Philosophy of Science III*, 161–178, North-Holland, Amsterdam. 1968.

[Otterspeer and Schuller tot Peursum-Meijer 1997] W. Otterspeer and J. Schuller tot Peursum-Meijer. *Wetenschap en Wereldvrede. De Koninklijke Akademie van Wetenschappen en het herstel van de internationale wetenschap tijdens het Interbellum*. Koninklijke Akademie van Wetenschappen, Amsterdam, 1997.

[Urysohn 1922] P. Urysohn. Les multiplicités Cantoriennes. *Comptes Rendus*, 175:440–442, 1922.

[Urysohn 1925] P. Urysohn. Sur un espace métrique universel. *Comptes Rendus*, 180:803–806, 1925.

[Urysohn 1926] P. Urysohn. Sur les multiplicités Cantoriennes. Suite. *Fundamenta Mathematicae*, 8:225–359, 1926.

[Poincaré 1887] H. Poincaré. Les Groupes de Équations Linéaires. *Acta Mathematica*, 4:201–311, 1887.

[Poincaré 1895] H. Poincaré. Les fonctions fuchsiennes et l'équation $\Delta u = e^u$. *Journal des mathématiques pure et appliqée*, 5th series, 4:137–230, 1895.

[Rigby 1984] A. Rigby. *Initiation and Initiative. An Exploration of the Life and Ideas of Dimitrije Mitrinović*. East European Monographs, Boulder, 1984.

[Schmitz 1990] H.W. Schmitz. Frederik van Eeden and the introduction of significs into the Netherlands: from Lady Welby to Mannoury. In H.W. Schmitz, editor, *Essays on significs. Papers presented on the occasion of the 150th birthday of Victoria Lady Welby (1837–1920)*, volume 23, pages 219–246, John Benjamins, Amsterdam/Philadelphia, 1990.

[Schmitz 1990] W.H. Schmitz. *De Hollandse Significa*. Van Gorcum, Assen, 1990.

[Schoenflies 1900] A. Schoenflies. Die Entwickelung der Lehre von den Punktmannigfaltigkeiten. *Jahresbericht der Deutschen Mathematiker-Vereinigung*, 82:1–250, 1900.

[Schoenflies 1908] A. Schoenflies. *Die Entwicklung der Lehre von den Punktmannigfaltigkeiten. II.* Teubner, Leipzig, 1908.

[Schoenflies 1913] A. Schoenflies. *Entwickelung der Mengenlehre und ihrer Anwendungen, I, 1. Hälfte.* Teubner, Leipzig, Berlin, 1913.

[Schoenflies 1920] A. Schoenflies. Zur Axiomatik der Mengenlehre. *Koninklijke Nederlandse Akademie van Wetenschappen. Proceedings*, 22, 1920.

[Schouten 1918] J.A. Schouten. Zur Klassifikation der assoziatieven Zahlensystemen. *Mathematische Annalen*, 72:1–66, 1918.

[Schroeder-Gudehus 1966] B. Schroeder-Gudehus. *Deutsche Wissenschaft und internationale Zusammenarbeit, 1914–1928. Ein Beitrag zum Studium kultureller Beziehungen in politischen Krisenzeiten.* Université de Genève. Imprimerie Dumaret & Golay, Geneve, 1966.

[Shatunovsky 1920] S.O. Shatunovsky. *Algebra as a study of congruences with respect to functional moduli.* Tekhnik. Contractor for the printing of notices of Novorossiysky University, Odessa, 1920. Algebra kak ucheniye o sravneniyah po funktsional'nym modulyam.

[Van Dalen 1984] D. van Dalen. *Droeve snaar, vriend van mij. De correspondentie tussen Brouwer en Adama van Scheltema.* De Arbeiderspers, Amsterdam, 1984. 2nd ed. 2002.

[Van Dalen 1995] D. van Dalen. Hermann Weyl's Intuitionistic Mathematics. *Bull. Symb. Logic*, 1:145–169, 1995.

[Van Dalen 1999] D. van Dalen. *Mystic, Geometer, and Intuitionist: The Life of L.E.J. Brouwer. Volume 1: The dawning Revolution.* Oxford University Press, Oxford, 1999. 2nd ed. 2002.

[Van Dalen 2000] D. van Dalen. Brouwer and Fraenkel on Intuitionism. *Bull. Ass. Symb. Logic*, 6:284–310, 2000.

[Van Dalen 2001] D. van Dalen. *L.E.J. Brouwer en De Grondslagen van de wiskunde.* Epsilon, Utrecht, 2001.

[Van Dalen 2005] D. van Dalen. *Mystic, Geometer, and Intuitionist: The Life of L.E.J. Brouwer. Volume 2: Hope and Disillusion.* Oxford University Press, Oxford, 2005.

[Van Dalen and Remmert 2006] V.R. Remmert and D. van Dalen. The birth and youth of Compositio Mathematica: 'Ce périodique foncièrement international'. *Compositio Mathematica*, 142:1083–1102, 2006.

[Van Stigt 1990] W.P. van Stigt. *Brouwer's Intuitionism*. North Holland, Amsterdam, 1990.

[Weyl 1921] H. Weyl. Über die neue Grundlagenkrise der Mathematik. *Mathematische Zeitschrift*, 10:39–79, 1921.

[Wilson 1928] W. Wilson. Representation of Manifolds. *Mathematische Annalen*, 98:552–578, 1928.

[Wolff 1920] J. Wolff. Over de stelling van Picard. *Koninklijke Nederlandse Akademie van Wetenschappen Verhandelingen*, 29:171–174, 1920.

[Zorin 1972] V.K. Zorin. On Poincaré's letter to Brouwer. *Russian Mathematical Surveys*, 27:166–168, 1972.

Index

D. van Dalen, *The Selected Correspondence of L.E.J. Brouwer*,
Sources and Studies in the History of Mathematics and Physical Sciences,
DOI 10.1007/978-0-85729-537-8, © Springer-Verlag London Limited 2011

Authors and Recipients of the Letters

L.E.J. Brouwer is for obvious reasons not included in the list of correspondents

D. van Dalen, *The Selected Correspondence of L.E.J. Brouwer,*
Sources and Studies in the History of Mathematics and Physical Sciences,
DOI 10.1007/978-0-85729-537-8, © Springer-Verlag London Limited 2011